This volume provides the first comprehensive, botan
survey of this beautiful group of plants since publica.... o. *The Genus Iris* by
W.R. Dykes early this century. Following the pattern of the original Dykes
monograph, botanical details, cultivation suggestions and general comments are
supplied, and the work is generously illustrated with accurate line drawings, colour
pictures of unusual species, and distribution maps. This new survey includes all of
the species that have been described so far this century and takes in to account the
many changes in classification that have taken place in the group. Information
currently scattered in the literature is brought together in one volume to provide an
authoritative reference for professional botanists and growers, and a mine of useful
information for amateur gardeners and iris enthusiasts.

A Guide to Species Irises

Their Identification and Cultivation

A Guide to Species Irises
Their Identification and Cultivation

Edited by The Species Group of the British Iris Society

Line drawings by CHRISTABEL KING
and maps by WILLIAM R. KILLENS

CAMBRIDGE
UNIVERSITY PRESS

CAMBRIDGE UNIVERSITY PRESS
Cambridge, New York, Melbourne, Madrid, Cape Town,
Singapore, São Paulo, Delhi, Tokyo, Mexico City

Cambridge University Press
The Edinburgh Building, Cambridge CB2 8RU, UK

Published in the United States of America by Cambridge University Press, New York

www.cambridge.org
Information on this title: www.cambridge.org/9780521206433

© Cambridge University Press 1997

This publication is in copyright. Subject to statutory exception
and to the provisions of relevant collective licensing agreements,
no reproduction of any part may take place without the written
permission of Cambridge University Press.

First published 1997
Reprinted 1997
First paperback edition 2011

A catalogue record for this publication is available from the British Library

Library of Congress Cataloguing in Publication data
A Guide to species irises: their identification and cultivation/
   edited by the Species Group of the British Iris Society;
   illustrations by Christabel King.
      p. cm.
   Includes bibliographical references (p.) and index.
   ISBN 0 521 44074 2 (hc)
   1. Iris (Plant) – Identification.   2. Flower gardening.
   I. British Iris Society.   Species Group.
   QK495.175G85   1997
   584'.24–DC20   95-23954 CIP

ISBN 978-0-521-44074-5 Hardback
ISBN 978-0-521-20643-3 Paperback

Additional resources for this publication at www.cambridge.org/9780521206433

Cambridge University Press has no responsibility for the persistence or
accuracy of URLs for external or third-party internet websites referred to in
this publication, and does not guarantee that any content on such websites is,
or will remain, accurate or appropriate.

Contents

Contributors	ix
Preface	xi
Illustration Acknowledgements	xvii
The Iris in History CHARLES LYTE	1
Cultivation and the Great Genus PETER MAYNARD	4
Chromosomes and the Genus Iris J.R. ELLIS	8
Identification Guide for Plants as Received	11
Subgenus Iris: the Bearded Irises	17
Section Iris NIGEL SERVICE	17
Section Psammiris (Spach) J. Taylor	58
Section Oncocyclus (Siemssen) Baker MARTYN RIX	62
Section Regelia Lynch	90
Section Hexapogon (Bunge) Baker	97
Section Pseudoregelia Dykes	98
Subgenus Limniris: the Beardless Irises	109
Section Lophiris (Tausch) Tausch (the Evansia Irises) CHRISTOPHER BREARLEY AND J.R. ELLIS	109
Section Limniris	121
Series Chinenses (Diels) Lawrence	121
Series Vernae (Diels) Lawrence	127
Series Ruthenicae (Diels) Lawrence	128
Series Tripetalae (Diels) Lawrence	131
Series Sibericae (Diels) Lawrence CHRISTOPHER GREY-WILSON	133
Series Californicae (Diels) Lawrence V.A. DICKSON-COHEN	144
Series Longipetalae (Diels) Lawrence	158

Series Laevigatae (Diels) Lawrence SIDNEY LINNEGAR	160
Series Hexagonae (Diels) Lawrence	167
Series Prismaticae (Diels) Lawrence	171
Series Spuriae (Diels) Lawrence MARIAN BOWLEY	172
Series Foetidissimae ANNE BLANCO WHITE	194
Series Tenuifoliae (Diels) Lawrence	195
Series Ensatae (Diels) Lawrence	202
Series Syriacae (Diels) Lawrence OFER COHEN	203
Series Unguiculares (Diels) Lawrence AARON DAVIS AND STEPHEN JURY	209

Subgenus Nepalensis (Dykes) 215
CLIVE INNES

Subgenus Xiphium (Miller) Spach 220
HENNING CHRISTIANSEN

Subgenus Scorpiris Spach 225
BRIAN MATHEW

Subgenus Hermodactyloides Spach (the Reticulata Irises) 279
WILLIAM. R. KILLENS

I. pariensis Welsh 1986 293
J.W. WADDICK

References	294
Select Bibliography	297
Glossary	299
Maps WILLIAM R. KILLENS	305
Line drawings CHRISTABEL KING	333
Colour plates	facing p.360*
Index of *Iris* species	361
Index of species names from other classifications	370

*A colour version of these plates is available for download from www.cambridge.org/9780521206433

Contributors

MRS A. BLANCO WHITE,
72 SOUTH HILL PARK,
LONDON NW3 2SN, UK

PROF. M.E.A. BOWLEY,
BROOK ORCHARD,
GRAFFHAM, PETWORTH,
WEST SUSSEX GU28 0PY, UK

MR C. BREARLEY,
31 FERNHEAD ROAD,
LONDON W9 3EX, UK

MR H. CHRISTIANSEN,
QUINTA DAS FLORES,
CASAL DO BORRALHO,
2665 MALVEIRA, PORTUGAL

DR. O. COHEN
HAR-GILLO FIELD SCHOOL CENTRE,
DOAR-NA ZEFON JEHUDA 90907,
ISRAEL

DR A.P. DAVIS,
c/o THE HERBARIUM,
ROYAL BOTANIC GARDENS,
KEW, RICHMOND,
SURREY, TW9 3AB, UK

MR V.A. DICKSON-COHEN,
42B BELSIZE PARK GARDENS,
LONDON NW3 4LY, UK

DR J.R. ELLIS,
THE GROVE,
WELL END, BOURNE END,
BUCKS SL8 5NY, UK

DR C. GREY-WILSON,
THE BLACK HOUSE,
FENSTEAD END,
Nr. HAWKEDON, BURY ST. EDMUNDS,
SUFFOLK IP29 4LH, UK

MR C.F. INNES,
HOLLY GATE NURSERIES,
BILLINGSHURST LANE,
ASHINGTON,
WEST SUSSEX RH20 3BA, UK

DR S.L. JURY,
DEPARTMENT OF BOTANY,
PLANT SCIENCES LABORATORIES,
THE UNIVERSITY OF READING,
WHITEKNIGHTS, PO BOX 221,
READING RG6 2AS, UK

MR W.R. KILLENS,
35 CLARKS LANE,
HALSTEAD, SEVENOAKS,
KENT TN14 7DG, UK

MS C. KING,
149 FULWELL PARK AVENUE,
TWICKENHAM,
MIDDLESEX TW2 5HG, UK

MR S. LINNEGAR,
6 OBAN GARDENS,
WOODLEY, READING,
BERKS RG5 3RG, UK

MR C. LYTE,
CARTERS CORNER PLACE,
COWBEECH ROAD, HAILSHAM,
EAST SUSSEX BN27 4HX, UK

MR B. MATHEW, V.M.H.,
'SAFFRON',
90 FOLEY ROAD,
CLAYGATE, ESHER,
SURREY KT10 0NB, UK

Contributors

MR P.R. MAYNARD,
43 SEA LANE,
GORING-BY-SEA,
WEST SUSSEX BN12 4QD, UK

DR E.M. RIX,
QUINCE HOUSE,
ROSE ASH,
SOUTH MOLTON,
N. DEVON EX36 4PW, UK

MR N. SERVICE,
BAZALGUE,
BALADOU,
46600 MARTEL, LOT,
FRANCE

DR J.W. WADDICK
8871 N.W. BROSTROM ROAD,
KANSAS CITY,
MO 64152, U.S.A.

Preface

Irises have been known and used for millennia. Linnaeus described and classified them in the eighteenth century; then in 1913 W. R. Dykes wrote the first modern survey which described the plants in detail, reclassified a number and offered some guidance towards their successful cultivation. It was published by Cambridge University Press and has remained the standard work, but is now totally inadequate. Since that time there has been one reclassification of Dykes by Prof. Ludwig Diels in 1930 and another by Prof. G.H.M. Lawrence in 1953. These were followed by Dr G.I. Rodionenko's totally different classification based principally on the characteristics of germinating seeds in 1967. Finally Brian Mathew reverted to updating the Dykes classification again in 1981.

Naturally, in the time between the various classifications, new irises were discovered and this has continued since Brian Mathew issued the second edition of his book. With vast areas of the northern hemisphere still to be explored there are certain to be more plants to come. Meanwhile, Brian Mathew has modified the classification of Hermodactyloides while Aaron Davis and Stephen Jury have reorganised the Unguiculares.

Apart from the original Dykes, there are three major books on irises circulating in the English reading world at present; *Growing Irises* by G.E. Cassidy and S. Linnegar (1982), which is a cultivation guide to irises reasonably likely to be available to gardeners and which is out of print; *Iris* by Fritz Köhlein (1981), again a gardener's guide to the many varieties of irises he is able to grow in central European conditions and which has superb photographs (and both these books contain material on iris cultivars); and *The Iris* by Brian Mathew, which was reprinted in 1989 to include notes on some of the new Chinese work. And there is *Iris of China* (1992) by James W. Waddick and Zhao Yu-tang with a more limited range. All those books are of value, but we, in the B.I.S. Species Group, found that we wanted more detail than they offered so that we could distinguish more easily both between the various sections and the various species.

Now we have produced this book in an attempt to assemble detailed descriptions of as many irises as possible between two covers where they will be easily accessible instead of being spread around many botanical publications in distant libraries. We have used the same format as the original Dykes to give full botanical details where possible and also some guidance on cultivation. The Guide is not a replacement for the books mentioned above; they are sure to have details we have not been able to include, but it is intended as a working handbook for a very wide and knowledgeable readership. It is not meant for beginners. Continuing

Preface

sales of the Dover facsimile of the original Dykes suggest that amateurs can make as good use of it as curators and staff of botanical and other gardens. We even hope that this may become a standard reference book.

Inevitably, as soon as this book has gone to the printers, new irises will appear. There will be errors and omissions and we shall be glad to hear of them in the hope that they can be corrected in another edition; at least they can be publicised somewhere.

For some sections we have been fortunate in having either an expert or an enthusiast to do all the work; their names feature in the contents list and the appropriate sections. For other sections there was no particular person available and there the work has been done by the editors and so no author is named.

Many users will regret that we have not included references and type locations. This was partly a matter of space and partly that references are now often misleading because so many irises have reverted to earlier names. As to type locations, these can be inaccurate, but we are more concerned that collectors should explore the wider environs of a plant's possible distribution than concentrate on areas that have been worked out. Many 'species' are in fact natural hybrids and it is important that areas of overlap between true species should be identified if possible. For potential Ph.D. candidates, we would point out that chromosome counts do not exist for a number of well-known species and that a lot more work on karyotypes would be useful.

For gardeners in general there are surprising gaps in knowledge of capsules and seeds. Much more work on the pollination of plants is needed; this emphasises that it is really much more useful to collect seeds rather than plants from the wild in order to enhance the likelihood that cross-pollination will be successful, since many plants are self-sterile. It should also be remembered that real life sizes of plants can vary considerably: the figures we give are a general guide to expectations. A plant with a flower stalk around 10 cm tall is unlikely to produce a form with a stalk of 60 cm, but can often achieve 16 cm.

Brian Mathew's classification has been followed in the main although it is clear that the work of Dr Rodionenko and Prof. Zhao will be of immense importance when it is finally possible to undertake a serious reclassification, not just of the Genus *Iris*, but of all the Iridaceae.

We hope that this *Guide* will be a useful garden book when it comes to identifying irises. All too often a longed-for plant turns out to be a disappointment. Quite apart from actually looking at the flowers, the way a plant grows and the shape of its seeds are guides to its identity and readers should be as observant of the actual plant material in their hands as they are of the growing conditions needed by exotic rarities. And far more thought needs to be given to growing conditions and concepts such as 'tender', 'hardy', 'drainage' and so on.

We are immensely grateful to those writers who have given so much time and effort to sorting out the various sections and series; to the Stanley Smith Horticultural Trust and British Iris Society for funding the colour plates and to

Preface

the B.I.S. Species Group, together with the Reg Usher Legacy, for the line drawings. And, above all, to Cambridge University Press for making this publication possible at all.

Anne Blanco White
Bill Killens
Charles Lyte

Note added in proof

As we go to press, two new and important pieces of information have reached us.

Firstly, recent Japanese work on the chemotaxonomy of *I. setosa* and its forms by Tsukasa Iwashina and Shunji Ootani published in the *Annals of Tsukuba Botanical Garden* in 1995 (14: 35–41) shows clearly that what we once knew as *I. setosa* var., or subsp., *canadensis* is clearly a separate species in its own right. It lacks several chemicals present in other plants of the species while possessing others peculiar to itself.

Secondly, Andrew Wheeler of the American Iris Society has done some comparative work on *I. pseudacorus*, *I. chrysographes* and *I.* 'Holden Clough' which shows conclusively that *I. chrysographes* is not one of the parents of 'Holden Clough'.

Preface to the 2011 Reprint

Irises have been known and used for millennia. Linnaeus described and classified them in the eighteenth century; then in 1913 W.R. Dykes wrote the first modern survey which described the plants in detail, reclassified a number and offered some guidance towards their successful cultivation. It was published by Cambridge University Press and has remained the standard work, but is now totally inadequate. Since that time there has been one reclassification of Dykes by Prof. Ludwig Diels in 1930 and another by Prof. G.H.M. Lawrence in 1953. These were followed by Dr G.I. Rodionenko's *The Genus Iris* (1961); a totally different classification based principally on the characteristics of germinating seeds in 1967. Finally Brian Mathew reverted to updating the Dykes classification again in 1981 and it was his classification on which we based the work though the format is that of the original *Genus Iris*.

We produced this book in an attempt to assemble detailed descriptions of as many irises as possible between two covers where they would be easily accessible instead of being spread around many botanical publications in distant libraries. We have used the same format as the original Dykes to give full botanical details where possible and also some guidance on cultivation. The *Guide* is not a replacement for the books mentioned below; they are sure to have details we have not been able to include, but it is intended as a working handbook for a very wide and knowledgeable readership. It is not meant for beginners. Continuing sales of the Dover facsimile of the original Dykes suggest that amateurs can make as good use of it as curators and staff of botanical and other gardens. We even hope that this may become a standard reference book.

When this book was originally published we predicted that many new irises were still waiting to be found. What we did not anticipate were the analytical advances such as molecular studies which would lead to more precise classifications and reclassifications. In addition there were errors and omissions which it is not possible for us to correct in this reprint, but since Cambridge University Press have generously permitted a new preface we can draw attention to some of the more invidious.

The worst error was the omission of *I. qinghanica* Y.T. Zhao from the Series Tenuifoliae. Then *I. narynensis* was misspelt and a number of page references in the index were inaccurate.

The plant we used to know as *I. setosa* var., or subsp., *hookeri* is now *I. hookeri* in its own right. Since 1997 *I. aschersonii* has been found again in the type location; *I. falcifolia* has grown at Kew for several years and even flowered in 2010. So, too, has *I. staintonii* although its introduction into cultivation was sadly short-lived. *I. phragmitetorum* has finally been abolished and proved to be a form of *I. laevigata*; *I. thompsonii*, after yet further investigations has been conclusively shown to be a colour form of *I. innominata*.

Indeed the Californicae group is now considered to be a complex: each named species

Preface to the 2011 Reprint

is still accepted as a species in its own right, but since they are genetically identical, they actually form one large species. It seems likely that other Sections or Series may be similar.

To Subgenus Iris we can add *I. bicapitata* Colas and *I. acutiloba* ssp. *longipetala* Mathew and Zarrei. DNA researches among the *Hexagonae* have tidied up specific relationships. *I. rutherfordii* J. Mart. Rodr., P. Vargas, Carine and Jury joins the Xiphiums which are being reorganised. *Ii. nezahatiae* Guner & Duman, *pseudocapnoides* Ruksans, and *pskemensis* Ruksans should be added to the Subgenus Scorpiris while *Ii. celicki* Akpulat and K.I. Chr., *kopetdhagense* Rodionenko, *zagrica* Mathew and Zarrei and *reticulata* var. *kurdica* Ruksans join Subgenus Hermodactyloides.

Additionally, *Hermodactylus tuberosus*, *Belamcanda chinensis* and *Pardanthopsis dichotoma* have been restored to the irises proper again. The first is now *I. tuberosa* (Lin.) Salisbury and classed with the Hermodactyloides; the others are now *I. domestica* (Kaempfer) Goldblatt & Mabberley and *I. dichotoma* Pallas. The trouble is that the last two have not been assigned to a specific section or series which makes life difficult for those compiling seed lists. Since Dr. Rodionenko is of the opinion that *I. dichotoma* is one of the oldest forms of iris, perhaps it could be temporarily assigned to a its own Subgenus.

Sadly, for technical reasons, it has not been possible to reproduce the original plates in colour [which are now available for download from www.cambridge.org/9780521206433], but much useful information about these plants can be obtained from the various websites spreading over the internet while books such as *Iris* – Fritz Kohlein (1981), *The Iris* – Brian Mathew (1989), *Iris of China* – James Waddick & Zhao Yu-tang (1992), *The Iris Family* – P. Goldblatt & J.C. Manning (2008) have generous illustrations. It is important to remember that work done in various places growing irises from collected seed has shown that many species previously believed to be of only one colour do in fact have several colour forms. Not only that, much work has been done to show that plants which previously failed to survive in temperate climates can be successfully cultivated there and in these days when conservation of rare forms is so important that work should be continued and expanded.

Many users will regret that we have not included references and type locations. This was partly a matter of space and partly that references are now often misleading because so many irises have reverted to earlier names. As to type locations, these can be inaccurate, and we are more concerned that collectors should explore the wider environs of a plant's possible distribution than concentrate on areas that have been worked out: for instance, *I. tectorum* has recently been found in a location far west of the expected areas. Many 'species' are in fact natural hybrids and it is important that areas of overlap between true species should be identified if possible. Similarly, there is no key to identification. Such keys are not, in practice, easy to use and are available from other sources. For potential Ph.D. candidates, we would point out that chromosome counts do not exist for a number of well-known species and that a lot more work on karyotypes would be useful.

For gardeners in general there are still surprising gaps in knowledge of capsules and seeds. Much more work on the pollination of plants is needed; this emphasises that it is really much more useful to collect seeds rather than plants from the wild in order to enhance the likelihood that cross-pollination will be successful, since many plants are self-sterile or appear to be. Little is known about the necessary conditions for ensuring that pollination will be successful. It should also be remembered that real life sizes of plants can vary considerably: the figures we give are a general guide to expectations. A plant with a flower stalk around 10 cm tall is unlikely to produce a form with a stalk of 60 cm, but can often achieve 16 cm. and a plant which is small in the wild may well double its size in cultivation.

We hope that this *Guide* will be a useful garden book when it comes to identifying irises. All too often a longed-for plant turns out to be a disappointment. Quite apart from actually looking at the flowers, the way a plant grows and the shape of its seeds are guides to its identity and readers should be as observant of the actual plant material in their hands as they are of the growing conditions needed by exotic rarities. Far more thought needs to be given to growing conditions and concepts such as 'tender', 'hardy', 'drainage' and so on.

We are immensely grateful to those writers who have given so much time and effort to sorting out the various sections and series; to the Stanley Smith Horticultural Trust and British Iris Society for funding the colour plates and to the B.I.S. Species Group, together with the Reg Usher Legacy, for the line drawings. And, above all, to Cambridge University Press for making this publication possible at all. Sadly, Bill Killens, who was responsible for the maps and the Subgenus Hermodactyloides has died, but he would have been delighted to see this reprint issued and to know that new plants had been added.

ABW

CL

Illustration Acknowledgements

We are extremely grateful to the following people and institutions for allowing us to use their colour photographs as illustrations in this book. Where possible we have not shown the type, but a different colour form; as it is not always possible to find a flowering plant in the wild, a number have had to be photographed in cultivation.

Stephen Anderton, Britain; 58
Bill Baker, Britain; 44
Anne Blanco White, Britain; 13, 39, 57, 75, 90, 97
Maurice Boussard, France; 22, 40, 73, 86, 98, 105, 111, 122, 126
P.L. Carter, Britain; 100
James Compton, Britain; 88
Aaron Davis, Britain; 87, 89
Vic Dickson-Cohen, Britain; 53, 54, 55, 56
Jack Ellis, Britain; 38, 42
Paul Furse, Britain (1904–1978); 19, 24, 25, 26, 99, 114
Christopher Grey-Wilson, Britain; 27, 28, 29, 45, 49, 50, 51, 78, 80, 81, 91, 102, 108, 120
Jennifer Hewitt, Britain; 68, 74
Clive Innes, Britain; 48
Stephen Jury, Britain; 94
Brian Mathew, Britain; 17, 23, 107, 109, 115, 117, 118, 119
Peter Maynard, Britain; 95, 96, 127
Martyn Rix, Britain; 15, 20, 33, 36
Jan Sacks, USA; 16, 37, 70
Nigel Service, France; 1, 2, 3, 4, 5, 6, 7, 8, 9, 10, 11, 12, 14, 72, 110
Yonatan Shkedi, Israel; 85
Tomas Tamberg, Germany; 69
Bernard Venables, Britain; 71
James Waddick, USA; 32, 34, 76, 77, 79, 82, 83, 84, 92, 128
Royal Botanic Garden, Kew; 18, 21, 30, 35, 41, 47, 52, 93, 101, 103, 104, 112, 113, 116, 119, 120, 125

Illustration Acknowledgements

Royal Botanic Garden, Edinburgh (Ken Grant), Britain; 31
Gothenburg Botanical Garden (Henrick Zetterland), Sweden; 106, 121, 123, 124
Species Iris Group of North America, USA; 46, 60, 61, 62, 63
The Society for Louisiana Irises, USA; 64, 65, 66, 67
The Clematis Nursery, Guernsey (Raymond Evison); 43
Wychwood Carp Farm, Britain (R.J. Henley); 59

The Iris in History

CHARLES LYTE

The iris is among the most ancient and valued of cultivated plants. From the days of the Egyptian Pharaohs, through pagan rites to Christianity it has been accorded a special reverence. The Egyptians believed it to be the eye of heaven and the symbol of eloquence and for this reason it was carved on the brow of the Sphinx. Thutmosis I brought irises home with him in the spoils of war from his expeditions against the Syrians in 1950 BC, and Thutmosis III had irises, probably *I. florentina*, carved on the walls of his temple. Another iris, believed to be an *Oncocyclus*, is engraved on a marble panel in the Temple of the Theban Ammon at Karnak.

Ancient Greek culture almost deified the iris, naming it after Iris, described equally as a nymph and a goddess, who was the messenger of the gods, travelling between the Olympian heights and earth on rainbows which were seen in replica in the multi-coloured falls of the flower. Greek women had a particular devotion to irises and had them planted on their graves in the belief that they would escort their souls safely to paradise.

Pagan Slavs dedicated the plant to the Slavic Jove, Perun, and thus it became known as Perun's Flower.

Like so many plants, trees and festivals that were the objects of pagan adoration, the iris was adopted by Christianity and accorded an affinity with the Virgin Mary as a symbol of purity along with *Lilium candidum*. It appears in many paintings of the Holy Family, such as the *Adoration of the Shepherds* by Hugo van Goes.

It has been of particular importance in French history, starting with the story of Clotile, the wife of the pagan Clovis I of France, being given a shield bearing three golden irises on a blue field by a hermit who had received it from an angel. Not only did it protect him in battle – his previous badge of three black toads had singled him out for special attention from his enemies, although why three golden irises should have been less distinctive is hard to understand – it also led to his conversion to Christianity. There is also the story of the ignorant French knight who could only remember the first two words of the Ave Maria. He retired to a monastery where the monks mocked him for a fool, but when he died an iris grew at the head of his grave and flowered to reveal the words Ave Maria on every petal. The monks dug down and found the roots touching the lips of the foolish knight.

The Fleur de Lis, the iris, became the royal badge of France and, until the Act of Union of 1801 when Britain's claim to French territory was finally laid to rest, it was also part of the British royal arms. During the reigns of the French kings

Philip, Charles VII and Louis XII, their kingdom was known as Lilium and its people as the Liliarts. The Burghers of Ghent were bound by oath not to make war on the 'Lillies'.

During the French Revolution the iris became an object of hatred because of its association with the monarchy; it was proscribed and anyone found wearing the flower went to the guillotine. Napoleon substituted a bee for the iris. Prior to that the revolutionaries had adopted the violet as their floral symbol. As well as the French and the English, the Emperors of Constantinople used the iris as their emblem.

From early times irises were valued in medicine, being credited with both curative and magical properties. Pliny says that in Northern Europe the harvesting of iris rhizomes involved elaborate rituals. Three months before harvesting, the soil around the plants was soaked with mead or honeyed water. A triple circle was drawn around each plant with the point of a sword, and when they were dug from the soil they were elevated as though offering a sacrifice to a deity.

In Austria and Hungary, and throughout the Alps, the iris was sought 'as if it were some precious metal', according to Pliny.

Its use in medicine goes back many centuries. Bancke's *Herbal*, published in 1525, recommends it for aches and pains, cramp, coughs and respiratory complaints, poisoned bites, and for aborting stillborn babies. Gerard prescribed it for bruising. Theophrastus described an ointment for curing scrofula. Both Culpeper and Lyte regarded it as a virtual cure-all, useful in the treatment of problems of the liver, spleen and bladder, gynaecological ailments, and dropsy, and for cleansing sores and ulcers, drawing splinters, easing chilblains, removing freckles, pimples and spots, and easing headaches.

Early settlers in North America learnt from the Native Americans the medicinal virtues of *I. versicolor*, which they used, among other things, as a cathartic and emetic. During the latter half of the nineteenth century this iris was regarded as one of North America's most useful medicinal plants, listed in the pharmacopoeia as an alterative, cathartic, sialogue, vermifuge and diuretic, and prescribed for hepatic, renal and lentitic complaints, and venereal diseases. Dried and ground iris seeds were even used as a coffee substitute in the United States.

Apart from the iris's commercial value as a decorative garden plant and increasingly as an important cut flower crop, Orris Root, or Root of Violets, derived from the rhizomes of *I. florentina*, and sometimes as an alternative from those of *I. germanica* and *I. pallida*, is still produced and harvested for the perfume industry. The sun-dried rhizomes, which are grown in the neighbourhood of Florence and Sienna, and also at the village of San Polo near Parma, develop a lovely fragrance of violets. The majority are sent to Grasse in France, where they are processed.

In the past there was a considerable production of sachets of powdered orris root, known as 'Violet Powder', which were used to scent linen and clothes. It was also mixed with anise for the same purpose. Until relatively recent times the dried rhizomes were fashioned into strings of beads, which were hung round babies' necks for them to chew on to relieve teething pains. At one time twenty million beads were exported annually from Leghorn and Paris. In western North America the peeled roots of *I. missouriensis* were used to relieve toothache. *I. germanica* was employed in the manufacture of toothpaste. Medicinally the iris is a plant that should be treated with caution: it contains a chemical, iridin, which can produce violent reactions in some people.

Few plants have attracted more local and common names. *I. foetidissima* is known as Gladden, Gladwyn and Gladwin, names derived from the Anglo-Saxon *laeded*, Gladens and Gladine. In Devon its seeds are called Adder's Meat or Adder's Berries, being regarded as poisonous. *I. pseudacorus* has a string of aliases: False Acorus, Segg, Cegg, Jacob's Sword, Flaggon, Mecklin, Myrtle Flower, Myrtle Grass, Myrtle Root, Yellow Iris, Fliggers, Daggers, Gladdyn, Levers, Shalder, Dragon Flower and Fleur de Luce. *I. germanica* rejoices as Blue Flag, Garden Flag, Flag Iris, Common German Iris, Blue Flower de Luce and Blue Flower de Lys, while *I. versicolor* is variously known as American Blue Flag, Water Flag, Flag Lily, Snake Lily, Dagger Flower and Dragon Flower.

As well as generating a most varied nomenclature, the iris has inspired poets and painters. Virgil, Milton, Shakespeare, Spenser, Byron and Tennyson have drawn upon this lovely flower for inspiration.

It appears repeatedly in paintings. Ophelia carries them in Millais's painting of her drifting to her death. They are in Gerard David's *Virgin and Child*; the Wilton Diptych; Constable's *Scene on a Navigable River*; Perugino's *Virgin and Child*; Lockey's charming study of *Sir Thomas More and his Family*; and the work of great Dutch flower painters.

Cultivation and the Great Genus

PETER MAYNARD

None has the right to lay his word at the foot of this august race unless he is prepared to say nothing about any other matter.

Reginald Farrer, writing on irises in The English Rock Garden.

Although confined to the northern hemisphere, the geographic and climatic distribution of the members of the iris family is remarkably diverse. They occur at sea level and in alpine meadows at altitudes of 4000 m, or more in favoured places; you can find irises growing in the Arctic tundra or in the semi-tropical parts of Asia, in marshes and in stony deserts. In all there are close to three hundred species and as many more recognised forms and varieties; the hybrids and cultivars are numbered like pebbles on a shore. This versatility means that whatever the situation of a garden there will always be an iris that will be at home. The iris species is very adaptable too, so with a modicum of thought and some simple artificial aids many of them can be accommodated within the bounds of an average suburban plot despite their differing natural habitats. Thus irises from the marshes can be grown in the dry gardens of East Anglia or Texas, while irises from the hot, dry deserts of Syria will flourish in the balmy air of Torquay or Vancouver. Although it is clearly easier to grow the plants that enjoy your own environment, the challenge of the 'ungrowable' is often irresistible. The key point is to know the plant and how it is adapted to its own environment and then to ascertain by enquiry and experiment what range of divergence will be tolerated.

With such a wide range of plants available, the 'Weekly Iris' is a realisable objective, but the whole gamut will need to be grown: herbaceous, bulbous and rhizomatous, hardy and half-hardy types, those suitable for the rock garden or for a pond margin, from tiny tots to giants of 2 m tall or more; even then you will need to work harder in August. However, there is no other genus of plants that can provide flowers in the garden for so many days of the year: when the sweet peas are on the compost heap and the rose bushes are bare sticks, irises are still flowering.

There are several good abecedaries now in print which deal satisfactorily with the iris species and their cultivation, but it would be as well to underline a few points. First and foremost, plants need to be contented if they are to thrive, and this means a healthy root system. Like you and me, plants suffer stress when they are moved and have their roots disturbed. Root action is also adversely affected if the soil is waterlogged, or conversely during drought conditions, and, like us again, when stressed, plants become more susceptible to bacterial or virus infections. A healthy plant is equally liable to attack by aphids and other insects, but is much more likely to survive. The usual soil improvement techniques will help towards an efficient root system with the herbaceous types: incorporation of well composted organic matter will make your soil more friable, better drained, yet moisture-retentive, and increase the general fertility. This is the real magic of muck. Note, however, that your compost should not display old cabbage stalks and orange peel. Mulching with compost will also be beneficial in areas of low spring rainfall, but remember that pogons and bulbous irises require at least a partial summer dormancy and favour soils that are dry and warm in summer, so good drainage is necessary; rotting organic matter around the resting plant may lead to fungal infection and death.

Irises have different requirements for soil pH and, although each is tolerant to a considerable degree, will not perform well if grown in soils outside their natural pH range. As a very broad generalisation the bulbous irises and bearded types prefer a soil pH near to neutral or above, whereas the beardless apogons such as the Californians, and particularly the Ensatas, require to be grown in acidic soils, that is at a pH lower than 7. This, however, is only a guide, so do check the specific habitat conditions for each species. Bearded irises will appreciate a top dressing of crushed limestone or dolomite chalk on marginal soils, but the old shibboleth of lime mortar rubble should now be discarded, primarily because lime mortar is rarely used nowadays. Conversely, the soil pH can be lowered if required by the application of elemental sulphur some months before planting; any top dressing or mulching should then be of ericaceous quality.

Most irises enjoy full sun when healthy, but there are some exceptions such as *Ii. gracilipes* and *verna*, and probably *Ii. sintenesii* and *forrestii*, which prefer the cooler growing conditions in the light shade of open woodlands where the leafy soil will still be moist in early spring. In gardens this means deciduous trees and shrubs, but don't confuse the friendly state of dappled shade with the dark, dry, infertile conditions that pertain under dense conifers or at the foot of a privet hedge.

Do remember that a plant is a sophisticated chemical factory powered by solar radiation. The raw materials are carbon dioxide from the atmosphere and the inorganic chemicals based on nitrogen, potassium, phosphorus, calcium, iron, magnesium, etc., which the plant takes up through its roots in aqueous solution from the soil. The products are plant tissue, the flowers and a catalogue of complex organic molecules. Not for nothing is a chemical factory colloquially known as a 'plant'! The leaves are there to absorb sunlight, so don't cut them down to 'tidy up' after flowering if you want more flowers next year. Cutting down the leaves of pogons is a prevalent fallacy among those who really should know better: remove the dead and damaged leaves certainly, but otherwise leave them alone. The one occasion when it is right to trim back healthy leaves is after transplanting to reduce the stress on the root system, but do leave those of evergreens such as *I. foetidissima* and *I. unguicularis* as long as possible: if their leaves are cut too short it will be some years before they flower again. And if disaster strikes a clump of evergreen irises through frost or drought, leave them alone for another season. They probably have their rhizomes in several layers; shoots will appear from deep down and respond to encouragement. Finally, when you read that some irises prefer poor soil this means low in nitrogenous compounds and organic matter, but high in mineral food from decomposed rock: poor for dahlias, but just right for the Algerian *I. unguicularis*. No living thing will thrive on a starvation diet.

Talking of replanting, the best time is after the flowering season and when the plant is beginning to make new roots for next year. You will need to get a feel for this, but as a guide the appearance of new basal growth is a clear sign that root action has commenced. Every year I am questioned about Californian irises dying; this is most often due to transplanting at the wrong time. After a hot, dry summer, the new root growth is often delayed into the period of autumn rain; the plants will not survive very long without roots. Do take a reasonably large piece of the plant to the new site – about the size of two clenched fists – for it will re-establish faster and flower sooner than one miserable growing point. As a rule, the herbaceous species can be left for many years to their own devices, like any other border plant, but the pogons move outward each year from the original planting site, leaving a patch of tangled, dormant rhizomes in the centre. If you find the cat sleeping comfortably there it is certainly time to replant the active rhizomes in fresh soil. The smaller pogons, which make excellent plants for the rock garden, will need replanting more frequently as they soon exhaust the soil around them and then degenerate. Grown in pots, the smaller species will need top dressing, new soil, or even replanting, annually.

Most iris species produce seed in the garden; for the amateur grower this is the best way of producing vigorous, virus-free new plants. I find that *I. foetidissima*, for example, is showing virus infection, in the form of badly streaked leaves, after two years in my garden. Germination of iris seed can be easy or maddeningly slow and erratic. Note that even with self-fertilised flowers, raising from seed is likely to produce variable offspring and for this reason it is incorrect to give to such seed-raised plants any cultivar name accorded to their parent. Further, even taking divisions from an herbaceous iris grown under a cultivar name can lead to problems unless scrupulous care has been taken to remove all ripe seed capsules. This is a particular difficulty with *I. unguicularis* cultivars as the capsules are formed at ground level and hence are rarely observed. The ripe seeds fall into the leaves and germinate, producing new and different plants. These new plants are usually formed at the outside of the clump, from where gifts are often taken; sadly, the seedlings from the best cultivars are rarely as fine as their parents. I have seen more strange irises named 'Mary Barnard' than would fill a book of registrations.

Those of you wishing to succeed with the more challenging species from the juno (Scorpiris) or Oncocyclus groups, or the reticulate bulbs, will require in most countries some means of keeping them from the summer rain during their dormant period. The traditional method of achieving this is a raised frame with the plants either planted directly into a bed of well-drained soil or in pots plunged in gravel. Both methods have their advantages and drawbacks; a final decision may well be a compromise based on available space and desired flexibility. Go easy on the organics with these plants, but be generous with low-nitrogen fertiliser or liquid feed for good growth and flower production, particularly with pot culture. Keep a careful watch for aphids, too, as all these plants are very susceptible to virus diseases, which are rapidly spread from plant to plant as the aphids bite into the leaves; the resulting infection is debilitating and often fatal. Seedlings are particularly susceptible. If you spray inside a frame, or use a fumigant smoke, ventilate the frame as soon as possible afterwards: in still moist air junos can rot with amazing speed, and one renowned grower lost a large collection this way. The reticulate bulbs are plagued by a fungal infection of the bulb called ink spot, so in addition to dusting the bulbs with a systemic fungicide before planting, do separate the bulbs if they clump up for you, otherwise the Black Death will quickly run through the whole population. There is a possibility that this disease is very liable to strike where a naturally non-alkaline soil needs extra calcium in the form of chalk or limestone.

Reading other articles on irises may already have told you that there exists almost an *embarras de richesses* with the species and you may have been lulled into a false sense of security through the taxonomic division of the genus into just six subgenera and eight sections, thus making the task of grasping the cultivation techniques less tedious. The relief is truly apparent rather than real since on arrival at the Section Limniris, the beardless ones, you are faced with a further division into sixteen series, the plants of which occur in substantially different habitats. It is therefore necessary to repeat once more the golden rule for successful cultivation of any species: read about the PLANT and its environment and determine which of the conditions are paramount for survival and which are optional. Then you will have a good chance of growing your choice of irises wherever you dwell.

Chromosomes and the Genus Iris

DR J.R. ELLIS

The somatic (non-sexual) chromosome number is the most usual cytological information reported for a plant species and is generally derived by the microscopic examination of cells in actively growing root tips. For each species, the chromosome number is usually constant in all tissues of the individual and in all individuals of that species. This constancy is maintained by the precise regularity with which the chromosomes duplicate prior to cell division during somatic development and by the regular halving of the chromosome number in the gametes prior to sexual reproduction.

The chromosome number is represented as $2n = \ldots$ and in the genus *Iris* ranges from a low of $2n = 16$ as found, for instance, in *I. attica*, to a high of $2n = 108$ as occurs in *I. versicolor*. Different species often have different chromosome numbers as is illustrated in the series Laevigatae, where the five species are characterised as follows: *I. ensata* ($2n = 24$), *I. laevigata* ($2n = 32$), *I. pseudacorus* ($2n = 34$), *I. virginica* ($2n = 70$) and *I. versicolor* ($2n = 108$). In other groups such as the series Californicae, all species have a common chromosome number of $2n = 40$ and the various species may be cytologically distinguished by differences in the shapes and sizes of the individual chromosomes.

When the chromosomes from a cell have either been photographed, or drawn accurately, the chromosome complement can be displayed in pairs in a linear sequence of gradually decreasing sizes. Such an arrangement is called a karyotype and a diagrammatic representation of the karyotype is referred to as an idiogram. Karyotypes, or idiograms, of related species can be compared to show the extent of chromosome similarities and differences and, in some instances, evolutionary relationships can be revealed. Using this type of information, Randolph and Mitra (1959a) showed that *I. pumila* had evolved from hybridisation between *I. attica* and *I. pseudopumila* followed by a doubling of the chromosome number.

For most species only the chromosome number is reported. This information is usually found in the species description and, as it is a species characteristic, can be considered as having some diagnostic value. For instance, the Laevigatae cultigen 'Rose Queen' is often listed as either a form of *I. laevigata* or an *I. ensata* × *I. laevigata* hybrid, but with a chromosome number of $2n = 24$ it clearly belongs to the taxon *I. ensata*. Chromosome numbers also have value in confirming or suggesting the parentage of natural or spontaneous hybrids. For example, the population of blue flag irises along a northern region of Lake Windermere could casually be assumed to be *I. versicolor* on the basis of floral morphology, but chromosomally this population is distinctive in having a somatic chromosome number of $2n = 89$ (Ellis 1975). This chromosome number strongly suggests that the population represents a hybrid between *I. virginica* ($2n = 70$) and *I. versicolor* ($2n = 108$) and the high sterility of the plants supports this suggestion. Such hybrids are common in the USA, where the two species overlap and where the hybrids have been given the binomial *I.* × *robusta* (Anderson 1928).

In other instances, chromosome numbers can only be a guide to the identity of the parentage of hybrids as, for instance, in the recently discovered hybrid 'Holden Clough'. This hybrid has a chromosome number of $2n = 37$ and most probably represents a hybrid between the yellow flag *I. pseudacorus* ($2n = 34$) and another iris with $2n = 40$. Possible contenders for the $2n = 40$ parent are *I. chrysographes* and related forms, *I. foetidissima*, or a member of the Californian or spuria groups. To establish the identity of the second parent and to confirm the involvement of *I. pseudacorus* will require the preparation of karyotypes of the different species and the hybrid, although in the meantime it would be possible to resolve the parentage without question by re-synthesising the hybrid from controlled interspecific pollinations. The recent report of a yellow sport in I. 'Holden Clough' (Collins 1986) has been interpreted as a reversion and proof that *I. pseudacorus* was definitely one of the parent species. If this has occurred, the simplest cytological explanation for the apparent reversion would be the elimination of the unknown parental genome, leaving intact the haploid genome of *I. pseudacorus*. In this situation the yellow sport would represent a haploid form of *I. pseudacorus* with $2n = 17$ and, like haploid forms known in other genera, should be characterised by having a smaller growth habit and complete sterility. However, in a recent instance of reversion observed by the author, the reverted spike, although showing morphological characteristics of haploidy, was nevertheless highly male- and female-fertile and yielded progeny indistinguishable from *I. pseudacorus*. In this instance it would seem that chromosome elimination was followed in the germ-line tissues by chromosome doubling to give the high fertility. The extraction of *I. pseudacorus*-like progeny gives confirmation of the involvement of this species in the origin of 'Holden Clough'.

Chromosome numbers are also important when evolutionary relationships between closely allied species are being investigated. One mechanism associated with the evolution of a new species is known as allopolyploidy (amphiploidy) and occurs by the spontaneous doubling of the chromosome number in a sterile interspecific hybrid. Natural hybrids are frequently sterile because of problems associated with the halving of the chromosome number during gamete formation. However, if the chromosome number by chance becomes doubled, gamete production can proceed normally and the hybrid becomes fertile. Such fertile hybrids are usually morphologically distinct from the parental or progenital species and are recognised as distinct species. Well-documented instances in the genus *Iris* are *I. versicolor* ($2n = 108$) from *I. virginica* ($2n = 70$) × *I. setosa* ($2n = 38$) (Anderson

1928) and *I. pumila* ($2n = 32$) from *I. attica* ($2n = 16$) × *I. pseudopumila* ($2n = 16$) (Randolph & Mitra 1959a).

Experimentally, chromosome numbers can be manipulated and in some species they can be doubled with relative ease. Plants with a doubled chromosome number are known as autotetraploids and are referred to as $4x$ types. They are usually characterised by an increase in size and robustness and in some instances such as $4x$ I. sibirica ($2n = 56$) and $4x$ I. *pseudacorus* ($2n = 68$) have been commercialised. In some species autotetraploids may occur naturally so that a species has two reported chromosome numbers, e.g. *I. sintenesii* ($2n = 16, 32$) (Randolph & Mitra 1959a). In other species, an intermediate number of chromosomes may also be present, giving rise to autotriploidy. In such instances the species will have three different chromosome numbers, as occurs for instance in *I. korolkowii*, where chromosome numbers of $2n = 22, 33$ and 44 have been reported (Simonet 1932a). Autotriploidy in this instance has presumably arisen from hybridisation between diploid and tetraploid forms. In *I. danfordiae* the form in commerce has an autotriploid chromosome number ($2n = 3x = 27$) whereas the wild forms are diploid with $2n = 18$. Since autotetraploidy has not been recorded in this species, the triploid form has presumably arisen from an unreduced gamete (i.e. a gamete in which the chromosome number had not been reduced (halved)) in a normal diploid population. All triploid plants are usually highly sterile.

In all instances where autopolyploidy occurs, the autoploids do not contain any new (different) genes to differentiate them from the undoubled forms; taxonomically, they are designated as forms (cytotypes) of the species involved. The significance of autopolyploidy within a species means that a species may have two or more chromosome numbers forming a polyploid series of $2n = 2x, 3x, 4x$, etc., where x is the degree of polyploidy and represents the gametic chromosome number of the diploid forms.

Finally, but on rare occasions, individuals may arise from gametes without fertilisation. Such individuals have half the chromosome number of the parent so that the somatic chromosome number is represented by $2n = x$. These plants are called 'haploids'. They have been found in many genera – particularly those which contain economic members – but in the genus *Iris* the only example known to the author is a 'haploid' form of *I. pseudacorus* with $2n = 17$. This plant is, of course, completely sterile, but can be maintained and increased by vegetative reproduction. Morphologically, like all known haploids in other genera, the *I. pseudacorus* haploid reflects the opposite trends of induced polyploidy in being smaller and less robust than the diploid form.

Identification Guide for Plants as Received

Full identification must depend on flowering and fruiting, but plants are often received in a dormant or cut-down state; sometimes there is no indication at all of the species, or the attribution may be doubtful. Even so, the plant must be grown because it might be very interesting. A brief guide to the identification of the vegetative parts is therefore given here.

All irises will thrive initially in a medium such as leaf mould or coir, with the addition of 25% of coarse grit, which will assist in good rooting and rapid growth. For very rare specimens a preparation of medium-grade charcoal helps to keep compost sweet. During growth, the compost should be kept damp, but not wet, and the situation should provide ample light, but not full sun in high summer. Low-nitrogen fertilisers should be used in moderation. Clean away any damp, dying, hollow roots, but not dry, wiry ones since these may only be waiting for damp conditions to start into growth, as with the Unguicularis group; take care because apparently dead roots can be healthy. If a fan arrives virtually or totally rootless, it is worth dipping it in hormone rooting powder, potting up and placing in an unheated propagation case.

Some irises will appear under more than one of the headings below; when a guess has been made about the identity of a plant, it should be compared with the line drawing for that group, with particular attention to the rhizome and leaf base arrangement where it has been possible to illustrate them.

1. Bulbs

(a) Reticulatas (Subgenus Hermodactyloides) (p.360)

Generally small bulbs with loose outer coats that look like nets when stretched out. Any leaves present are usually notable for their squarish cross section except for *I. kolpakowskiana*.

(b) Xiphiums (Subgenus Xiphium) (p.358)

Bulb size depends on the species, but they all have smooth, tough outer coats. Leaves tend to have a semi-circular cross section.

(c) **Junos (Subgenus Scorpiris) (p.359)**

The bulb is characterised by several long, fairly thick roots, but they may have been broken off. The outer coats are loose and thin. The leaves tend to a semi-circular cross section and when several have developed there is a resemblance to sweet corn. Bulbs will survive when they have lost the storage roots and it is sometimes possible to grow a new bulb from a broken-off root. Great care is needed for most of the species in this group so refer to detailed cultivation notes.

2. Pseudo-bulbous

These plants are not bulbs at all, but their root-base structure may give that impression at first sight.

(a) **Nepalenses (Subgenus Nepalensis) (p.357)**

Several storage roots are joined by a vestigial rhizome and they do look a little like a juno, but totally lack any 'bulb'.

(b) **Series Tenuifoliae (p.353)**

The tough remains of leaf bases build up until they resemble a bulb, but they will feel 'wrong' and it will not be possible to separate them as a clump of bulbs can be divided. These plants come from arid areas; the remains of the dead leaves are usually very sharp and tough to deter predators.

(c) **Series Syriacae (p.355)**

The wet conditions in which these plants grow in spring enables some of the dead leaves to rot, and so the clump looks quite different from the Tenuifoliae, but should be handled with equal care for the same reason. The rhizomes grow almost vertically.

3. Subgenus Iris

(a) **Pogon or bearded irises (Section Iris) (p.334)**

These are the easiest to recognise with the relatively long rhizomes regardless of actual size and frequency of offsets.

(b) Section Psammiris (p.335)

The commonest is *I. humilis,* which can be confused with the smallest pogons, but the rhizomes have a stoloniferous appearance and the 'ruff' of persistent dead leaves is distinctive.

(c) Section Oncocyclus (p.336)

The rhizomes of any year are generally short, compact and lumped close together. Very young leaves tend to be sharply curved away from each other.

(d) Section Regelia (p.337)

Old growing points tend to be one on top of another, almost like corms, with offsets out on stolons.

(e) Section Hexapogon (p.338)

The rhizomes are very lumpy and compact, usually with very thick roots.

(f) Section Pseudoregelia (p.339)

Rhizomes are similar to those of the hexapogons, but the leaf tips are much more rounded. If there is any doubt, it is best to treat the plant as a hexapogon.

4. Evergreens

These plants should never have the leaves cut hard back.

(a) Series Foetidissimae (p.352)

Cut a sliver off the end of a leaf and the quite strong smell is diagnostic.

(b) Evansias (Section Lophiris) (p.340)

Not all of these are evergreen, but the whole 'cane' group is. A basic fan looks rather like a pogon, but there may also be a long or short rhizome-like 'stem' with a fan at the tip, which is characteristic.

(c) Series Californicae (p.346)

Leaves are extremely varied in width; texture is tough, and orange-brown old leaves are usually present in the clump; rhizomes tend to make large clumps fairly quickly and are seldom more than two or three layers deep.

(d) Series Unguiculares (p.356)

Leaves generally narrow, but *I. lazica* has wide glossy ones, resembling *I. foetidissima* without the smell. In an old clump the rhizomes may be six, or more, deep. These plants do not spread fast, so clumps are very compact. If the plant is very dry, place on a damp surface for a few days until root hairs appear, then pot up.

5. Water irises

The outstanding detail here is the curious internal veining of the leaves, which is obvious when they are looked at against the light.

(a) Series Laevigatae (p.348)

Rhizomes have disintegrating remains of old leaf bases and leaves have a heavy mid-vein, except for *I. laevigata* itself, which has lopsided leaf tips: the inner edge is straight, the outer a long curve.

(b) Series Hexagonae (p.349)

Rhizomes usually greenish, heavily segmented and long between the growth points; rather similar to *I. milesii*, but tubular and not conical.

6. Herbaceous

The only common characteristic is that they lose the present year's leaves at the end of the season, but may have young green leaves through the winter. The very small ones are described in a separate section.

(a) Section Lophiris (p.340)

I. milesii and *I. tectorum*. The first has long, green, heavily segmented, cone-shaped rhizomes; treat for slugs, which aim for the flowering shoots. The second has short conical rhizomes; the leaves are 'pleated'; watch for greenfly.

(b) Series Tripetalae (p.344)

The rhizomes are usually thick and the leaves fairly wide and often yellowish when very young.

(c) Series Longipetalae (p.347)

Rhizomes are short in themselves, but can make very large clumps; new season's leaves can be quite tall by winter and much growing is done during the colder months. Narrow leaves.

(d) Series Sibericae (p.345)

Clumps are compact and the tips of new leaves can be found around the edges of the clumps in early autumn; *I. clarkei* is distinguished by a solid stem and *I. wilsonii* by a mid-rib on the leaves. Leaves fairly smooth, clumps compact, small new tips among old leaves in winter.

(e) Series Prismaticae (p.350)

Plants that are all too frequently confused with sibiricas. Clumps are lax to the point of disintegration; differing from the sibiricas, the slender rhizomes are unlikely to be stacked on top of each other, although there may be some interweaving.

(f) Series Spuria (p.351)

All this group have the same characteristics in growth: the rhizome has a bulbous extension beyond the leaf bases; young leaves are very upright and old ones have an odd trick of collapsing at the base after transplanting. In most cases if a leaf is damaged there is a faint smell similar to that of *I. foetidissima*.

(g) Series Ensatae (p.354)

Leaves distinctively and narrowly ribbed across whole width; roots may be extremely long and extensively branched.

7. Miniature plants

These need just as rich a compost as larger ones. Leaves at flowering time will be short, but continue to grow through the summer. Since a miniature pogon is so similar to a big one they are not included here, but the rhizomes are much thicker than those of the groups below.

Identification guide

(a) Section Lophiris (p.340)

Ii. cristata and *lacustris* can be mistaken for tiny pogons, but the rhizomes are proportionately much thinner and longer. *I. gracilipes* tends to form tight clumps and the leaves arch outwards and downwards, quite unlike those of the rest of this group.

(b) Series Chinenses (p.340)

The only likely member of this group to turn up for some time yet is *I. minutoaurea*. Thin, wiry rhizomes make a compact clump. The leaves are very narrow. Nodules may be found on the roots.

(c) Series Vernae (p.342)

Slender, wandering rhizomes with leaf fans which can be mistaken for tiny pogons. The flowering spike does not bear a fan of leaves; this characteristic helps to differentiate the plant from the lophirises. It must have an acid compost.

(d) Series Ruthenicae (p.343)

Forms very variable in appearance: similar to *I. verna*, but with compact clumps.

Subgenus Iris: the Bearded Irises

Section Iris   NIGEL SERVICE

Map 1, Figure 1.

Section Iris, popularly known as the pogon or bearded irises, consists of those species with bearded falls and that produce non-arillate seeds. They tend to have heavy, rather short-branching rhizomes and flowers which usually have standards and falls of different colours, or different shades of one colour, the falls being darker, brighter and more strongly patterned. The leaves are sword-shaped, though often more or less curved, frequently glaucous and usually quite broad for their length. The stems of the shortest species rarely, or never, branch; those of the taller usually do so. The size, colour and texture of the bracts and, in particular, of the spathes which surround and enfold the buds and lower part of the flowers are important features in differentiating between species.

Pogon irises occur from the Atlantic coast of southern Europe eastwards into central Asia, growing commonly in fairly open country and in reasonably dry situations, although this is generalising, perhaps unforgivably, for I have seen *I. lutescens* flowering well in running water in the south of France. But dampness is seasonal; what you can be certain of is that all situations will become dry once the spring inundations are over.

Bearded species are rare as true mountain plants, although some occur at quite high altitudes. The range of a number of species is quite extraordinarily limited; in a couple of cases the range consists of only one known station.

A high degree of sterility, at least under garden conditions, is a curious feature of some species; attempts to effect fertilisation can continually fail. However, pods are formed and viable seeds too on occasion and it is clear that insects, in this case ideally bumblebees (*Bombus* sp.), have an ability we have not to set seed. It seems that a large mass of flowers of similar general attractiveness, be they of one species, or several of close colouring, can set up a situation where the possibilities of fertilisation are enhanced: the pollinating insects seem attracted in a way that concentrates their efforts. A few spikes of *I. germanica* may be wholly sterile; a few score spikes may produce a few pods. To try to rationalise this is not easy, but any observer of such a scene will note what appears to be a complete lack of method in the working of the insects. The same part of a flower may be visited time and again by the same individual, other blooms often being similarly worked over during the process while others again, close by, are ignored. Some breeders recommend that hand pollination should be repeated – preferably using different flowers – at intervals on several days. Mixed pollen may be the key. At the same time, many species are thought to have flowers that are virtually self-sterile; this makes it more important that plants should be increased, with breeding taking second place.

As well as containing the type species of the genus *Iris*, the pogons are notable in providing the material for two of the economic uses of a genus which is otherwise commercially important only to the florist. True, other irises increasingly used in breeding have growing popularity, but they cannot rival the sheer magnificence of the Tall Bearded hybrids in their season, and this is reflected in the market for such plants.

All through the descriptions and comments, the degree of variability is a constantly recurring subject. Some species seem to display this much more than others; these usually have wide distribution and populations isolated from each other. Measurements given in the descriptions are indications of what can be expected, but extremes may easily be found. This does not detract in any way from the value of such measurements, as they have been found to be typical of the species concerned.

Leaf lengths are given as at anthesis. In many cases the leaves further elongate after flowering and may well reach double the length given.

A further complication arises in the matter of stem heights. In some of the shorter species, increasingly long stems are produced as the season progresses, and I have noticed that in cultivation stems tend to be noticeably shorter than in the wild. It seems to be a very common feature in this section that where a form has a well-developed stem the leaves will tend to be less sickle-shaped than those of shorter-stemmed forms.

Variation, indeed, extends to all parts of the plant and even seeds can, and do, depart from the normal.

Species are usually described as evergreen or herbaceous. In fact, a complete state of either is not common and all degrees between exist: partly new growth, partly the lower parts of old leaves still green.

Cultivation

The requirements of this section are basically so similar that a few general words now will save constant repetition later. In their ease of cultivation these species form a sharp contrast with some other sections of subgenus *Iris*.

Planting should be in a well-drained, open situation with maximum exposure to sun. Extremely light soils will not be conducive in most cases to free flowering and may also, I suspect, inhibit increase. In Britain, rhizomes should have their upper parts exposed above the soil, but these species will tend to find their own level as they multiply.

Transplanting is best done soon after flowering, the ideal apparently being to seize the time when the new roots are well into growth, but before they have started to branch. If left too long, the following year's flowering will be affected.

There is no reason I can think of for cutting back the leaves of established, healthy plants except a spirit of unnatural tidiness, but with newly planted rhizomes it cuts down the leverage of the wind and so aids stability until the root system is established and reduces the chance of fragile new roots being broken. Some other method might do as well, though, and so let the leaves do their work, but fans are difficult to stake. When the leaves become limp and the worms begin to drag them into the ground it is as well to cut them back, or pull them away from the base, so as to reduce the risk of rot in the leaves spreading back to the fan base and so to the rhizome. This is especially needed in damp autumn and winter climates as it also reduces slug and snail attacks. Dying foliage can also encourage persistent fungus attacks.

The dwarf species are, on the whole, more demanding. Because of their shallow roots, young plants in particular are prone to frost heaving. A winter mulch will help, but it is advisable to watch them as well. Winter damp is a problem, too, and alpine house culture is favoured by many specialists in the UK.

There is a possibility that further species have still to be discovered. Another dwarf Russian iris as yet unnamed, from what used to be the USSR, has been postulated as a species and, indeed, brought to flowering in Britain by a specialist alpine grower.

I. albertii Regel

Distribution
Foothills of the Tien Shan and Fergana Mountains.

Description
Rhizome stout, compact. **Leaves** about 40 cm tall, 4 cm wide, erect, sub-glaucous, rather obtuse, purple at base, tips often dead at flowering. **Stem** 35–60 cm, 1–2 stem leaves, 2–3 branches, lower not longer than the inflated bract, 2–3-flowered, upper shorter, 1-flowered. Spathe green, membranous, inflated, rounded, scarious at apex, with 2–3 flowers. Ovary rounded; tube about 1.5 cm long, funnelform towards apex. **Flower**: falls 2.5 cm wide, 5 cm long, long-cuneate, bright lilac-violet, haft yellowish-white veined reddish-brown extending out to level with the end of beard, which is pale bluish-white tipped yellow; standards suborbicular, coloured as falls, narrowing sharply to a grooved yellowish haft veined reddish-bronze; style arms broad, pale, about 3 cm long, crests short, serrate, reflexed; stigma entire; filament pale mauve, longer than anther; anther white; pollen cream. **Capsule** about 5 cm long, rounded, tapering gradually at either end. **Seeds** irregular, compressed, rather pyriform to D-shaped, brown, wrinkled. **Flowering** in May. $2n = 24$.

Cultivation
It does best in a warm, sheltered and sunny position, fairly dry in winter. In such conditions it is very hardy and disease-resistant.

Observations
A feature of this species is said to be the abrupt ending of the veining running up from the hafts of the falls onto the base of the blade in a line with the extremity of the beard and not fading into the ground colour of the blade as in other species. This characteristic does seem to be common, though not universal. Flower colour varies from violet to lavender and, rarely, yellow.

f. *erythrocarpa* **Rodionenko**
This is a taller variant with a stem up to 70 cm where the immature capsule takes on a cherry-red coloration.

I. albicans **Lange**

Synonyms
I. alba Savi, *I. florentina* Kunze, *I. florentina* Kohler, *I. florentina* Redouté, *I. florentina* Trat., *I. majoricensis* Barc.

Distribution
Yemen and south Arabia.

Description
Rhizome stout, nodular, pale. **Leaves** evergreen, ensiform, glaucous, about as tall as stem, 1.5–2.5 cm wide. **Stem** 30–45 cm, occasionally taller, rarely with 2 branches confined within green bracts; spathe broad, rounded, green, scarious at apex, about 4 cm long; 1–3 flowered at apex, 1 flowered on branch; pedicel very short; ovary trigonal, 2 cm long; tube 2–2.5 cm. **Flower**: falls about 9 cm long, 4.5 cm wide, obovate–cuneate, white with fine olive-green veins in the haft running up onto the very base of the blade; beard white tipped yellow, but wholly yellow in haft; standards obovate, 9 cm long, 4.5 cm wide, white with a yellowish-green veined haft; style arms white; crests white, acute-tipped, serrate, 1.2 cm; stigma entire; filament white, 1.7 cm long, anther cream, 1.7 cm long, pollen cream. **Capsule** ovate, roundedly triangular with pronounced grooves at the angles and shallower ones on the side. **Seeds** oval, brown, wrinkled. **Flowering** May–June. $2n = 44$.

Cultivation
I. albicans has been in cultivation for a very long time, spread wide by the Moslem custom of planting it on graves. By its survival it has shown its tolerance of quite

a wide range of conditions. Any well-drained, sunny position will suit this handsome garden plant, but the foliage may suffer in any but a mild winter and buds will be lost in a harsh one. It may be that the north European climate is to blame, but it does not generally set seed easily.

Observations
W.R. Dykes (1924) wrote that the leaves are often twisted when fully grown. This may be so, but I do not think too much reliance should be placed on this for identification purposes, as forms of *I. germanica* show the same characteristic.

Subspecies and varieties
The fine violet form cultivated under the name 'Madonna' (*I. madonna* Sprenger) is sometimes reported to be the more common variant in the wild. Conversely, it has elsewhere been suggested that this is a hybrid. In 1992, however, I had part of a Madonna rhizome revert to the white-flowered form and this appears to be purely *I. albicans*.

I. alexeenkoi Grossheim

Distribution
Transcaucasus, Baku Province.

Description
Leaves 11–12 cm long, 0.8–1.2 cm wide, larger than those of *I. pumila*, ensiform, more or less falcate, grey-green. **Stem** very short, sometimes scarcely developed; spathe about 10 cm long, acute, narrow, single-flowered; tube long and slender, much exceeding spathe. **Flower**: falls 4–5 cm long, 1.3 cm wide, narrow, spathulate, oblong, reflexed, violet, beard light violet or yellow; standards slightly longer than falls, about 1.5 cm wide, violet, ovate, narrowing gradually towards the retuse apex and abruptly into the haft; style arms about 3 cm long, whitish with a violet keel; crests about 0.5 cm long, acute, dentate, reflexed; anther light blue, equal to filament. **Capsule** up to 5 cm long, oblong–ovoid, triangular. **Flowering** April. $2n = 32$.

Cultivation
I know nothing of its requirements in cultivation, but Brian Mathew has grown it and persuaded it to flower. He found that plants always became quickly infected with virus.

Observations
Only dark purple and black-violet flowers are otherwise reported. Little additional information is available. The karyotype is distinct, showing certain similarities to

I. pseudopumila, but morphologically *I. alexeenkoi* is undoubtedly similar to a rather large *I. pumila*.

The illustration accompanying Grossheim's description shows a very nodular rhizome and both the standards and falls having rather oblong blades, the falls abruptly obtuse at the apex, the standards somewhat more acute. The only photograph I have seen confirms it as a desirable iris with a very long and slender perianth tube bearing the flower well above the somewhat falcate leaves and with somewhat bowed, connivent standards and falls abruptly reflexed nearly as far out as the tip of the beard.

I. aphylla Linn.

Synonyms
I. benacensis Kerner ex Stapf, *I. biflora* L., *I. biflora* Pallas, *I. biflora* Reich., *I. biflora* Sweet, *I. bifurca* Steven ex Baker, *I. bisflorens* Host, *I. bohemica* Fiek, *I. bohemica* Schmidt, *I. breviscapa* Opiz, *I. clusiana* Tausch, *I. dacica* Beldie, *I. extrafoliacea* Mikan, *I. falcata* Tausch, *I. fieberi* Reich, *I. fieberi* Seidl., *I. furcata* Bot. Mag., *I. furcata* Lindley, *I. gracilis* hort., *I. hungarica* Waldst. & Kit., *I. melzeri* Prodan, *I. nudicaulis* Hooker fil., *I. nudicaulis* Lam., *I. nudicaulis* Reich., *I. polonica* Blocki ex Aschers, *I. reflexa* E. Berg, *I. rigida* Fieber, *I. scariosa* Wild. ex Link, *I. schmidtii* hort. ex Baker, *I. subtriflora* Fieber ex Klatt.

Distribution
Widespread from central and southern Germany to European Russia.

Description
Rhizome compact, stout. **Leaves** 15–30 cm long, 2–3 cm wide, entirely herbaceous, outer sometimes subfalcate, ensiform, glaucescent, often flushed purple at base. **Stem** 6.5–25 cm tall, slender with 1–2, rarely 3, long branches, the lowest usually at or near the base; spathe green, often purple-stained, inflated, rounded, outer longer, apex sometimes scarious, 5–6.5 cm long; 2 flowers at apex, 1 on branches; pedicel 0.5 cm; ovary oblong, hexagonal, 1 cm long; tube green, marked purple, 1.6–2.5 cm long. **Flower**: falls 5–6 cm long, 2–3 cm wide, obovate, often retuse, dark violet, narrowing to a pale, cuneate haft veined darker, beard whitish to pale blue, yellow in haft, rather sparse; standards paler, rather oval, narrowing sharply to a short canaliculate brownish-marked haft; style arms pale, keeled, with crest deltoid, dentate, acute-tipped, 0.6–1 cm long; stigma entire; filament about 1.5 cm long, very pale violet; anther blue-edged, same length; pollen white or bluish. **Capsule** oblong, hexagonal, 6-grooved, 3–6.5 cm long, 1.3–2.3 cm wide. **Seeds** broadly ovoid to pyriform, brown, wrinkled. **Flowering** April–May and may rebloom in autumn. $2n = 48$.

Cultivation
Growing this species is not regarded as difficult, although it may be rather prone to virus infection. Depending on size, the rock garden might be the best place, but drainage is not the problem it is with some of the smaller species.

Observations
As well as the impressive list of synonyms, numerous forms have been given some varietal ranking at one time or another, although none of these has been generally supported. Variation lies mainly in the length of stem and perianth tube and in the dimensions of the leaves, which have been reputed to be as narrow as 0.5 cm. Shorter forms do exist, even very dwarf ones; the only good distinction seems to be the shape of the ovary/capsule. At least trigonal ovaries are not unique in the species to var. *hungarica*.

Subspecies and varieties
Var. *hungarica* Waldst. & Kit. is set apart by some and seems to be generally regarded as distinct to one degree or another by eastern European botanists. It is often taller than typical forms with a stem of 20–30 cm; the ovary is trigonal, elongate, deeply 3-grooved; capsule cylindrical, elongate or obovate with ridges in pairs separated by grooves and a thickened neck at the base.

I. attica Boiss. & Heldr.

Synonyms
I. ochridana Hayek, *I. pumila* L., *I. pumila* Sibthorp & Smith, *I. pumila* subsp. *attica* (Boiss. & Heldr.) Hayek.

Distribution
South Yugoslavia through Greece to north-western Asia Minor.

Description
Rhizome very close-growing. **Leaves** 0.4–0.9 cm wide, evergreen, falcate, slender. **Stem** normally extremely short; spathes green, membranous, acute, sheathing the tube, the inner very insubstantial; 1 flower; ovary 1–2 cm; tube long, 4–6 cm. **Flower**: falls 4.3 cm long, 1.8 cm wide, pale yellow with a browny-purple blotch on the blade, sharply reflexed, beard whitish; standards usually somewhat longer and wider, pale yellow, oblong–ovate, narrowing abruptly to brown haft; style arms whitish with pale yellow crest, narrow acute-pointed; stigma rounded; filament colourless, anther white. **Seeds** rounded–oval to obtusely pyriform, wrinkled. **Flowering** April or May. $2n = 16$.

Cultivation
In Britain, the surest way to grow, flower and keep *I. attica* seems to be in the alpine house where a gritty loam can be concocted to suit it. Grown outside, cold dampness seems to undermine its constitution.

Feeding and quite frequent transplanting will be necessary as, like *I. pumila*, it seems to exhaust its available soil. If it is happy, it is very free-flowering.

Observations
Flower colour varies through shades of light yellow, often with a contrasting blotch of plum or dusky-violet covering much of the fall, to purple with a bright blue beard. The exact status of this attractive dwarf species and its relationship to *I. pumila* has always been a matter for debate. There are quite close morphological similarities to small forms of this latter species and this has in the past led to *I. attica* often being classified as a subspecies; in the *Flora Europea* (Webb & Chater 1978) it is still so treated. Karyotype analysis, however, shows it to be quite distinct (Randolph & Mitra 1959b; Bini Maleci & Maugini 1981) and it is probably one of the species which, perhaps with *I. pseudopumila*, gave rise through hybridisation to *I. pumila*.

Subspecies and varieties
The Turkish forms are distinct in having the outer valve of the spathe often keeled. Previously it was not questioned that *I. attica* was a species with a stem so abbreviated as not to signify in any descriptions: 'more or less stemless' and 'near sessile' being the sort of terms used. But quite recently on the island of Euboea, in a coastal area only separated from Attica by a strait some five miles wide, I found quite an extensive colony where some plants had stems as long as 13 cm. Coupled with this curious feature, some other characteristics of this colony have undergone some enlargement and exceed the limits generally recognised for the species: the leaves can be larger and the perianth tube may reach at least 8 cm in length. The spathes seem to have slightly more substance than is normal. Perhaps the most extraordinary thing, though, was the discovery of a plant in which the flowers had beards on the hafts of the standards.

It is possible that this is the result of some ancient incursion by some other species. The most probable is *I. reichenbachii*, although its range today seems to come no nearer than the Khalkidiki peninsula, the nearest point of which is about 200 km away, although this would not account for the beard. For any sort of answer we will have to await at least a chromosome count.

I. babadagica Rzazade & Goln

Distribution
East Caucasus mountains.

Description
Stem 10–15 cm, 2-flowered. **Flowers** purple-violet. $2n = ?$

Cultivation
Requirements unknown.

Observations
Unfortunately nothing more can be written about this species; it is said to exist with a type location near Mount Babadag and that, with what is set out above, is about all we know. The brief information came from Dr Rodionenko, who mentioned it in the *Iris Year Book* (1967) where he noted that it was approaching *I. furcata*, another species on which information is lacking.

I. clausii O. Schwarz

Distribution
Endemic to the area around Mt Kasbek, Georgia.

Description
Leaves sickle-shaped. **Stem** up to 10 cm in cultivation; 1-flowered. **Flowers** pale green with velvety falls. $2n = ?$

Cultivation
I. clausii is in cultivation in England and has been flowered by Tony Willis under alpine house conditions. Even so, it did not prove easy initially.

Observations
This recently described species is said to be close to *I. pumila* and how distinct it will prove to be remains to be seen. Further hoped-for information had unfortunately not come to hand when the publication deadline was reached.

I. furcata Bieb.

Synonym
I. pumila Linn.

Distribution
Widespread between the Black Sea and the Caspian Sea.

Description
Leaves herbaceous, more slender than those of *I. aphylla*. **Stem** shorter than normal in *I. aphylla* and more slender, branching usually from above the middle.

Flowers smaller and of deeper colour; style crests about 1.2 cm, deltoid, dentate, acute. **Seeds** said to be smaller, reddish-brown, oval, wrinkled, or rather compressed. $2n = 24$.

Cultivation
The only plant I have raised of this species flowered itself to death in the first year, as it made no offsets.

Observations
Mathew clearly regards this species with some reserve, but L.F. Randolph and J. Mitra (1961) found the karyotype distinct from that of *I. aphylla*. *I. furcata* has a diploid chromosome count, but they felt it improbable that this species represented an ancestral form of *I. aphylla*.

I. germanica

Synonyms
I. skamnili ined., *I. spectabilis* Salisb., *I. suaveolens* N. Terr., *I. varbossiana* Maly, *I. violacea* Savi, *I. vulgaris* Pohl.

Distribution
Unknown in the wild.

Description
Rhizome segmented, stout. **Leaves** 30–45 cm long, 2–3.5 cm wide, variably evergreen, ensiform, glaucous. **Stem** strong, 30–60 cm or more with up to 3 branches, lowest spreading and 8–17 cm long; bract keeled, sometimes foliaceous, uppermost near sessile with bract short; spathes green, purple-marked, upper part scarious; usually two-flowered at apex; pedicel short; ovary roundedly trigonal, about 1.6 cm long; tube variably marked purple, about 2.5 cm long. **Flower**: falls 7.5 cm long, 3.8–5.5 cm wide, obovate–cuneate, whitish in haft, veined reddish-brown extending onto base of blade, which is variably bright purple-violet, beard white or pale blue tipped yellow, yellow in haft; standards obovate, usually broader, paler and more blue than falls; style arms very pale lilac with darker keel; crests around 1.3 cm, somewhat dentate, coloured as standards; filament pale purple, about 1.8 cm long, anther white, subequal; pollen white, poorly produced. **Capsule** rarely formed, narrow-oblong triangular, usually concave-sided. **Seeds** few, oval acute-tipped to bluntly pyriform, wrinkled. **Flowering** in mid-April and well into May, the date depending on the form. $2n = 44$.

Cultivation
I. germanica is, in general, very easy to grow, but forms vary in their susceptibility to cold and damp. The form Nepalensis seems particularly prone to the latter; even after quite mild, wet winters there is much rot in my plants.

Observations
It is still not possible to have a clear picture of *I. germanica*; it remains an enigma despite being so widely grown. This is a highly involved complex of more or less closely related irises and, as envisaged by Mathew (1989), whom I have mainly followed here, includes a number of what I suppose I must loosely term 'near-species' at present, which were previously regarded as distinct.

Many problems attend *I. germanica*. For instance, owing to a certain lack of fertility, seed is rarely set and seedlings, when such things have been produced, have in the past been said to come up more like *I. aphylla*. This has not been my experience, however, with the fair number of seedlings that have flowered for me. The refusal of these plants to be pollinated save by insects makes methodical investigation nearly impossible, but a chance seedling for Wessel Marais, in an area where only one form of the species appears to grow, is nearly identical to its parents.

In *Flora of Turkey* (1984), Brian Mathew wrote that in that country he had seen mixed populations including numerous forms of which some were identifiable as named cultivars, some not. What does this indicate? A keen horticultural bent in earlier human populations?

Stems seem capable of being produced, regardless of season, at any mild moment or in any favoured spot. I saw a very fine and full flowering of Vulgaris in Belsize Park, north-west London, at Christmas in 1986.

Subspecies and varieties
Mathew's (1989) way of listing the 'near-species' within the *germanica* complex, but under their species names – while specifying that he did not necessarily regard them as species – indicates a point of view not yet fully arrived at. That his thoughts on these irises continue to develop is shown by the fact that when he wrote about the genus for *Flora of Turkey* (1984), which cannot have been long after the first edition of *The Iris* (1981), he had concluded that *I. junonia*, previously included in the complex, was a good species.

The 'near-species' differ to a material extent from *I. germanica* in the narrow sense and will therefore, if eventually they are included within that species, presumably be given some varietal ranking; possibly, for instance, as subspecies with *cypriana* having a var. *mesopotamica*. This is a taxonomic problem, however, with the possibility that none of them arose in the wild. All have good fertility, having, with one reported exception, higher chromosome counts than *I. germanica*.

I. belouini D. Bois & Cornault

Distribution
Morocco and Andalucia, but not wild.

Description
Leaves 40–58 cm tall, 3.5–4.5 cm wide, ensiform, somewhat glaucous, pale green, widest in upper 1/3. **Stem** 89–137 cm tall, 2–3-branched, these being up to 30 cm long and each bearing 2–3 flowers with 3 flowers at the apex; spathe pale green with a pale-scarious apex, sometimes slightly stained a faint purple and somewhat keeled; tube longer than ovary, 2–2.5 cm. **Flower**: falls about 9.5 cm long, 4.5 cm wide, obovate–cuneate, bright violet, haft whitish veined brown, beard white tipped yellow; standards nearly as long as falls and up to 6 cm wide, suborbicular–obtuse, short hafted, pale violet, style pale, crests somewhat more purple and about 1.8 cm long; filament and anther subequal, about 1.8 cm long. **Capsule** about 6.5 cm long, 2 cm broad. **Seeds** irregular, D-shaped to pyriform, dark red-brown, coarsely wrinkled. $2n = 48$.

Observations
Unique in the section in that it loses its leaves after flowering and becomes quite dormant until the autumn. In this respect it is more like the arillate species.

Originally introduced from Morocco, I have found it at two locations north of Ronda in Spain. Not in cultivation as far as I know.

I. biliottii Foster

Distribution
Asia Minor, but not known as a wild plant.

Description
Leaves about 50 cm tall, 4 cm wide, darkish green, ribbed, erect, broadest above the middle. **Stem** 50–90 cm tall with 2, sometimes 3 branches; spathe narrow, green, not purple-stained, not scarious except at extreme apex, divergent, about 7 cm long; ovary 2.5 cm, rounded trigonal, 6-grooved; tube 2– 2.5 cm, green, faintly marked purple below standards. **Flower**: falls 7–9 cm long, 3–3.5 cm wide, cuneate, somewhat retuse, bright violet with a broad white yellow-edged haft, veined brown, which barely continues onto base of blade; standards 6.5–8 cm long, 4–5 cm wide, obovate, short-hafted, rather bluer than falls, haft pale yellow, marked brownish; style crests obtusely deltoid, violet (composed of darker blue and lilac stripes), 1.0 cm long; filament white, 1.8 cm; anther cream, 1.5 cm. **Capsule** elongated–oval, sometimes very narrow, about 6.9 cm long, 6-sided with a deep groove between each rounded face. **Seeds** brown, rounded, wrinkled. $2n = 44$.

Observations
W.R. Dykes (1924) stated that it was evergreen, but I find it much more herbaceous than forms of *I. germanica*. The distribution of this species is very much wider than previously supposed; I have seen it at Hattusas and in the Taurus mountains near Pozanti, Adana Province (see *I. junonia*). Despite the reported chromosome count, it seems quite fertile.

I. croatica J. & M. Horvath

Distribution
North Croatia.

Description
Leaves up to 60 cm long, 3.8 cm wide, green-glaucous, prominently nerved, sometimes falcate, more or less herbaceous. **Stem** 26–70 cm, often 3-branched starting below the middle; spathe inflated, outer valve keeled, green, upper 1/4 scarious, 3–4 cm long; ovary 3-ribbed, about 1.3 cm long; tube 1.7–2.5 cm long. **Flower**: falls 5–7 cm long, 2.1–3.5 cm wide, obovate narrowing gradually, blade violet, haft pale yellow or white veined purplish, beard pale blue tipped yellow; standards 5.3–7.8 cm long, 2.7–3.7 cm wide, erect, elliptico-spatulate narrowing abruptly to the haft, violet, often paler than the falls, veined dull brownish-yellow at base; style arms cuneate with a darker keel, crests deltoid, about 0.8 cm, finely denticulate; filament much longer than anther. **Capsule** 6-grooved, 3 being deeper, about 4 cm long, 1.5 cm wide. **Seeds** red-brown, globular, short-beaked to bluntly pyriform, wrinkled. $2n = 48$.

Observations
Mathew (1989) regards this as a synonym of *I. germanica*, but besides its possibly natural occurrence in the mountains north of Zagreb, it possesses an interesting feature: if the illustrations accompanying the original description are reasonably accurate, it seems the style crests may be closer to those typical of *I. aphylla* and it is to that species that the Horvaths likened it.

I. cypriana Baker and Foster

Description
Leaves glaucous, narrow, not notably wider in the middle, evergreen. **Stem** 90 cm tall, somewhat weak, 2–3-branched; spathe to 5 cm long, broad, navicular, outer nearly wholly scarious, 3-flowered at apex and sometimes on lowest branch, otherwise 2-flowered; ovary elongatedly oval, unevenly hexagonal with 3 sides wider; tube 2.5–3 cm long. **Flower**: falls 12 cm long, 6.5 cm wide, obovate–cuneate, reddish-lilac with thin darker veins, haft cream-white veined greenish-brown

continued onto blade to a level with the end of beard, beard white tipped orange; standards obovate–unguiculate, lighter and bluer than falls, haft with reddish-brown markings; style crests rather quadrate, divergent, serrate; filament white; anther cream. **Capsule** elongate–ellipsoid. **Seed** pyriform, wrinkled, not compressed. **Flowering** June–July. $2n = 48$.

Observations
The flowers are very large with a diameter of up to 17.5 cm. M. Foster noted it still in flower in mid-July. Introduced from Cyprus, it is not regarded as being known in the wild, but Dr K.H. Rechinger found a similar plant in the mountains of north-west Iraq in what he reported as an undeniably wild situation.

I. mesopotamica Dykes

Distribution
Found widely in the Middle East, but not as a wild plant.

Description
Leaves 45–60 cm long, 5 cm wide green, only slightly glaucous, widest in middle. **Stem** 90–120 cm tall, sturdy, erect, with 2–3 branches; spathe broad, membranous, scarious in the upper 1/3, 3–4-flowered at apex, 2-flowered on branches, occasionally 3-flowered on lowest; ovary roundedly and concavely triangular with a groove down each angle; tube about 1.2–2 cm. **Flower**: falls obovate–cuneate, light violet, more purple in centre of blade, haft whitish with bronzy-purple veins, which continue some way onto the base of the blade, beard white, increasingly yellow-orange towards haft; standards obovate–unguiculate, light violet, paler than falls, the short, pale haft marked bronzy-purple; style arms with a blue-purple keel, crests violet; filament white; anther cream. **Capsule** oblong–trigonal, 5–6.5 cm long. **Seeds** large, pyriform, brown, wrinkled. $2n = 48$.

Observations
The colour described is like the flowers I saw in southern Turkey, but Maurice Boussard found a much darker plant in the Lebanon. Père Mouterde found what he thought to be wild populations in the mountains of northern Syria.

I. trojana Kerner ex Stapf

Distribution
Western Asia Minor.

Description
Leaves long, ensiform, erect, glaucescent, narrow, acute-tipped. **Stem** about 1 m, robust; branching starts low and secondary laterals may be produced from lower

branches; lower subtending bracts wholly herbaceous, but upper becoming partly scarious; spathe scarious in upper part, herbaceous below, purple-stained with up to 3 flowers; ovary short, oblong, rounded–trigonal, 6-grooved; tube slightly longer than ovary, about 2 cm. **Flower**: falls obovate–cuneate, blade purple-violet, haft with whitish edges, yellow veined coppery-purple, beard white tipped yellow becoming wholly yellow in haft; standards elliptic, emarginate, violet, narrowing abruptly to a paler haft veined red-brown; style crests barely divergent, bluish-violet, reflexed; filament white; anther white, subequal or longer. Capsule about 6 cm long, obovate to ellipsoid, roundedly triangular. **Seeds** wedge-shaped, light brown, compressed. $2n = 48$.

Observations
The secondary heads, produced from the lower branches on occasion, are something I have only seen otherwise on *I. kashmiriana* and on one form of *I. germanica* proper.

Another intriguing tetraploid ($2n = 48$) *germanica*, seemingly wild, was apparently collected by J. & M. Horvath and by L.F. Randolph & K.H. Rechinger on separate occasions in south-west Yugoslavia and in north-west Greece, with stems varying between about 40 and 80 cm and flowers reported as being lavender-purple with paler standards. It was tentatively called *I. skamnilii*, but never published. Further, on Euboea, to the east of Greece, I have seen another tall-stemmed (90–110 cm) 1–2-branched iris with standards bluer and not much paler than the purple-violet falls, and with distinctively angled green bracts.

Forms of *I. germanica* proper

These are, with the exception of Amas, all 44-chromosome plants. Several were originally described as species. When brief descriptions do not appear in *The Genus Iris* (Dykes 1913a), I have marked where reference to them will be found. The probability is that the species originated in the natural hybridisation of parents belonging to two species, possibly themselves now extinct, later introduced into cultivation and transported around in migrations and other human movements until their true origins were lost, probably beyond all hope of resolution.

Alba
This is close to being an albino form of Nepalensis with broad falls and a yellow beard. W.R. Dykes noted that it produced 2-flowered branches, a feature almost unknown in this species *sensu stricto*. It often has some bluish markings on it and slightly bearded standards, (Dykes 1924, p. 224).

Amas
More herbaceous than most forms. Stem 56 cm, sturdy; spathes almost wholly scarious by flowering; falls blue-purple, hafts veined golden-brown, beard blue

tipped yellowish-orange; standards orbicular, light blue-purple, often with a few isolated hairs in the haft. Early flowering. $2n = 48$.

Synonyms: I. amas Foster; Macrantha (however, Dr Lee Lenz lists them separately, holding Macrantha to be a taller plant, to 70 cm).

Askabadensis
Stem 56 cm. Falls reddish-violet with a white haft veined ochre becoming light golden-brown on lower blade, beard white tipped bright yellow; standards obovate, lavender with yellowy-white hafts marked brown. This is the lightest-flowered form and very distinctive. $2n = 48$.

Australis
Falls violet-purple, more intense towards the centre, lower third dirty white veined violet-purple, beard white tipped pale gold; standards the same colour as falls and slightly longer, rounded–obovate or obovate–elliptic, yellowish towards the edge at base and whitish in haft.

Synonyms: I. australis Tod., Atroviolacea.
W.R. Dykes saw the possibility that this was in fact *I. kochii*, but A. Todaro's description of the colour of the veins on a white ground 'evident at the base of the falls' seems to rule this possibility out.

Caerulea
Only a little darker and rather bluer than Askabadensis, with the veining at the base of the falls rather browner and more diffuse. This plant came to me from the collection of the Jardin des Plantes in Paris.

Cephalonian form
Flowers pinkish-purple (Dykes 1915c).

Cretan
Stem about 90 cm, sturdy, free-flowering; falls a dark rich violet-purple; standards bright-violet blue (Dykes 1924).

Deflexa
Falls dark purple, obovate-cuneate, reflexing half-way down; standards obovate–unguiculate, bright lilac (Knowles & Westcott 1838).

Synonym: I. deflexa Knowles & Westcott.
An Illustrated Dictionary of Gardening (Nicholson & Upcott Gill 1889), equates this with Nepalensis, Dykes (1913a) with Kharput; what I see most in the illustration is the extraordinarily undulate stem, not unlike one French form.

However, Deflexa was sent from northern India and required greenhouse protection.

Florentina
Spathe very membranous by flowering and soon becoming wholly scarious; a semi-albino, both standards and falls variably marked with faint blue; the hafts of the standards are always lightly bearded.

Synonyms: I. florentina L., *I. pallida* Kohler.
W.R. Dykes (1915a) mentioned a dark black-purple form; later, in America, C. Mahan observed that part of his plant had mutated – or possibly reverted – to produce a fine violet sport (Mahan 1988), so it seems there may be more than one clone involved.

Fontarabie
A short form with comparatively large flowers, standards blue-violet, beard a deeper yellow than that of Vulgaris.

This was one of W.R. Dykes's favourites, but it does not seem still to be in cultivation and the description is hardly detailed enough to enable it to be identified. Perhaps some fuller description appears in the French literature.

Gypsea
Pale pearly white (Nicholson & Upcott Gill 1900).

Istria
White-flowered with greenish veins on the haft and a white beard. Larger-flowered than Florentina. W.R. Dykes compared it to Nepalensis, but it is not the albino which he named Alba (Dykes 1924, p. 216).

I have never seen it, but J. Flintoff of Seattle has a plant that answers to this description.

Kharput
Young leaves edged red. Stem about 60 cm. Falls deep red-purple with a bluer margin, haft white veined reddish brown-purple, beard white tipped yellow; standards light violet with a purple infusion or sheen, haft whitish, yellow towards edge.

Synonym: Asiatica.

Metkevic form
Leaves 90 cm long, over 5 cm wide. The albino of the Mostar form (Dykes 1913b).

Mostar form
Flowers reddish purple (Dykes 1913b).

Nepalensis
Stem 40–50 cm. Flowers dark reddish-purple, falls much blacker in tone, haft becoming pale purple on base of blade veined brown-purple, beard white tipped yellow; standards with a very pale yellow-fawn haft marked purple-brown.

Synonym: *I. nepalensis* Wallich; Atropurpurea.
This form has some reputation for not being fully hardy in that the developing buds can easily be damaged by frost.

Sivas
Leaves weak, yellow-green, narrow. Falls dark indigo-purple, haft palest blue veined with faint brown, beard only slightly yellow-tipped on blade, yellow in haft; standards a very blue violet, often with a few hairs on the hafts; style crests divergent.

Veglia
A light blue-purple self (Dykes 1924, p. 184).

Vulgaris
Falls a reddish violet, beard almost white, slight blue tone, becoming more yellow in the haft; standards blue-violet.

Said by W.R. Dykes to be the commonest form in England and also, apparently, in the USA.

There are other forms without names, probably numerous others.

I. glaucescens Bunge ex Ledeb.

Synonyms
I. biflora Falk, *I. eulefeldii* Regel, *I. scariosa* Willd. ex Link.

Distribution
Widespread from north of Aral Sea to east of Lake Balkash.

Description
Rhizomes thick, yellowish. **Leaves** 12.8–20.6 cm long, 6.9 cm wide, ensiform, falcate, glaucous. **Stem** 7.7–17.8 cm long, simple; spathe membranous, acuminate, sheathing the tube; tube purple, 2 cm. **Flower**: falls spathulate, entire, blue-violet; standards erect, emarginate, coloured as falls with a purple, yellow-marked haft; style crests narrowly acute, serrate at apex, reflexed; pollen blue. $2n = ?$

Cultivation
Not much can be suggested. R.I. Lynch (1904) found it shy-flowering and said that it required the treatment of an Oncocyclus, but was less difficult to grow. This, of course, referred to *I. eulefeldii*.

Observations

This species has only quite recently been re-established (Shevchenko 1979) after being included for a long time in *I. scariosa*. Most of the literature is in Russian so it is not easy to work out the reasoning.

From herbarium material, the more easterly plants certainly appear to be slighter, but a more robust iris was described by Regel under the name *I. eulefeldii* from within the *I. glaucescens* range; herbarium specimens thus labelled seem as large as *I. scariosa*. If my understanding is correct, it would perhaps be right to give *I. eulefeldii* the varietal status sometimes allowed to it under *I. scariosa*.

W.R. Dykes's description (see observations under *I. scariosa*) does not seem to equate very closely either with Ledebour's original or with the only other brief information we have to go on. However, *I. eulefeldii* is reported to come from the Zailiyskiy Alatau, a range of mountains south of Alma Ata, so Foster's material might equally refer to that or to an intermediate form for, from Dykes's remarks, he also had access to Foster's notes on *I. eulefeldii*.

It seems worth giving an outline of W.R. Dykes's *I. scariosa* description as it may show the degree of variation within *I. glaucescens*: **Rhizome** somewhat slender. **Leaves** 15.2–30.5 cm long, 1.2–1.9 cm wide, erect, or only slightly falcate, very glaucous. **Stem** 5–15.2 cm long, usually 2-flowered; spathe 5–6.3 cm, membranous, scarious; tube 2.5–3.8 cm, brown-purple. **Flower** falls 4.3 cm long, 1.6 cm wide, purple with a yellow haft, blade much reflexed, obovate–cuneate, beard white tipped purple; standards about 3.7 cm long, 1.3 cm wide, obovate, unguiculate, red-purple, darker in haft; styles with small lanceolate deltoid crests; pollen white. **Capsule** trigonal, pointed. **Seeds** pyriform, brown, wrinkled.

He further remarks on the curious red-purple of the flowers, veined brownish. R.I. Lynch (1904), making no mention of either *I. glaucescens* or *I. scariosa*, gives a description of *I. eulefeldii* which, sadly, we cannot take as typical of our species. Obviously a lot more needs to be known about all this.

I. griffithii Baker

Distribution

Uncertain, but probably limited to north-east Afghanistan, E. Nuristan (Kafiristan).

Description

Rhizome stout; leaves of non-flowering fan much wider (if from same plant), inner erect, outer somewhat falcate, 15–30 cm tall, 2.7 cm wide. **Stem** simple, normally probably 8–20 cm with one short stem leaf below the middle; spathe firm-textured, probably green, narrow, divergent, sometimes somewhat separated, the inner somewhat shorter and less substantial; tube long, 2–5 cm. **Flower** probably purple, although that on one herbarium sheet seems much paler; falls about 5 cm long, beard probably white, dense; standards about 4 cm; style arms around 3 cm; anther 0.8 cm. $2n = ?$

Observations

There is only herbarium material to go on and this has not been added to since W.R. Dykes's time. All colour has faded and there is only one sterile leaf-fan. One stem measures an extreme 39 cm and bears two stem-leaves, the upper above the middle point.

I. imbricata Lindley

Synonyms
I. flavescens Sweet, *I. obtusifolia* Baker, *I. sulphurea* Koch, *I. talischii* Foster ex Spreng.

Distribution
North Persia, Transcaucasia.

Description
Rhizome stout, compact. **Leaves** broad, ensiform, pale green-glaucescent, outermost often very obtuse-tipped, shorter than stem, 2–3 cm wide. **Stem** 30–50 cm tall with one stem-leaf and 2–4 branches contained within long, inflated, pale green bracts; spathe inflated, light green, membranous, the outer sharply keeled, 4–6 cm long; pedicel very short; ovary rounded, 6-grooved; tube about 2.5 cm long, green mottled brown-purple. **Flower**: falls 5–6.5 cm long, 2.5–3.7 cm wide, obovate–cuneate, pale yellow veined brown-purple in haft, barely extending onto blade, often sharply reflexed with a whitish, yellow-tipped beard; standards 5–6.2 cm long, 3–3.8 cm wide, orbicular–oblong, coloured as fall, tapering sharply to a short, reddish marked haft; style arms about 3.5 cm, crests subquadrate, serate, about 1.2 cm; anther white, usually longer than filament, 1.6 cm; pollen cream. **Capsule** long, oval, roundedly hexagonal. **Seeds** large, pyriform, dark brown, wrinkled, compressed. $2n = 24$.

Cultivation
A good garden plant; it is perfectly hardy, but in damp areas a bulb-frame is recommended. In fact the finest display of this species I have seen, and that certainly not in a damp area, was in Jim Bingley's raised and protected Aril bed at Flatford Mill.

Observations
A purple-flowered form is often mentioned but its existence has never been verified and must be doubted. It is said, however, that if insufficient lime is present in the soil the flower may be blotched purple and this may be behind such reports. Alternatively, they may be due to confusion with *I. albertii*.

I. junonia Schott & Kotschy

Distribution
Given as endemic in the Cilician Taurus.

An unsatisfactory situation prevails with this species at present and makes it impossible to give a description without some examination first. The original description is so unspecific that it could apply to almost any iris you cared to present: W.R. Dykes described a tall, 4-branched, blue-purple-flowered plant with paler standards and leaves much shorter than the stem. Recently the name has been taken to apply to a smaller iris, usually with a stem below 30 cm and 2–3 branches; the leaves ordinarily exceed the stem.

First a brief outline of W.R. Dykes' description (1913a, p.174), of the iris then known as *I. junonia*: **Leaves** short in comparison to stem, 30–36 cm long, 4.5 cm wide. **Stem** 51–61 cm tall, 4-branched, the lowest branch 9 cm long, the uppermost very short; spathe about 4 cm long, 2-flowered, pale green, scarious in upper part; ovary trigonal with concave sides; tube 2.5 cm long, green. **Flower**: falls 9 cm long, 4.5 cm wide, obovate–cuneate, haft white veined yellow-brown, blade light blue-purple suffused redder just beyond end of beard, beard white tipped orange; standards obovate–unguiculate, pale lavender blue, haft spotted redbrown on white; styles with large triangular crests; filaments, anthers and pollen all cream. **Capsule** 6.5 cm, trigonal with slightly hollow sides, grooved at angles. **Seeds** large, pyriform, dark-brown, wrinkled.

Notwithstanding some distinctions and the fact that both were included in Dykes (1913a), I very much wonder if this description is not of a form of the iris named *I. billiottii* Foster, 33 years after the original description of *I. junonia* in 1854, which I found in several places around Pozanti, just north of the Cilician Gates, though not previously recorded in the Taurus Mountains. This feeling is compounded by the inclusion of *I. junonia* under the umbrella of the *germanica* complex with which the second candidate – described below – has nothing in common.

The iris at present considered to be *I. junonia* is a species appearing much closer to *I. purpureobractea* than to *I. germanica*.

Description
Rhizome short, nodose. **Leaves** 20–30 cm long, 2.5–4.5 cm wide, outer falcate, somewhat glaucous. **Stem** about 19–35 cm tall with up to 2 or more very short branches each with one flower enclosed in wholly green bracts, 2-flowered at apex; spathe 5 cm long, inflated, barely keeled, green, lightly stained at margin, extreme apex membranous; pedicel very short, ovary about 1 cm long; tube about 1.2 cm long, green, finely speckled. **Flower**: falls about 5.5 cm long, 3.3 cm wide, blade light blue in centre, paling towards the edges, sharply reflexed with grey veins running up from the short, white brown-veined haft, beard white tipped yellow; standards about 5.7 cm long, 3.7 cm wide, pale blue, yellowed at base,

obovate, retuse, narrowing abruptly to the short haft; style arms nearly white, crest 1.1 cm, pale blue, rather narrow and acute; filament 1.7 cm long, white; anther 1.3 cm long, white; pollen white. $2n = 48$.

Flower colour varies from shades of yellow and whitish to a rather deeper blue-violet and the beard may be tipped orange.

In so far as it tells us anything, the original description could be said to lean somewhat towards the large species, but no more than that. It would therefore seem prudent to acquiesce to the *status quo* and accept this delightful small species as *I. junonia*, since it seems unlikely the matter can be resolved unless another iris is found in the Taurus.

I. kashmiriana Baker

Distribution
Endemic to Kashmir.

Description
Rhizome stout. **Leaves** around 60 cm long, 3.5–4.5 cm wide, somewhat glaucous, ribbed, broad. **Stem** thick, 50–70 cm tall, occasionally taller with 1–2 branches from long green bracts often exceeding the branch, the lowest from above the middle, and often 1–2 lower stem leaves; spathe 7–10 cm, green, slightly inflated, long-acute, the outer keeled, 3-flowered at apex, 2–3 on branches; pedicel usually under 1.0 cm; ovary 1.7–2 cm long, unequally 6-grooved; perianth tube green, 2.5 cm or more. **Flower**: falls 7.2–8.8 cm long, 3–3.9 cm wide, obovate–cuneate, spreading, white often with some blue marking and yellow-green veining extending from haft roughly to end of beard, which is white, tipped yellow, continuing white at base in haft; standards 6.9–8 cm long, 3.2–4 cm wide, obovate to oblong–elliptic, shorter and broader, but coloured as falls, greenish-veined, haft often sparsely bearded; style arms about 5 cm, crests large, triangular, acute, barely dentate, 1.8 cm; filament cream, 1.7–2 cm; anther cream 1.4–1.7 cm; pollen cream. **Seeds** globular, dark red-brown, wrinkled. $2n = 44, 48$.

Cultivation
In cultivation, *I. kashmiriana* has a reputation of wasting away after a good initial flowering; it is certainly not in general cultivation. Other reports, however, speak of its becoming well established in gardens. It may well be a matter of damp winters and lack of sun again. Flowering probably occurs in the second half of May.

Observations
I. kashmiriana is the albino form of a rather less common *I. kashmiriana* purpurea, which differs from it in various ways. Commonly in Kashmir this is granted the honorific noun 'var.' although it does not ever seem to have been described as

such; perhaps this ought to be so as it is quite distinct from the type in more than colour. Purpurea is taller, with a stem up to 1 m and more and with often 3 or 4 branches, possibly even more, and can produce secondary flowers, even secondary branches, of some length. The greatest number of flowers I counted was 14. It has a shorter perianth tube, rarely reaching 2 cm, and the flowers are larger: falls 8.5–10.9cm long, 3.4–5.2 cm wide and rather less flared, standards 6.9–9.0 cm long and 3.5–5.5 cm wide; a flower diameter of up to 11.5 cm.

Most of these features could perhaps be set down to the extra vigour of a non-albino, but not the significantly shorter perianth tube, nor the extraordinarily irregular pyriform to oval seeds it produces. Any great degree of colour variation is rare: I saw only one plant with strikingly darker flowers and one much paler.

I. kashmiriana presents something of a mystery. With the exception of *I. glaucescens* much further north, it is the most easterly of the species in this section and is exceptionally isolated from the other tall, or even quite tall, branching ones. Apart from obvious escapes, it seems to occur only in Kashmir and even there appears to grow only in connection with human settlement, usually graveyards. Most sites are in the Vale of Kashmir, the bed of a not so very ancient lake, geologically speaking. After draining, the vegetation, apart from what was round its rim, must have migrated or been introduced by people. Kashmir is surrounded by lofty mountains with alpine flora; the only migration route for plants is up from the Indian plains. Whether it is natural there, but extinct now in the wild, or whether it was introduced by humans and from where cannot really even be guessed at.

Even in Kashmir it forms few seeds, but I do not know how much significance should be attached to this; such irises really need bumblebees for fertilisation and such bees are not common there.

I. kochii Kerner ex Stapf

Distribution
Around Rovigno and Trieste. Questionably wild.

Description
Rhizome large. **Leaves** 30–40 cm long, near-herbaceous, ensiform, erect or slightly curved, glaucescent. **Stem** 30–45 cm, 1–2 branches, the upper very short; spathe lanceolate–acute, purple-stained around margin, outer green, scarious above, inner nearly wholly scarious; pedicel short; ovary oval–cylindrical, 6-grooved; tube about 2.5 cm long. **Flower**: falls about 8 cm long, 5 cm wide, obovate–cuneate, a fine purple-violet with a whitish haft with brown veins running out onto only the very bottom of the blade on a light violet ground, beard white tipped yellow; standards about 8 cm long, 5.5 cm wide, obovate, rounded, coloured only very slightly paler, narrowing abruptly to yellow haft with reddish markings; style arms with violet keel, crests slightly bluer than standards, rather

broad, quadrate, coarsely dentate; filament white or pale violet; anther white. **Capsule** oblong, trigonal, grooved at the angles with an obscure ridge down each face. **Seeds** elongated, pyriform, brown, wrinkled. **Flowering** late April and May. $2n = 44$.

Cultivation
I. kochii seems to present few problems and its shortish stature and tidy habit, as well as the striking dark purple-violet of the flowers, make it worth while to grow.

Observations
This is not unlike yet another form of *I. germanica* and shares with that species the characteristic of an apparent low degree of fertility. The theory is that it may have originated as a natural hybrid of *I. germanica* and *I. pallida* subsp. *cengialti*.

All plants I have seen are very uniform in colour and patterning. It may be distinguished from the forms of *I. germanica* most similar to it by its lack of clearly visible patterning at the base of the falls; what there is is notably brown in colour, but rather obscure and quickly lost. Direct comparison will reveal its bluer tone and its smaller size.

I. lutescens Lam.

Synonyms
I. benacensis Kerner ex Stapf, *I. burnatii* Baker, *I. chameiris* Bert., *I. erratica* Tod., *I. gracilis* hort., *I. italica* Parl., *I. neglecta* Parl., *I. olbiensis* Henon, *I. pumila* Linn. (in part), *I. pumila* Savi, *I. pumila* Vill., *I. statellae* (Tod) Baker, *I. virescens* Delarb.

Distribution
East Spain, along French Mediterranean, north-west Italy.

Description
Rhizome compact, branches crowded. **Leaves** ensiform, subglaucous, 6–20 cm long. **Stem** 1–30 cm, simple, spathe insubstantial, uninflated, green or scarious above, more or less divergent in upper 1/3, 1- or rarely 2-flowered; tube 1.75–5 cm. **Flower**: falls obovate–cuneate, often retuse, haft with diffuse veins on a pale ground, beard yellow; standards elliptic–oblong, usually wider than falls; style crests triangular, acute-tipped, dentate; stigma entire; filament white or palest purple; anther cream or pale purple; pollen cream. **Capsule** roughly 4–6.5 cm long, oval or oblong, rounded–trigonal, 6-grooved, variable. **Seed** red-brown, pyriform to irregular, sometimes somewhat compressed. **Flowering** March and April. $2n = 40$.

Colour varies widely: purple, violet, bright to pale yellow often with greenish veining or greyish suffusion, yellow blotched purple, white often tinged blue on falls. Other more exotic colours may be found in limited areas: subtle blends which defy description.

Cultivation
I. lutescens is reputed not to be entirely hardy, but I have not found it to suffer in extreme winters with around 34 °F of frost; not even flowering seemed impaired, so any lack of hardiness must be due to some combination of factors. Very good drainage is obviously necessary as is maximum sun. Also, replanting every two or three years must be done without fail.

Observations
I. lutescens has been established (Webb & Chater 1978; Maugini & Bini Maleci 1981) as the correct name not only for the previously well known *I. chameiris* Bert., but also for the often postulated species *I. italica* Amb. and *I. olbiensis* Henon. The fact that these names have been variably supported for so long as separate entities is indicative of the wide diversity of the species and, indeed, some recent authorities would go directly to the other extreme, recognising the two species *I. chameiris* and *I. lutescens*, the former having a var. *italica* and the latter divided into two subspecies: subsp. *lutescens* and subsp. *olbiensis*.

My own experience indicates that there seems to be no geographical basis in the distinctions which gave rise to these epithets and no large concentrations of them either. They are spread through the populations in a haphazard manner. Variation extends to all aspects of the plant, not least the shape of the floral parts.

I. lutescens is very rarely 2-flowered, although this may occur more frequently towards the east end of its range. Branching is also recorded, but it is so uncommon as to be of no real significance.

My own researches have convinced me of one thing: such maxims as 'the stem always exceeds the tube', 'tube not more than twice as long as the ovary', 'standards shorter than falls', 'tube approximately 1" long', 'spathe divergent, revealing ovary', 'visible stem always produced', just do not stand up when wild populations are examined. Some are usually so, some are frequently so, some are almost never so. They are the sort of inaccurate rule-of-thumb precept which we can well do without.

There have been reports of occurrences further south in Italy. It is hard to know how much to credit these, but clearly they cannot be ignored. However, reports of *I. pumila* in France and Spain must refer to short-stemmed forms of *I. lutescens*.

I. marsica Ricci & Colasante

Distribution
Central Apennines, endemic.

Description
Rhizome nodular, compact. **Leaves** about 50 cm long, 3.5 cm wide, herbaceous, glaucescent, falcate, acute, broadest in the middle. **Stem** 30–40 cm, rarely to 65 cm, usually with a short sheathing stem-leaf and 1 or 2 branches, the lower subequal to the green, slightly inflated, keeled, purple-marked bract, the upper shorter; spathes variable to 6.5 cm long, often exceeding perianth tube, slightly inflated, rounded, acute-tipped, green, purple-marked; pedicel short; ovary elongate–oblong, bluntly trigonal; tube 2–3.5 cm, green with violet stripes. **Flower**: falls 6–8 cm long, 2–4 cm wide, obovate, narrowing gradually to the whitish haft veined purple with a white beard often tipped yellow; standards 5.8–9.5 cm long, 2.3–4.5 cm wide, violet, elliptic with a narrow haft about 1/2 as long as blade; style arms whitish with a violet keel, crests acute, dentate; filament blue or white, anther white, subequal; pollen cream. **Capsule** oblong–ovate with 6 grooves, three more accentuated. **Seed** oval, somewhat beaked, reddish brown, finely wrinkled. **Flowering** early May. $2n = 40$.

Cultivation
This should not prove difficult. Coming from a mountain area, *I. marsica* is perfectly hardy. Good drainage in full sun will probably satisfy its needs.

Observations
I. marsica is recorded from only a very limited number of stations, all but two of them lying within the area of the Abruzzi National Park, east of Rome. Its appearance is similar to *I. germanica*, but the leaves are more curved and greener and the stem is less glaucous. The plant is almost wholly herbaceous.

I. pallida Lam.

subsp. *pallida*

Synonyms
I. dalmatica hort., *I. fulgida* E. Berg, *I. germanica* Sibthorp, *I. hortensis* Tausch, *I. illyrica* Tomasini, *I. mandraliscae* Tod., *I. marchesettii* Pamp., *I. odoratissima* Jacq., *I. pallido-caerulea* Pers., *I. plicata* Lam., *I. sicula* Tod.

Distribution
North Italy; Slovenia; Croatia; Bosnia.

Description
Rhizome massive, ridged. **Leaves** 45 cm or more long, variably herbaceous, broad, glaucous. **Stem** 40–110 cm, or more, with usually 1 stem-leaf and several short branches from whitish, scarious bracts; spathe pallid, scarious; ovary short, 6-grooved; tube 0.8–1.2 cm long. **Flower**: falls around 8.3 cm long, 5.3 cm wide,

rounded–obovate–cuneate, pale blue-violet haft, pale, brownish veining, beard yellow, becoming white tipped yellow on blade; standards 8 cm long, 5 cm wide, broad, obovate, coloured as fall, haft veined red-brown; style arms broad, 2–2.4 cm long, crests rounded, dentate, about 1 cm long; filament pale violet, 2 cm; anther cream, 1.5 cm; pollen cream. **Capsule** oblong, trigonal to hexagonal, 6-grooved, 5 cm long. **Seeds** large, wrinkled and very irregular, compressed, somewhat cubical, greyish-brown, dark red-brown or light golden-brown. **Flowering** May. $2n = 24$.

Cultivation
A fine and striking garden plant, very easy to grow in any reasonable situation and requiring only to be divided when growth becomes congested. The two variegated forms, though not as free-flowering, are both handsome.

Observations
It seems that variation can be much wider than is indicated by those plants in cultivation and that not even the range of colour variants is represented. Dykes (1913b) wrote of the wide variation he found in one small area near Ragusa (Dubrovnik): deep violet, blue-purple, through pink and reddish-purple shades to pale mauve and white. I have one form, rather dark-flowered, collected by Martyn Rix, which can have four branches and up to thirteen flowers, four at the apex. The appearance of the seed pods varies markedly from that described.

Two decorative and popular variegated forms exist, but whether these originated in the wild is not clear. The distinction lies in the colour of the variegated part of the leaves: in one case this is a creamy white, in the other a soft yellow. This condition is always an unstable one and some variation in the amount and nature of the leaf markings will be found.

I. pallida is the basis of the orris industry in Tuscany, apparently only this species has the necessary quality of rhizome to give the fine extract required (Sani 1963). In other countries where orris is produced, a wider range of species is apparently employed. Bini Maleci & Maugini (1977) have shown that the majority of these Tuscan *I. pallida* are male-sterile with a distinct karyotype, a shorter stem and paler-coloured flowers appearing rather earlier than the full-fertile minority.

subsp. *cengialtii* (Ambr.) Foster

Synonyms
I. cengialtii Ambr., *I. portae* Foster.

Distribution
North-east Italy and Slovenia.

Description
Leaves 15–45 cm tall, 2–3 cm wide, fully herbaceous, green, barely glaucescent, ensiform with the outer often falcate. **Stem** 30–45 cm tall, slender, branched; spathes brown-scarious; tube 0.6–0.9 cm. **Flower** violet-blue with standards slightly paler; falls 7.7 cm long, 5.2 cm wide; standards 7 cm long, 4.5 cm wide; style arms 2 cm long; crests acute-tipped, about 1.4 cm long. **Seeds** rounded to pyriform. $2n = 24$.

Observations
Colour variation is not great, but variation in other characteristics is so wide that var. *loppio* Foster cannot be regarded as in any way distinct. Intergrades with subsp. *pallida* apparently occur: I have a form collected by P.J. Christian with spathes nearly as pale as those of subsp. *pallida*.

Plants grown under the name var. *dalmatica* seem to lack real distinction and it seems possible that in a search for scientific accuracy 'dalmatica' has been applied indiscriminately and would be better kept for plants collected from that geographical area.

I. perrieri Simonet ex P. Fournier

Synonym
I. aphylla L.

Distribution
Endemic to Savoy, Dent d'Arclusaz.

Description
Rhizome nodular, compact. **Leaves** 22–28 cm long, 2–2.4 cm wide, fully herbaceous, green, falcate, very heavily ribbed. **Stem** around 17–27 cm long, usually branched from around the middle or, if unbranched, with one inflated stem-leaf, rarely 2-branched or branching from near base; spathe 3.5–4.6 cm, green, inflated, valves widely separated, remaining green for an extended period after flowering, 2-flowered at apex; pedicel very short; ovary rounded, 6-grooved, 1.2–1.4 cm; tube 1.7–2 cm. **Flower**: falls about 5–7 cm long, 2.2–3 cm wide, obovate–cuneate to clavate, apex rounded or retuse, violet, more or less heavily veined up to end of beard, which is white or pale bluish becoming yellow-tipped towards the haft; standards 5.4–6.7 cm long, 2.3–3 cm wide, oblong, violet, style arms pale violet with a darker keel; crest acute-pointed, dentate, 1–1.6 cm long; filament pale violet, 1.2–1.7 cm; anther 1–1.2 cm. **Capsule** broadly oblong, obtuse at apex, pronouncedly 6-grooved. **Seeds** reddish-brown, pyriform, wrinkled, somewhat compressed. $2n = 24$.

Cultivation
There is no real problem in a sunny, well-drained spot, although *I. perrieri* seems rather slow-growing and is not often free-flowering.

Observations
There is no great variation in colour, which is bright violet of various tones within quite a limited range.

The identification of this species with *I. aphylla* has a long history going back to P.E. Perrier de la Bathie and A. Songeon's original report of its discovery in 1890. It does not, however, appear to stand up to scrutiny: significant differences lie in the leaves, the stems, the branching, the spathes, the falls, the style crests, the seeds and the karyotype (Randolph 1959).

The production of a second branch and of low branching (what presumably led Perrier to assume it was *I. aphylla*) seems to be dependent on some factor operating over a small area as both features occur in only one limited part of an already very limited range.

It may seem surprising that, all the above being so, both Alain Richert's and Jean Peyrard's clones, the only two in circulation as far as I know, produce rather long laterals from the base. The reason for this lies perhaps in the comparative accessibility of the area where this type of branching occurs. Because of the nature of the slope, it is around here that *I. perrieri*'s range most closely approaches the path running between two huts used by herdsmen up for the summer pasture and leading from both to the only available source of water. Few would clamber about on that mountain slope for fun. They would walk along this path, see these irises and gather them from the nearest impressive clump.

I. pseudopumila Tineo

Synonyms
I. lutescens Gussone, *I. panormitana* Tod., *I. pumila* Bivona, *I. violacea* Parl.

Distribution
Malta, Sicily, Gargano, west Croatia.

Description
Rhizome compact, with crowded growths and rather fleshy roots. **Leaves** 10–30 cm long, 1–2.2 cm wide, evergreen, ensiform, glaucescent, erect, the outer often narrowing abruptly to an oblique tip. **Stem** simple, 3–25 cm, largely concealed by rather inflated stem and basal leaves; spathes green, membranous at apex, usually slightly keeled and closely sheathing tube, 1-flowered; pedicel to 1.5 cm; ovary narrow, around 1.5 cm; tube 5–8 cm, green faintly marked purple below standards. **Flower**: falls around 6.2 cm long, 2.5 cm wide, obovate–cuneate, purple becoming

yellow-veined, darker towards the haft, around 6 cm long, 2 cm wide, beard white tipped blue; standards oblong to elliptic, paler; style arms keeled with triangular, dentate, long-pointed crests of about 1.5 cm. **Capsule** roundedly hexagonal, fusiform to ovate acuminate tipped, 6.5 cm long, 1.8 cm wide. $2n = 16$.

Cultivation
This is not difficult to grow, though harder to persuade to flower much and prone to rhizome rot.

Observations
Flower colour can vary through violet, purple, yellow and variegata; occasionally white. Beards white, yellow or bluish. The Croatian and Sicilian forms are reputed to be dwarfer, but if this is so the species overall height in south-east Italy must far exceed that usually given, for in Sicily stems of over 20 cm are not uncommon. The form from Zadar, however, is exceedingly small.

Very dwarf forms can be distinguished from *I. pumila* by a perianth tube that is comparatively shorter, a spathe that is more herbaceous and less membranous, and a capsule of a different shape that deshisces apexially.

A recently published *Flora of France* holds that *I. pseudopumila* occurs very rarely in Provence; other French floras have also stated this. If this is so, there seems no doubt that those occurrences are escapes from cultivation and not spontaneous.

I. pumila Linn.

subsp. *pumila*

Synonyms
I. aequiloba Ledeb., *I. angustifolia* Miller, *I. biflora* L., *I. binata* Schur., *I. clusiana* Reich., *I. coerulea* Spach, *I. diantha* Koch., *I. gracilis* E. Berg, *I. guertlerii* Prodan, *I. longiflora* Ledeb., *I. lutea* Ker Gawl., *I. lutescens* Sprun. ex Nyman, *I. napocae* Prodan, *I. pluriscapia* Prodan, *I. pseudopumila* Janka, *I. pseudopumilaeoides* Prodan, *I. sarajevoensis* Prodan, *I. steniloba* D.C. ex Baker, *I. transylvanica* Schur., *I. tristis* Reich., *I. violacea* Ker Gawl., *I. violacea* Sweet.

Distribution
Austria to the Urals.

Description
Rhizome short-branching, compact. **Leaves** 7–10 cm long, 0.8–1.5 cm wide, linear–ensiform, subglaucous, variably deciduous and variably falcate. **Stem** very short, only rarely visible; spathes narrow, lanceolate, green, scarious at tips, closely investing tube, the inner membranous, the outer more substantial, occasionally

slightly keeled, 1-flowered; ovary rounded–trigonal, 1 cm or less; tube slender, 5–10 cm or more. **Flower**: falls about 4.5–7 cm long, 1.5–2.5 cm wide, oblong–cuneate to spathulate, rounded at apex, often sharply reflexed, beard pale yellow or bluish, yellow in haft; standards usually wider, rounded-elliptic to ovate, narrowing abruptly to the haft; style arms pale at edges with a darker keel, crests deltoid, serrate, acute- or obtuse-tipped; filament and anther subequal, 1.5–1.8 cm long, variably coloured; pollen blue or cream. **Capsule** trigonal, pointed, about 4 cm long, dehiscing below apex. **Flowering** in March or April. **Seeds** usually small, subspherical, wrinkled, light brown. $2n = 30, 32$.

Cultivation
A sunny, well-drained site with rather alkaline, loamy soil with the addition of some grit, particularly if it is heavy, is probably best. Too rich conditions, W.R. Dykes observed, tend to blur the distinctions between the forms; leaves become lush and the plants less compact. However, the short-branched rhizomes make for a very densely packed growth and the soil available is rather quickly exhausted. Frequent transplanting is the answer, best done shortly after flowering. Seed does not often set in cultivation, but germination is likely to be good; seedlings can develop rapidly and may flower at 18 months.

Observations
Every aspect of this widespread species is variable, from the rhizome and the leaf shape to the shape of the seeds. The range of flower colours is uniquely wide, blue-violet to very red-purple, yellow, cream, white, bronze, often with a darker blotch covering most or all of the fall blade. Even the scent is variable. A stem, if visible at all, is never normally of any real length; 1.5 cm is normal, 5 cm a possible suggested maximum. Branching is virtually unknown, although there are herbarium specimens with this feature.

Although I have a plant collected by Peter Davis which is 2-flowered, *I. pumila* is, in principle, a single-flowered species. It is, however, extremely floriferous. Dr Blazek has observed that more than one flower can spring from a single rhizome and that offsets, themselves without any root system, can produce additional flowers.

The capsule is peculiar in not being fully divided internally. This characteristic is variable in different forms.

Subspecies and varieties
Numerous botanists, notably the Rumanian Julius Prodan, have worked to make some sense of order in this very variable group of plants. Variants of differing rank have been described, geographical groupings suggested and a number of related species split off from time to time, but so far no really illuminating way of regarding the complex has been put forward.

subsp. *taurica* (Loddiges) Rodionenko & Shevchenko

Synonym
I. taurica Loddiges.

Distribution
Widespread in the former USSR, north Caucasus, Crimea, Volga.

Description
Leaves prominently nerved, shorter, broader and blunter-tipped than those of subsp. *pumila*, not falcate and broadest around 2/3 up. **Stem** under-developed, 2.5 cm or less; flowers usually pale yellow, but also purple and variegated. **Flower**: falls usually horizontal or flaring, oblique, reflexed at end of blade; standards obovate–oblanceolate, connivent. **Capsule** with sharp-pointed apex. $2n = 32$.

Observations
Although in appearance subsp. *taurica* is closer to the Austrian forms of subsp. *pumila*, this entity is possibly only distinct in its karyotype and morphologically speaking may be impossible to set apart. Dr. J.H. Leep grew it and describes it as having more foliaceous spathes and floral segments of a distinct shape. The anthers have also been reported as being twice as long as in the type, but this would make them very long indeed.

var. *elongata* Lipsky
This variant is recorded in *Flora of the USSR* (Komarov 1935) as producing stems up to 12 cm.

In *The World of Irises* (1978), Barbara Whitehouse and Bee Warburton observed that in hybridising there was no apparent difference from other *pumila* crosses.

I. *purpureobractea* Mathew & Baytop

Distribution
North, north-west and south-west Turkey.

Description
Rhizome nodular. **Leaves** 10–20 cm long, herbaceous, straight or slightly falcate, grey-green. **Stem** 20–35 cm with 1–4 short branches, the uppermost near-sessile, produced from purple membranous, short, inflated bracts; spathe inflated, keeled, obtuse, totally infused purple, 2 cm, rarely to 5 cm, long, becoming scarious by end of flowering, 2-flowered at apex, 1, rarely 2, on branches. **Flower**: falls 5–6 cm long, 2.5–3.5 cm wide, obovate–cuneate, pale blue suffused darker, veined deep blue in haft, or pale yellow with greenish-brown veining, beard yellow; standards

5–6 cm long, 2.5–3.3 cm wide, obovate with a narrow haft, pale blue or pale yellow; style arms about 3–4 cm long, crests about 1 cm long; anther 1.4 cm, shorter than filament; pollen white. **Capsule** oblong with an acute-pointed apex, about 5 cm long, 1.5 cm wide. $2n = ?$

Cultivation
I find some difficulty in persuading this species to flower regularly, but otherwise there seems no problem. However, it is only just coming into general cultivation so it is too early to say how adaptable it is likely to prove. May blooming.

Observations
At its best, *I. purpureobractea* is really a most striking plant with almost ice-blue flowers and bracts of quite dramatic darkness. In the yellow-flowered forms the bracts are not so intensely stained.

I. reichenbachii Heuffel

Synonyms
I. ambertellon ined., *I. athoa* Foster, *I. balkana* Janka, *I. bosniaca* Beck, *I. chalcidice* ined., *I. kobasensis* Prodan (probably), *I. macedonica* Nadji, *I. reichenbachiana* Baker, *I. reichenbachiana* Heuff., *I. serbica* Panc., *I. skorpilii* Velen., *I. straussii* Lynch, *I. tenuifolia* (Vel.) Prodan.

Distribution
Serbia and Macedonia to north-east Greece.

Description
Rhizome compact, close growing. **Leaves** near-herbaceous, ensiform, subglaucous, more or less falcate. **Stem** 10–30 cm, simple with usually several short stem-leaves from near the base, rarely taller; spathe, green, sharply keeled, broad, closely enclosing tube, 1–2-flowered; pedicel to 1 cm; ovary cylindrical, 6-grooved, 1.3–1.5 cm long; tube about 2.5 cm. **Flower:** falls 5–6 cm long, 2–3 cm wide, obovate–cuneate, yellow or violet, sharply reflexed with a dense yellow beard; standards oblong–elliptical to obovate, coloured as falls, shorter and wider and narrowing sharply to a short haft; style arms near-colourless with a deeper keel, crests short, about 0.5 cm, serrate; filament 1.5–1.8 cm; anther 0.9–1.0 cm. **Capsule** elongate–elliptic, cylindrical, shallowly 6-grooved. **Seeds** light brown, pyriform, wrinkled. **Flowering** late April and May. $2n = 24, 48$.

Cultivation
Should be in light, rich, well-drained soil. This is another species that seems to require quite frequent transplantation. Reputed to be rather susceptible to fungal and bacterial infection.

Observations
The texture of the flowers of *I. reichenbachii* is somewhat slight and insubstantial; colours include smoky and brownish-purple shades as well as yellow and violet. A number of forms have been described, often as separate species, but there really seems little to set them apart. A short branch may very occasionally be produced; Randolph came upon a tetraploid plant in northern Greece.

I. revoluta Colasante

Distribution
Endemic to one small islet off south Italian coast near Lecce.

Description
Rhizome thick, nodose. **Leaves** 40 cm long, 3 cm wide, ensiform, falcate, acute, glaucescent. **Stem** longer than foliage to nearly 65 cm with 1–3 short stem-leaves, not sheathing, and rarely 1 branch about 5 cm long; spathes 5–6 cm long, rather inflated, subacute, green, scarious at apex; 2–4 flowers at apex, 1–2 on branch; pedicel short; ovary oblong–subtrigonal, around 2 cm; tube funnel-shaped with violet stripes below standards, 3.8 cm. **Flower:** falls 7.6 cm long, 4.8 cm wide, obovate, blue-violet, margins slightly and irregularly toothed, hafts pale veined violet, beard white or pale violet, yellow-tipped; standards 7.8 cm long, 4 cm wide, elliptic, arcuate with a short canaliculate haft, paler than falls; style arms 3 cm, crests acuminate, somewhat dentate; filament 1.2 cm; anther 1.5 cm. **Capsule** oblong–ovate, roundedly hexagonal, obscurely 6-grooved, 8.6 cm long, 3 cm wide. **Seeds** suboval. $2n = 40$.

Cultivation
From the appearance of the plant at Kew, *I. revoluta* does not seem happy in London and blooms poorly.

Observations
This seems an individual species limited to its minute habitat and, from the description, bearing its flowers on a normally uncomplex inflorescence for such a tall-stemmed iris.

I. scariosa Willd. ex Link

Synonyms
I. astrachanica Rodionenko, *I. elongata* Fischer, *I. longiflora* Herbert ex Baker, *I. ventricosa* ined.

Distribution
On both sides of the river Volga.

Description
Leaves longer than stem to around 33 cm, 1.8 cm wide, ensiform. **Stem** about 19 cm long, partly clothed with sheathing leaves; spathe scarious at tips, very membranous, so much so as to be semi-transparent, 2-flowered; ovary about 2.8 cm. **Flower**: falls bearded. $2n = 40$.[1]

Colours mentioned in connection with this species are red-violet, dark violet, light violet-blue, light blue often with light violet shading, near-white and yellow. These may, however, refer to *I. glaucescens*.

Cultivation
I. scariosa, if indeed it was that, has not persisted in cultivation and has proved slow in growth while it did survive, flowering sparsely. A well-drained sandy soil is suggested and it might flower in May.

Observations
Shevchenko (1979) divided this species, previously regarded as widespread, into two, reinstating Bunge's *I. glaucescens* and leaving *I. scariosa* occupying only a limited area to the west of the Caspian Sea. He found morphological and ecological distinctions, but the brief English summary included in the text does not enumerate these.

W.R. Dykes's description in *The Genus Iris* (1913a) is based on a note made by Foster from specimens collected near Lake Balkash and so refers to *I. glaucescens*. Rodionenko's description in the *Iris Year Book* (1967) could easily do the same.

I. schachtii Markgraf

Distribution
Asiatic Turkey.

Description
Rhizome creeping, nodular. **Leaves** about 22 cm long, 1.5 cm wide, semi-herbaceous, narrow, ensiform, glaucous. **Stem** about 20 cm with short semi-sheathing basal leaves and often one stem-leaf, 1–3 short branches more or less contained within the bracts; spathe 2.5–5.5 cm long, slightly inflated, green sometimes stained purple, membranous at apex; ovary broad, smooth, 1.5 cm long; tube 1.5–3 cm long. **Flower:** falls 4.5 cm long, 2.5 cm wide, obovate, obtuse, narrowing gently into the haft with dark veining extending onto the blade, beard white tipped yellow; standards broadly elliptic–obtuse narrowing abruptly to the short

---

[1] Randolph and Mitra made this count from material collected, apparently, near a railway station called Adjana-Arca in Kazakhstan. I have not been able to trace this place; although it looks as if it is possible that *I. scariosa* might just occur in the extreme west of that republic, it is much more probable that this also was *I. glaucescens*.

haft; style arms about 3 cm long, crests acute, dentate, about 0.8 cm long; stigma entire; anther white, 1 cm long. **Capsule** ellipsoid, 2.5–4.5 cm long, 2–2.5 cm wide. **Seeds** pyriform, reddish-brown, wrinkled. $2n = 48, 49$.

Cultivation
This species is, happily, just starting to become available in commerce, but not a great deal is known yet about its cultural requirements. Sun seems indicated and lots of it, and a well-drained open soil. Bulb frame cultivation has been suggested as the best method.

Observations
Colour varies from violet, through purple with bluish or yellow beards and shades of yellow with yellow beards and often rather purplish, greenish or cream markings on the falls, to white.

I. setina Colasante

Distribution
Probably endemic to Monte Trevi–Sezze Romano.

Description
Leaves 30 cm long, 1–1.5 cm wide, slender, widest in upper half. **Stem**: about 27.5 cm high, slender, 2-branched, the lower long, 9.5 cm, the upper shorter with long, slender green bracts exceeding the branches in length; spathes 4.5 cm, slender, scarious and rather insubstantial above, green in lower 1/3, valves about equal; pedicel short; ovary about 1.5 cm; tube about 2.2 cm, green. **Flower**: falls 8 cm long, 3 cm wide, violet, obovate and rather short-hafted, beard yellow; standards 7.7 cm long, 4 cm wide, oval with a rounded apex, narrowing quite gently to a slender haft; style crests about 1 cm long, narrow. $2n = ?$

Observations
This species has been quite recently described. I have only seen herbarium material and it is on this that the description is based.

I. suaveolens Bois. & Reut.

Synonyms
I. jugoslavica Prodan, *I. mellita* Janka, *I. rubromarginata* Baker, *I. straussii* Hansok, *I. straussii* Leicht. ex Micheli.

Distribution
From Albania and Macedonia to west Asia Minor.

Description
Rhizome small, compact. **Leaves** 7–13 cm long, 0.8–1.3 cm wide, herbaceous, narrow, ensiform, falcate. **Stem** simple, very short or up to 14 cm, not usually exceeding 5 cm with several leaves from, or near, the base; spathe 4–6 cm long, herbaceous, subinflated, lanceolate long-pointed, keeled, divergent and exposing the tube, appearing much like the leaves, 1–2-flowered; ovary cylindrical, 6-grooved; tube 3.5 – 5 cm or more. **Flower:** falls 4.5 cm long, 1.3 cm wide, oblong–cuneate, rounded or retuse, strongly reflexed, beard white or blue; standards oblong with a short haft, broader than falls; style arms narrow with deltoid, serrate crests about 0.6 cm long; anther longer than filament; pollen bluish-white. **Capsule** trigonal, tapering slightly to a pointed apex. **Seeds** small, subspherical, brown, wrinkled. $2n = 24$.

Cultivation
Like the other very dwarf species, *I. suaveolens* has a limited root system and can exhaust the available soil in a short time. Move every 2–3 years and top-dress in spring with limestone grit and leaf-mould, but do not bury the rhizomes deeply.

Observations
Colours are in many cases distinctive: pale fawny-violet, smoky brown-purple, mahogany, faded mulberry all occur, also yellow and clearer violets and purples. Variation is wide: a long-stemmed form has been described from Serbia under the name var. *jugoslavica* Prodan, which reaches up to 30 cm with 3–5 stem-leaves and longer spathes. Features such as the length of the perianth tube and length and openness of spathes are as variable as stem length and substance of spathes. This last feature, together with the width of the valves, which may be as narrow as 0.25 cm or as wide as a centimetre or more, may have some geographical connotation.

Variation in the shape of the seed pod and the thickness of its walls also occurs: that in the description is as given by W.R. Dykes, which agrees with that produced by my own plant, but the widest point can occur towards the top or the shape can oblong.

Forms where the stem is obvious can be distinguished from *I. reichenbachii* by their divergent, pointed, narrower spathes and proportionately longer tube and by their generally more deciduous habit.

I. subbiflora Brotero

Synonyms
I. biflora L., *I. fragrans* Salisb., *I. lisbonensis* Dykes, *I. longiflora* Vest, *I. nudicaulis* Hooker fil.

Distribution
South Portugal and Andalucia.

Description
Rhizome compact. **Leaves** 16–30 cm long, 1.5–2.5 cm wide, evergreen, ensiform, broad, often exceeding stem. **Stem** 20–30 cm long, simple, concealed in lower part by usually 2 short stem-leaves; spathes acuminate, green sometimes stained purple, often scarious above, the outer often slightly keeled, divergent towards apex, 1-flowered; pedicel short; ovary obscurely trigonal; tube 3.5–5 cm long, green marked purplish. **Flower**: falls around 7.5 cm long, 3.5 cm wide, obovate, violet, narrowing to a broad cuneate haft coarsely veined brown-purple on a white or greenish-white ground, beard bluish becoming dull yellow in haft; standards obovate, often retuse, paler than falls with a short, broad, grooved, pale haft veined red-brown; style arms colourless with a violet keel, crests deltoid to semi-ovate, 1.4 cm; stigma entire; filaments blue, 1.3–1.8 cm; anthers bluish or white outlined blue, 1.4–1.8 cm. **Seed** brown, wrinkled, broadly pyriform to oval. $2n = 40$.

Cultivation
A hot summer sun is necessary to ripen the rhizomes and a poor spring can abort flowering.

It has some reputation for not persisting in England, but Mathew has found it quite hardy. W.R. Dykes, however, obviously continued to have trouble with it right up to 1924. Clearly it must vary in this respect and raising it from seed might be advantageous.

Observations
Colour variation is not wide: violet, deep blue-black, dull reddish-purple, clear dark purple and a near albino of a rather yellowish cast is about the recorded range.

Webb and Chater (1976) reduced *I. subbiflora* to the rank of a subspecies of *I. lutescens,* in preparation for volume 5 of *Flora Europaea,* in which they typify it as having stems of usually more than 20 cm and spathes of up to 8 cm. My experience looking at it in the wild does not bear out these findings: in more than half the measurements I made the stem failed to reach 20 cm and spathes quite often exceeded the limit given, being up to 11 cm.

var. *lisbonensis* (Dykes) Dykes

This variant is distinguished by its bare stem, its longer perianth tube and its longer spathe valves. The spathes have also been held to be plain green without any purple infusion. This last feature does not hold good in the wild: purple coloration can occur in var. *lisbonensis* and, in fact, intermediate forms are probably quite common in many populations, failing to conform in one way or another.

I. taochia Woronow ex Grossheim

Distribution
Area north-east of Erzurum, Tortum.

Description
Leaves 1.5–2.5 cm wide, erect or slightly falcate, grey-green, as tall as, or taller than, the stem. **Stem** 15–30 cm tall with usually 2 branches, the upper sessile or shorter than its inflated bract; spathe 3–5 cm, navicular, inflated, pale green becoming scarious at apex; tube 1–1.8 cm. **Flower:** falls about 5–6 cm long, obovate–cuneate, rounded at apex, base of blade prominently veined with a yellow, or white tipped yellow, beard; standards 5–6.5 cm long, 3–3.5 cm wide, obovate–elliptic with a short brownish-marked haft; style arms 2.8–3.8 cm long, rather narrow with a short obtuse crest. **Capsule** about 3–6 cm long, 1.5–2 cm wide, ellipsoid. **Seed** brown, irregularly D-shaped, wrinkled. $2n = ?$

Cultivation
I have not, myself, cultivated this species, but it is reputed not to be difficult under normal conditions though, perhaps, rather sensitive to winter damp.

Observations
Colour varies from pale to bright yellow, dark, dull purple to violet. This species has noticeably longer foliage in comparison to the stem and this gives it a much leafier look than the other Turkish species.

I. timofejewii Woronow

Distribution
Endemic in east Caucasus.

Description
Rhizome short-branching, moderately thick. **Leaves** 6–15 cm long, 0.5–0.8 cm wide, fully herbaceous, slender, glaucous, falcate. **Stem** up to 20 cm, simple; spathe lanceolate, acute, sharply keeled, green, somewhat membranous, purple veined, 2-flowered; ovary around 1 cm long; tube about 3–5 cm long. **Flower** falls clavate, violet, narrowing to a yellowish, or white, haft veined purplish, beard white tipped violet, yellow in haft; standards erect, oblong–lanceolate, often notched, violet, narrowing gently to a narrow, dull golden-brown haft with this colour continued well up onto the base of the blade; style arms violet with narrow, acute, dentate crests, 0.7 cm long; anther and filament subequal or filament longer; pollen blue. **Capsule** rounded–ovate, beaked, obscurely trigonal with rounded angles and shallow grooves. **Seeds** dark red-brown, pyriform to oval, finely wrinkled. $2n = 24$.

Cultivation
I. timofejewii would obviously need sunny conditions with good drainage and dry summers. Cultivation in a bulb frame has been suggested to counteract the frequent dearth of fine weather in England. As this species thrives in the open in France, it seems that actual winter protection is not required.

Observations
My plants, grown from collected seed, are more dwarf than is normally indicated in descriptions so there may be some degree of variability, at least in this respect. Little is known about it.

G.I. Rodionenko (1971) has noted the flowers of *I. timofejewii* have a somewhat oncocyclus-like appearance and points out that its range lies close to that of the Caucasian oncocyclus species, but, in fact, it is related cytogenetically to *I. scariosa*. He further observes (1961) that it shares with *I. pumila* the peculiarity of possessing a capsule that is not fully divided internally when mature. So this attractive little iris shares characteristics with a curiously diverse selection of species.

I. variegata Linn.

Synonyms
I. amoena D.C. (possibly), *I. belgica* Spach, *I. corygeii* Lynch, *I. dragalz* Horvath, *I. flavescens* Delile (possibly), *I. flavescens* Kummer & Sendtner, *I. lepida* Heuffel, *I. leucographer* Kerner, *I. limbata* Besser ex Steud., *I. mangaliae* Prodan, *I. reginae* Horvath, *I. rudskyi* J. & M. Horvath, *I. squalens* E. Berg.

Distribution
Widespread from Bavaria to Moldavia and the Ukraine.

Description
Rhizome stout, ridged, compact. **Leaves** herbaceous, bright green, more or less purple-marked at base, strongly ribbed, often falcate, can exceed stem, width variable to 3 cm. **Stem** 20–45 cm with up to 4 branches, spathe green, membranous-edged, sometimes purple-marked, inflated, keeled, acute; pedicel very short; ovary 1–1.4 cm, 6-grooved; tube yellowish-green, about 2–2.5 cm. **Flower** falls 5–7 cm long, 2–2.5 cm wide, blade oval to obovate, heavily veined dark reddish-brown becoming solid at apex, beard bright yellow, dense, haft whitish with a pale yellow edge veined brownish, cuneate to obscurely panduriform; standards elliptic–rounded, bright yellow, narrowing gently to a slender greenish-marked haft; style arms yellow, crest deltoid, acuminate, acute-tipped, 1.2 cm; stigma entire; filament 1.4 cm; anther shorter than filament; pollen cream. **Capsule** ellipsoid to ovoid, concavely hexagonal, grooved along each ridge, about 4 cm long. **Seeds** pyriform to oval, colour and size variable, wrinkled. $2n = 24$.

The colour form described is typical enough of the species, but variation is wide: the ground colour of the falls can be whitish or yellow, the veining can be less dense at the apex and sometimes there can be a pale yellowish margin. Or this can be bluish, rather than brown, on a pale blue ground. The standards can be paler or brighter or browner-yellow with more or less veining extending up from the base. Yellow coloration can be entirely absent, even from the beard.

Altogether a very distinct species which you would be unlikely to take for any other with its unique flower colouring and flaring falls.

Cultivation
I. variegata is normally the latest-flowering of the section. It is not difficult in a sunny position, and it is perfectly hardy. Hardy, that is, by British standards: Dr Blazek found that it failed in Prague.

Observations
Dykes (1913a), describing the shape of the falls, wrote '. . . passes gradually without any constriction into the wedge-shaped haft'. All the three variants of this species I grow have rather noticeable constrictions at this point, not actually a very usual feature in the section.

I. variegata crosses freely with *I. pallida* wherever their ranges coincide, and a number of hybrids resulting from this have been named as species: *I. buiana* Prodan, *I. lurida* Ker Gawl., *I. lurida* Reich., *I. lurida* Solander, *I. lurida* Spach, *I. neglecta* Hernem, *I. nyaradiana* Prodan, *I. rhaetica* Bruger, *I. rosaliae* Prodan, *I. rothschildii* Degen, *I. sambucina* L., *I. squamata* Prodan & Buia., also probably *I. flavescens* D.C. and *I. squalens* L. Mathew suggests *I.* × *lurida* Aiten as a blanket name to cover all such hybrids.

A number of garden forms are still in circulation from past centuries and are sometimes confused with the species.

Subspecies and varieties
Several variants have been described as species, but differences, even where significant at all, are small. A number of others, or the same, have received descriptions as lesser taxa, although none seem widely accepted. It does seem worth describing one, however, as it appears somewhat distinct.

var. *pontica* Prodan

From Romania. **Leaves** rather longer and 1.8 cm wide. **Stem** 70 cm and may commence branching below the middle, bearing up to 9 flowers; spathe to 6 cm, more inflated; ovary 1.4–1.6 cm; tube 3–3.5 cm. **Capsule** 6.5–8 cm long, 1.4 cm wide.

The plant described as *I. mangaliae* is probably the same.

Iris species

Synonym
I. falcata Bablonas & Papanicolaou

Distribution
North-central and north-east Greece.

Description
Rhizome stout. **Leaves** 12–20 cm long, 1–2.5 cm wide, exceeding stem, strongly falcate, broadest above the middle. **Stem** simple, 3–5 cm; spathe 7–7.7 cm, greenish, keeled, the outer valve longer, with a long-acuminate, sharply tipped apex, 2–3-flowered; pedicel very short; tube about 2–3 cm. **Flower:** falls 4.5–5 cm long, 1.5–2 cm wide, obovate–spatulate, somewhat acute tipped, bearded, pale yellow; standards 4–6 cm long, 2.4–2.8 cm wide, elliptic narrowing quite abruptly into the haft. **Capsule** 4.5 cm long, 2.5 cm wide. **Seeds** subglobose to globose. $2n = 24$.

This iris was published as *I. falcata* (1984), but unfortunately this is *nomen illegitimum* as the name had already been used by Tausch in 1824 and is a synonym of *I. aphylla*.

It is described by the authors as being close to *I. reichenbachii* and *I. suaveolens* with the same chromosome count. The leaves are very pronouncedly sickle-shaped and the larger number of flowers have distinctive zig-zag veins on them.

Section Psammiris (Spach) J. Taylor

Map 2, Figure 2.

This is an odd little group of irises which seems to display characteristics from all the other members of the subgenus while excluding something which would make it possible to include them with another group.

I. bloudowii Bunge

Synonyms
I. flavissima Pall., *I. flavissima* var. *bloudowii* Baker, *I. flavissima* var. *umbrosa* Bunge.

Distribution
Siberia, Tien Shan mountains, Mongolia and China.

Description
Rhizome compact, thick, segmented, irregularly shaped, fibrous old leaf bases persistent. **Leaves** up to 30 cm long, 0.7–1.3 cm wide, strong, dark grey-green

with dark tips in clusters of 4 or 5, erect or slightly curving, bases sheathing, several parallel veins. **Stem** 10–35 cm, unbranched, leaves at base only; spathes ventricose, keeled, pointed, with brown veining and sometimes flushing, outer pointed; 2–3 flowers; pedicels 0.5–2 cm long; tube funnel-shaped, 1.0–1.5 cm long; ovary green with 6 purple stripes, 1.5 cm long, 0.3–0.5 cm diameter. **Flower:** bright yellow, to 5.5 cm diameter, falls to 5 cm long, 2.5 cm wide, obovate, narrow haft veined purple-brown on yellow; beard yellow, large hairs extending back beyond styles; standards to 4.5 cm long, 1.2 cm wide, oblong, unguiculate, erect, much smaller than falls and brighter yellow; styles flat, narrow, to 2.5 cm long; crests oblong or quadrate, narrow with toothed edge; stigma entire, rounded; stamens about 2 cm, filaments short, anthers large. **Capsule** 5 cm long, hexagonal, surface with reticulate veining, narrowing at both ends; dehisces below the top. **Seeds** to 0.5 cm long, 0.3 cm wide, pyriform, brown, wrinkled, small whitish aril. **Flowering:** May. $2n = 22, 26$.

Cultivation
It is a native of sandy, woodland edges on mountains and alpine meadows. Dykes found it shy-flowering in spite of moving it around, while Mathew finds it easier than *I. humilis*; Foster was successful. Good drainage, then, and full sunshine or very light shade on a loamy soil with plenty of leaf mould and dry winter conditions. It has overwintered in the open at Kew.

I. curvifolia Y.T. Zhao

Distribution
China: Xinjiang province.

Description
Rhizome about 2 cm diameter, yellow-brown; roots fleshy, little branching. **Leaves** 10–20 cm long, 1–1.5 cm wide, falcate, with fleshy sheathing bases, widest in middle, apex acuminate or abruptly pointed. **Stem** 8–10 cm wide, leafless; 2 flowers; spathes 3, to 6 cm long, 1.3–1.8 cm wide, lanceolate, apex acuminate, margins membranous, herbaceous; pedicel very short; tube funnel-shaped, to 3 cm long. **Flower** yellow with brown veining, 4.5–6 cm diameter; falls about 4.5 cm long, 1.5 cm wide, obovate, bearded; standards about 4 cm long, 1.3 cm wide, oblanceolate; style arm about 3 cm long, 0.4 cm wide, bilobed; lobes obliquely lanceolate; stamens about 2.2 cm long; anthers golden yellow; ovary 1.8–2.2 cm long, cylindrical. **Capsule** 4 cm long, 2 cm diameter, obovate, apex rounded, yellow-green, with short beak, smooth, 6-ribbed. **Seeds** pyriform. $2n = ?$

Cultivation
Probably similar to others of this group, but it is not yet in general cultivation.

Observations
I. curvifolia was first found in 1976 and little is really known about it. It is similar to *I. bloudowii,* but the leaves are sickle-shaped, the stems shorter and the spathes pointed; it is also close to *I. scariosa,* but the flowers are yellow and the seed pod ovate with a short beak.

I. humilis Georgi

Synonyms
I. arenaria Waldst. & Kit., *I. flavissima* Pall., *I. flavissima* subsp. *stolonifera, I. pineticola.*

Distribution
Extensive from Austria east through central Europe to south-east Russia, Siberia, Rumania and Mongolia.

Description
Rhizome slender, stoloniferous. **Leaves** to 10 cm long, 0.2–0.7 cm wide, erect, incurving at tips, light glaucous green, growing continuously. **Stem** short, with sheathing basal leaves; spathes 2–3, outer acutely lanceolate, with slightly scarious upper half, inner less pointed, 2–3 flowers; pedicel 0.75 cm long; ovary about 1 cm long, rounded trigonal, ridged on each face to give a nearly hexagonal section; tube 0.5 cm, funnel-shaped; ovary cylindrical. **Flower**: buds sometimes green touched with bronze; 3–4 cm diameter; falls to 3.5 cm long, 1.2 cm wide, bright yellow, oblong blade rather shorter than haft, which is veined brown-purple, haft horizontal; beard orange, club-shaped hairs tipped brown; standards to 3 cm long, about 0.3 cm wide, oblong, unguiculate, yellow with brown-purple edging on haft and shorter than falls; styles to 2.5 cm, narrow, crests short, triangular, acutely tipped; stigma entire; filaments colourless, same length as anthers; anthers cream with green-black edging; pollen greenish. **Capsule** 3 cm long, tapering to outer end, remains of flower persistent, rounded trigonal, dehiscing below apex. **Seeds** pyriform, long-necked, brown with flat, creamy-white aril. **Flowering** between April and June. $2n = 22$.

Cultivation
This is a plant from sandy or stony areas from altitudes of 200–1500 m, which may account for some of the difficulties of keeping it in cultivation. Dykes recommended planting in a 5 cm layer of sand over soil with a lot of added leaf mould. Replant after flowering or when the rhizomes show signs of overcrowding or weakening. This is probably better done too soon rather than too late. These small plants exhaust the soil relatively faster than larger ones. With hand pollination seed can be obtained and germinates easily, with the seedlings coming to flower in their second year after fast growth. They should be given plenty of space.

B. Mathew, though, has not found the plant easy. Much may depend on the actual seedlings.

I. mandschurica Maxim.

Distribution
Ussuri region of the former USSR, Manchuria and possibly Korea.

Description
Rhizome short, thick, creeping, bases of old leaves remaining entire; roots thick, scarcity of branching. **Leaves** 15 cm long, 0.8–1 cm wide, ensiform, green, lengthening throughout season. **Stem** to 20 cm tall, with a single stem-leaf; spathes 3, lanceolate, scarious at edges; 1–2 flowered; pedicel to 1 cm long; tube to 2.5 cm. Ovary to 1.2 cm long, spindle shaped. **Flower:** yellow, to 5 cm diameter; falls 4–5 cm long, 1.5 cm wide, with maroon veining, obovate, cuneate; standards to 3.5 cm, narrow; styles to 3 cm long, flat, narrow; crests obtuse, dentate, relatively wide; stamens to 2 cm, anthers yellow. **Capsule** to 6 cm long, 1.5 cm diameter, spindle-shaped with obvious ribbing, narrowing to long pointed end, dehiscing from below apex. **Flowering** May. $2n = 34$.

Cultivation
Not in general cultivation yet, but is a native of sunny areas among shrubs and on woodland edges.

I. potaninii Maxim.

Synonym
I. thoroldii Baker ex Hemsley.

Distribution
Central and eastern Tibet; western China.

Description
Rhizome compact, forming dense plants with strong, fleshy roots; curving remains of old leaves persist above soil surface in fibrous mass. **Leaves** 5–10 cm long, 0.2–0.3 cm wide, straight, upright, ends obtuse, growth persists through season. **Stem** subterranean, 1–2 sheathing leaflets; 1 flower, 2 spathes to 4.5 cm long, 0.6 cm wide, narrowly lanceolate; tube 1.5–4 cm long, slender, but splayed at top; pedicel very short; ovary to 0.7 cm, spindle-shaped. **Flowers** yellow, 3–4 cm diameter; falls to 3.5 cm long, 1.2 cm wide, edges curved up, beard yellow; standards 2.5 cm long, to 1 cm wide, erect, tip irregularly rounded; style 2.8 cm long, 0.6 cm wide, flat; crests curved and toothed; stamens about 1.5 cm long, anthers

compact, purple. **Capsule** to 3 cm long, 1.6 cm wide, chubby, 6-ribbed, obtuse at base, narrowing sharply to short beak carrying dried remains of flower, dehiscing from mid-zone. **Seed** about 0.3 cm diameter, reddish-brown, wrinkled, flattened, globular. **Flowering** May–June. $2n = ?$

Cultivation
No information available.

Observations
There are resemblances between this plant and *I. bloudowii* and *I. tigridia*. It differs from the first in having an obconical, not prostrate, rhizome and from the second in the erect, old leaf fibres.

Section Oncocyclus (Siemssen) Baker MARTYN RIX

Map 3, Figure 3.

Most of the members of this section are easily recognised by their large, delicately veined flowers, which are always borne singly on the stem. All have a fine beard and a dark patch, called a signal patch, on the falls. The seeds are large with a ring-like aril. The section occurs in semi-desert areas of the Middle East from the Sinai peninsula northwards through Israel, Jordan and Syria to central Turkey in the west and Georgia and north east Iran in the east. Those species found on the Anatolian plateau, northwards into Armenia and in Iran are generally smaller and hardier than those from the south.

The very similar Regelia section generally has two flowers per stem (although *I. afghanica* is similar to the Oncocyclus section in its solitary flowers) and a beard on the inside of the standards as well as on the falls; members of this section are found in central Asia from Samarkhand eastwards. The oncocyclus irises have a reputation for being difficult to cultivate because they require a period of dormancy in summer and are liable to be attacked by fungi and bacteria in humid weather. They are also exceptionally susceptible to virus infection spread by aphids.

Members of the Oncocyclus section have been crossed with various members of Section Iris (pogon irises) to form so-called Arilbred irises. 'Lady Mohr', raised by Salbach in 1943, was one of the earliest of these crosses. Hybrids were also made with members of the Regelia section, so-called regeliocyclus, notably by the firm of Van Tubergen, to try to produce-large flowered irises hardier and easier to cultivate than the pure oncocyclus.

Cultivation (especially of the northern group in an English climate)

General
Oncocyclus irises originate in dry, Mediterranean-type climates. These have hot dry summers, rain and often snow in winter, and sometimes very wet springs. Most

of the growth and flowering takes place during this warm and comparatively wet spring. Within this general pattern there is of course much local variation between the extremes of the hot Negev and Sinai deserts and the cold steppes of Erzurum and the high mountains of Hakkari and the Iranian Elburz. In climates with wet summers, oncocyclus irises need to be grown under glass, or otherwise kept hot and dry for the summer months (in the northern hemisphere from late June to September). In Mediterranean climates, which naturally have a dry summer, plants grown in the garden should be protected from irrigation.

The plants spend their summer dormant period with long (to 30 cm or more), unbranched fleshy roots attached to smallish rhizomes. In autumn, if watered, these long roots produce thin feeding roots and some leaf growth is generally made. New long fleshy roots are produced in spring ready for the summer dormant period, and last year's roots die back and shrivel. Renewed leaf growth starts in spring; new rhizomes are built up in late spring during and after flowering while the leaves are slowly dying back.

Containers

Cultivation is easiest in a large greenhouse. The plants should be grown in deep pots, or drainpipes, to allow the long roots to grow unhindered and find some moisture even when the surface is hot and dry. These pots are plunged into a bed of coarse sand up to half their depth or more. I have had good growth and flowering in clay-drain pipes: the best being 15–30 cm in diameter; smaller pipes dry out quickly and even a small plant soon becomes crowded. Larger pipes, such as chimney pots, are difficult to handle. Plastic pipes and extra deep plastic pots are probably equally good.

Alternatively, plants can be grown in raised beds either in the greenhouse or in the open, protected by a frame. Composts and watering can be the same as for plants in pots and there is the advantage that the plants are less likely to dry out during sunny spells in spring. The roots can also roam freely through the compost.

Compost

Satisfactory growth can be made in a large variety of composts provided they are alkaline, fertile, sterile and well drained. A mixture of two parts coarse sea sand, or crushed limestone, to one part burnt clay, or loam, or chalk soil and one part sterilised leaf-mould, or peat-based compost, makes a satisfactory start. Burnt clay was used by Paul Furse to good effect in bulb frames and its high potash content helps to produce good root growth. Similar clay-type minerals are also excellent to keep the compost open and well-drained; perlag is light and porous; montmorillonite is slightly heavier and usually sold in granules like instant coffee. Pieces of chalk rock, the size of walnuts and smaller, would also make an excellent addition to the compost. In the wild plants grow in widely different soil types from loose sand and pumice dust to sticky clay which becomes rock-like in summer; the compost used for individual species can reflect the soil of their wild habitat and frequency of watering can be varied to take soil differences into account.

Watering

The first watering of the growing season should be in late autumn. There is some dissent among growers as to how soon this should be, but mid-October is certainly early enough and November not too late. Established plants seem to develop perfectly well if left dry until December. The aim should be to start root growth without encouraging too much leaf, which is prone to die off and go mouldy in the dark days of mid-winter. Growers in sunny climates such as California and the south-western USA, South Africa and Australia do not have this problem and can allow the plants to grow slowly through the winter.

I visited north-eastern Turkey in October and there very little growth is made before the winter snows begin. Some areas had received rain, but in other places nearby the soil was quite dry, and a covering of snow likely any day.

After the first two or three waterings of autumn the compost should be kept rather dry until spring; dry enough to discourage root rot, but not so dry that the thin roots die. In spring, i.e. by late February, watering should become more frequent again, especially in sunny weather, and the leaves should begin to grow. The plants benefit from weekly feeds of high-potash fertiliser such as is sold for tomatoes at fruiting time, or from a generous scattering of similar slow-release fertiliser on the surface. During March and up to flowering the plants can take plenty of water, but I have found this is best put on in cold, windy weather if possible, not in a panic on a sudden hot day. Watering the plants when the soil is warm encourages root rot, which is shown by the sudden yellowing of the leaves. If the leaves do go yellow prematurely, the plant should be dug up, any dead roots cut off, the whole plant soaked in fungicide and then replanted in dry sand. In autumn it can be replanted in fresh compost and should have been saved.

After flowering, watering and feeding should be continued at somewhat longer intervals while the leaves are green, which may be as late as the end of June. It is at this time that the new rhizomes and resting roots, and the flower buds for next year, are formed.

B. Gavrilenko recommends the following liquid feeding. In autumn 1 or 2 feeds of 15 g ammonium nitrate, 15 g potassium nitrate, 30 g superphosphate in 10 l of water; during spring growth up to flowering, 3 feeds of 27 g ammonium nitrate, 36 g potassium nitrate, 80 g superphosphate in 10 l of water.

Light and air

At all times of the year, but especially in winter, it is important that the plants receive as much light and air as possible. Oncocyclus irises are plants of open hillsides and windy steppes, and short tough leaves are less susceptible to mould than lanky, etiolated ones. It is also important that the rhizomes receive sufficient heat in summer. When collecting plants in Turkey, I often noticed that the soil in *Iris* habitats remained warm even after sunset. A temperature of 23 °C (75 °F) is recommended for summer ripening in Fritz Köhlein's *The Iris* (1981). However, during growth, and especially before flowering, and particularly when grown in a greenhouse, the plants should be kept as cool as possible.

Re-potting
Plants are best re-potted or re-planted about every three years when the soil becomes exhausted or the rhizomes overcrowded in the pot. A good way is to re-pot a third of the collection each year so the task is less onerous and at any one time a third of the plants are at their optimal stage of growth.

Hardiness
The species from the mountains of Turkey and Iran are perfectly hardy and should be able to survive −20 °C, especially if any leaves formed in autumn are covered with snow, or with a light covering of dry peat. Species from southern Turkey, Syria, Lebanon and Israel are tenderer, but most will probably tolerate up to −5 °C for short periods, but not prolonged freezing of the the rhizomes. Of the southern species, those from higher altitudes, *I. gatesii, I. kirkwoodii, I. cedretii,* and *I. libanotica* should be the hardiest.

Pests and diseases
The worst ailment to which the oncocylus irises are prone is caused by a virus. For this reason it is vitally important to control aphids, which spread viruses, and ants, which spread aphids. Spray the leaves as soon as they appear with a persistent insecticide and repeat as recommended on the bottle. Watering with a systemic insecticide is an alternative. 'Organic' sprays such as oil washes and soap solutions will need to be applied regularly, too, and any potential source of aphids sprayed also. Equally important is the ruthless destruction of any plant that appears to be infected before it has a chance to pass on its virus. New acquisitions, especially if from a cultivated source, should also be checked for virus as soon as the leaves appear. Virus infection shows up as pale yellowish-green streaks on the usually bluish-green leaf and gets worse as the leaves get older and the virus multiplies. Old iris hybrids of garden origin are almost certain to be infected and should be kept well away from healthy stock, or avoided altogether.

Fungus diseases can also be a problem, particularly *Botrytis* and *Fusarium*. They can be controlled by suitable fungicides and by good cultivation, but preventative sprays in late autumn against *Botrytis* and in spring against *Fusarium* are probably advisable.

Seed sowing
Young, healthy plants can be raised from seed. This is not difficult, only slow and unreliable. Oncocyclus seed germinates erratically, there is evidence that inhibitors in the seed are lost gradually by leaching as well as over time (i.e. after 5 or 6

years). Sowing the seed while still green, as it is when the capsule begins to split, is advocated by Kenneth Bastow, who says that it gives a higher percentage of immediate germination. Dry seed should be sown in autumn in pots of sandy soil and stood outside to receive all available winter rain. As soon as the leaf of a young plant is seen the pot should be brought inside, fed and the young plants grown on as normal. Either after three leaves have appeared, or in autumn, the plantlets are dug out and re-potted and the pot put outside again so further seeds can germinate the next winter and so on until either all the seeds have germinated or the grower is fed up with waiting. Oncocyclus seed has been known to germinate after 8 years or more; there is even a record of 25-year-old seed germinating! Young seedlings do not have such a complete need to become dormant as mature plants and can be kept growing through the summer in cool conditions. Embryo culture gives quicker and more reliable results than seeds. Recently, Neville Watkins has found that treating seeds by soaking for one week, scraping, and chilling for two months, as if they were bearded iris cultivars, produces rapid and comprehensive germination.

Breeding

It is necessary to hand-pollinate two clones to be sure to get a good set of seed of a species, but if suitable flowers are not available, beautiful hybrids can be obtained between species and between oncocyclus and regelias. Some of these oncogelia hybrids are even fertile. Hybrids with pogoniris are harder to achieve until they are made at tetraploid level.

Chromosome numbers

All the species counted have $2n = 20$.

Synopsis of species

1. Found in Turkey, Iran and Transcauscasia. Leaves narrow, generally less than 0.6 cm wide; stems short, up to 30 cm.

Iris acutiloba, barnumae, camillae, iberica, meda, paradoxa, sari (small form), *sprengeri.*

2. Found in S Turkey and the Syrian desert and further south. Dwarf and medium-sized species; leaves mostly less than 1 cm wide; stems to 30 cm.

Iris assadiana, bostrensis, damascena, heylandiana, mairiae (Sinai), *nectarifera, nigricans* (S Jordan), *petrana* (S Jordan), *swensoniana* (S Syria), *yebrudii* (Syria).

3. Found in S Turkey, Syria, Iraq, Lebanon and Israel, usually in mountains. Large species; leaves around 1 cm or wider; stems 45 cm or more.

Iris antilibanotica, aurantiaca, basaltica, bismarkiana, cedretii, gatesii, haynei, hermona, kirkwoodii, lortetii, sari (large form; stems about 20 cm), *sofarana* (S Lebanon), *susiana, westii* (S Lebanon).

4. Coastal species; on the Mediterranean coastal plains of Israel.

Iris atropurpurea.

5. Desert species; in the Negev, Sinai and Jordan deserts.

Iris atrofusca.

I. acutiloba C.A. Meyer

subsp. *acutiloba*

Synonyms
I. fominii Woron. ex Grossh., *I. szovitsii* C.A. Meyer.

Distribution
Transcaucasia on the Apsheron peninsula north of Baku and elsewhere north of the Kura river.

Description
Flowers about 7 cm in height, veined with purple on a background of cream to pale violet; standards oblanceolate, acute, considerably larger than falls; falls lanceolate, recurved at the apex with a small signal patch just beneath the beard and a second dark patch at or near the apex; beard brown. **Flowering** May.

Observations
from low hills below 200 m. This subspecies has seldom been seen in cultivation in the West. It is apparently variable in colour and flower shape in the wild, although it is found only a small area, as five forms have been named.

subsp. *lineolata* (Trautv.) Mathew & Wendelbo

Synonyms
I. ewbankiana M. Foster; *I. helena* (Koch) Koch.

Distribution
in Russian Transcaucasia south of the Kura river; in Turkmenia in the Kopet Dag; in N Iran in Khorassan and Gorgan; in Azerbaijan south to Ispahan; 1300–3000 m.

Description
Rhizomes slender, to 1 cm wide densely tufted, without stolons. **Leaves** 0.2–0.6 cm wide often recurved and falcate. **Stems** 8–25 cm, usually about 10 cm tall. **Flowers** 5–7 cm across, closely and finely veined with brown or purple on a buff or cream ground; standards obovate to oblanceolate, up to 5–8 cm long, to 4 cm

wide, acute; falls lanceolate, 4–7 cm long, to 2.5 cm across, with a single spot, acute; beard of stiff, dark brown or blackish hairs. **Flowering** April–May.

Observations
From dry stony plains and rocky mountainsides, often on loose volcanic soils, but also in heavy clay. Variable in depth of colour, in the width of the standards and falls, and in the position of the falls. They are often horizontal or barely reflexed, especially in plants from the Kopet Dag in the east of the species' distribution where the plants named *I. ewbankiana* originated. I have found this the easiest species of all the oncocyclus irises to grow in Kent and it has proved very free-flowering, especially the clone Rix 889 from Kuh-i-Savalan in north-west Iran.

I. antilibanotica Dinsmore

Distribution
Central Syria, in the Antilebanon above Bluden.

Description
Rhizome small, compact. **Leaves** 7–8, up to 20 cm long, about 1 cm wide, falcate. Stem to 40 cm tall. **Flowers** dark purple, bicoloured, standards paler than falls; standards to 10 cm long, 8 cm wide, purple with darker veins; falls 6–8 cm long, 5 cm wide, dark maroon, reddish brown or purplish without veins or dots; signal patch small, blackish; beard yellow, the hair sometimes purple-tipped; style arms light brown, strongly keeled, with the lobes the same colour as the falls. **Capsule** not recorded. **Flowering** May–June.

Observations
Recorded from rocky mountainsides at 2000 m. The habitat of this species is described in detail by Peter Werckmeister in the *Iris Year Book* (1957).

I. assadiana Chaudhary, Kirkwood & Weymouth

Synonym
I. barnumae var. *zenobiae* Mouterde (in part).

Distribution
Syrian desert, Ain-al-Baida, Qarytein and Hafar, mainly to the west of Palmyra.

Description
Rhizome small, forming small clumps with stolons to 12 cm long. **Leaves** 6–8, 4–12 cm long, 1 cm wide or less, strongly falcate and reflexed. **Stem** to 15 cm tall. **Flowers** scented; standards 6–8 cm long, 4–5 cm wide, maroon-purple to dark

purple, same colour as falls, obovate; falls 5–5.6 cm long, 2.5–3.5 cm wide, often recurved; signal patch about 1 cm wide, black, velvety, oval, wider than long; beard a median band of long bright yellow hairs, with lateral short, purple hairs; style arms arched, not keeled, orange streaked with purple. **Capsule** about 4 cm long. **Flowering** April.

Observations
A small species growing on low chalky hills at 800–1000 m. Around Qarytein white, yellow and pale forms are recorded.

I. atrofusca Baker

Synonyms
I. hauranensis Dinsm.; *I. jordana* Dinsm.

Distribution
Israel and Jordan, from the N Negev east into the Judean desert, E Samaria, the Yarmook Valley, the S Golan and the Jordan Valley down to −250 m.

Description
Rhizome stout, compact. **Leaves** 5–8, 1 cm wide, erect or falcate. **Stem** 20–30 cm tall. **Flowers** fragrant, purplish-brown to dark purple; standards 7–9 cm long, 3–4.5 cm wide, incurved, with heavy veins and dense dots; falls 6–7.5 cm long, 3–4.5 cm wide, recurved, dark brown-purple to nearly black; signal patch broad, brownish-black; beard yellow, tipped with brown; style arms greenish yellow, spotted with purple. **Capsule** not recorded. **Flowering** April.

Observations
Found growing on loess and chalky hills with *Phlomis, Echinops* and *Eremostachys laciniata*. Yellow-flowered forms are not uncommon.

I. atropurpurea Baker

Distribution
Israel, on the Mediterranean coastal plains of Sharon and Philistea.

Description
Rhizome stoloniferous, with thin stolons, forming wide clumps. **Leaves** 7–11, 0.5–0.8 cm wide, glaucous, falcate. **Stem** 25–35 cm tall. **Flowers** usually concolorous, red-brown to dark purple, rarely reddish or early black; standards 5–5.8 cm long, 4.5–6 cm wide, erect, incurved, sometimes slightly paler than the falls, without dots and with obscure veins; falls 3.5–6 cm long, 2.5–4 cm wide,

recurved, uniform in colour; signal patch a black semi-circle, velvety around the mouth of the tunnel; beard of dark-tipped yellow hairs on a yellow ground; style arms with erect lobes. **Capsule** not recorded.

Observations
Found growing on chalky hills and in sandy loam. It is now very rare because of urbanisation of the habitat and is largely maintained in reserves.

I. aurantiaca Dinsmore

Distribution
S Syria, in the Djebel Druze at about 1600 m.

Description
Rhizome medium-sized, compact. **Leaves** up to 9, to 25 cm long, 1 cm wide, erect or slightly arched. **Stem** to 50 cm tall. **Flowers** scented, golden yellow to coppery-brown; standards 8.5 cm long, 5.5 cm wide, with fine purplish veins; falls 7 cm long, 4 cm wide, obovate, with minute purplish-red spots and very fine reddish veins; signal patch 1.5 cm across, orbiculate, dark maroon or reddish-yellow; beard yellow, dense, the hairs with minute purple tips; style arm keeled, golden yellow with fine purple to brownish-purple dots on a creamy ground. **Capsule** 8 cm long, rather narrow. **Flowering** April–June.

Observations
Found growing in volcanic lava and cinders. Forma *wilkiana* Chaudhary, syn. var. *unicolor* Mouterde, has bright yellow flowers without red-purple veins or dots; it is found with the type on Tell Jaffna.

I. barnumae Baker & Foster

Description
Rhizomes tufted, without stolons, up to 1 cm diameter. **Leaves** narrow, 3–5 cm wide, often recurved, falcate. **Stem** 10–25 cm. **Flowers** purple or yellow, without conspicuous veins, 7–12 cm high, standards usually larger than falls.

Two subspecies and three forms are recognised.

subsp. *barnumae* f. *barnumae*

Distribution
SE Turkey, east of Van and northeast of Yukşekova; NE Iraq at Hisar-i Rost; very rare; NW Iran around Khoi and Rezaiyeh; at 1300–2600 m.

Description
Flowers purple; falls elliptic to obovate, 2–3 cm wide, with a narrow median beard of white or purple hairs; signal patch small and black. **Flowering** May–June.

Observations
Found on dry, bare, stony hills, with *Acantholimon* and other dwarf shrubs. This is an attractive, dwarf and neat species, but I have found it one of the more difficult to grow and flower in cultivation.

f. *protonyma* (Stapf) Mathew & Wendelbo

Synonyms
I. polakii Stapf f. *protonyma* Stapf.

Distribution
NW Iran around Khoi and Rezaiyeh and as far south as Hamadan and Sanandaj.

Description
Flowers dark purple; falls small, around 2 cm wide, recurved and pointed with a fine black beard covering much of the upper surface. **Flowering** May–June.

Observations
Paul Furse reported a range of probable hybrids between this form and *I. meda* in the Hamadan–Sanandaj area and likened its fine dark beard to moleskin. The northern populations appear to be nearer to f. *barnumae*.

f. *urmiensis* (Hoog) Mathew & Wendelbo

Synonym
I. chrysantha Baker.

Distribution
NW Iran, west of Lake Rezaiyeh (also called Lake Urmia) and SE Turkey near the Iranian border in Hakkari.

Description
Flower pale to medium lemon yellow, unmarked, with a darker yellow signal patch; falls 2–4 cm wide. **Flowering** May.

Observations
This appears to be a yellow variant of f. *barnumae*, but Mathew records that the two do not grow together and that f. *urmiensis* is common in the mountains to the

west of the lake. It appears to be free-flowering in cultivation and grew and flowered well for some years at Wisley in the 1970s.

subsp. *demavendica* (Bornm.) Mathew & Wendelbo

Synonym
I. demavendica Bornm.

Distribution
N Iran, in the Elburz mountains from Kandavan to Firouzkuh.

Description
Leaves usually erect, not recurved. **Flowers** purple, standards and falls wider than in f. *barnumae;* beard white, narrow; signal patch inconspicuous. **Flowering** June–July.

Observations
Found on rocky mountain slopes, 2300–4200 m. In *The Iris* Mathew (1989) gives detailed recordings of soils and air temperatures in autumn at a locality where this iris grows in the Elburz at 3200 m. The striking feature is the warmth of the soils (ca. 13 °C) even after the first frosts and within about two weeks of the winter snow covering. This iris has proved difficult to cultivate for any number of years. I have grown a plant from seed, but it remained small and never flowered.

I. basaltica Dinsmore

Distribution
Syria, Tell Kalakh to Hadidia region, west of Homs. Found growing on stony basalt hillsides at 700 m.

Description
Rhizome stout, compact. **Leaves** 9–12, slightly falcate, about 24 cm long, 1.5–2 cm wide. **Stem** to 70cm. **Flowers** about 15 cm diameter, standards paler than the falls; standards 8.5–10.5 cm long, 7–7.5 cm wide, almost orbicular with embossed blue dots and veins on a pale greenish ground; falls 9 cm long, 5 cm wide, recurved, ovate to lanceolate with heavy, 'felty-thick' dark purple to almost black veins and dots in the lower parts on a pale greenish ground; signal patch orbicular or truncate triangular, 1.5 cm diameter; beard sparse, of very long hairs, maroon-purple tipped with rusty yellow. **Capsule** 6–11 cm long, inflated, 6-lobed. **Flowering** March–April.

Observations
This species is reported by S.A. Chaudhary to be in danger of extinction. It was also recorded in the past from Krak des Chevaliers. P. Mouterde regards this species as the probable source of the cultivated *I. susiana* L.

I. bismarkiana Regel ex E. Damman & C. Sprenger

Synonym
I. nazarena (Foster ex Herb.) Dinsm.

Distribution
N Israel and S Syria: on the southern slopes of Mount Hermon, in upper and lower Galilee and on Mount Tabor. Also recorded from Transjordan. Up to 1300 m.

Description
Rhizome medium to stout, stoloniferous with long, thin stolons. **Leaves** 6–8, 30–50 cm long, in a fan, erect, obtuse. **Stem** 30–50 cm tall. **Flowers** 15 cm diameter, standards much paler than the falls; standards 7–9 cm, orbicular, with fine blue veins and scattered purple dots on a white or bluish ground, yellowish at the base; falls 6–7 cm long, 4 cm wide, round–ovate with few embossed red-brown spots and maroon, purple or crimson veins on a creamy-yellow ground; signal patch large, orbicular, blackish-red to purple; beard dark purple. **Flowering** March–April.

Observations
Found in mountain scrub on heavy soil formed from basalt, on terra rossa and on soft chalky rock.

I. bostrensis Mouterde

Synonym
I. atropurpurea var. *purpurea* Dinsm.

Description
Rhizome short, compact. **Leaves** usually 8, 20 cm long, 1 cm wide, weakly falcate or erect. **Stem** to 40 cm tall. **Flowers** with standards slightly paler than the falls; standards 8–10 cm long, 5–7 cm wide, streaked brown-purple, often paler in parts on a yellowish or brownish ground; falls 6.5–7.5 cm long, 3–4.5 cm wide, reflexed and often folded back, densely spotted and veined with purple-brown on a yellow ground; signal patch to 2 cm wide, a semi-circle, often notched; beard dense, of bright yellow hairs minutely purple-tipped; style arms keeled, with dense dots on a yellow ground. **Flowering**: April.

Section Oncocyclus

Observations
Often found as a weed of cornfields on the Damascus to Jordan road.

I. cedretii Dinsmore ex Chaudhary

Distribution
North-west Lebanon, in the area of Les Cedres, near Bsherri, at about 2000 m.

Description
Rhizome of medium thickness, compact. **Leaves** 8–9, to 23 cm long, 1–2 cm wide; stem leaf none or single. **Stems** to 40 cm long. **Flowers** 18 cm diameter; standards 8.5–11 cm long, 6–7.5 cm wide, obovate, with very fine veins and dots on a white ground, pale inside, dark outside, strongly incurved; falls 6.5–9.5 cm long, 4.5–5.5 cm wide, ovate, narrowed to the tip, veins very fine, dense and embossed dark maroon to maroon-purple on a white ground, dots very fine and dense; signal patch 1.7–2 cm long, 1.5 cm wide, orbiculate, dark maroon-purple, in the middle of the fall; beard sparse, hairs rusty-brown, pink-purple or mottled; style arms strongly arched. **Capsule** 8 cm long, inflated. **Flowering** April.

Observations
This species was collected by S.D. Albury, M.J. Cheese and J.M. Watson in 1966 (ACW 925).

I. damascena Mouterde

Synonym
I. sofarana forma *qassioumensis* Werckmeister

Distribution
Syria, on the Jabl Qasyoun (Kassioum) near Damascus.

Description
Rhizome short, compact. **Leaves** 5–8, usually strongly falcate. **Stem** to 30 cm tall. **Flowers** 15 cm in diameter, standards much paler than the falls; standards 9 cm long, 6 cm wide, very pale, finely and delicately veined and dotted with purple on a creamy-white ground; falls 8 cm long, 5 cm wide, obovate to elliptic, densely dotted and veined with dark purple-brown on a creamy white ground; signal patch small, 1.5 cm long, 1 cm wide, elliptic, dark purple; beard sparse, purple hairs. **Flowering** March.

Observations
This species is reported by Chaudhary to be in danger of extinction. It is described by Werckmeister (1957) where there is a photograph of this species and several others from Syria.

I. gatesii Foster

Distribution
South-eastern Turkey in the districts of Urfa, Mardin and Siirt, growing on dry limestone slopes and in rock crevices, 1050–2000 m. Also in N Iraq in the district of Amadiya and Penjwin.

Description
Rhizome stout, compact, to 2 cm in diameter. **Leaves** 5–7, 0.5–1.1 cm wide, mostly upright, greyish-green. **Stem** 45–50 cm tall. **Flowers** 16–20 cm diameter, with a white, pinkish or creamy-yellow ground colour, minutely spotted and finely veined with reddish-brown, brownish-purple or nearly black; standards slightly paler than the falls; standards 8–9.5 (–13) cm long, 7–9 cm wide, broadly ovate to orbicular; falls 8–8.5 (–11) cm long, 5–8 cm wide, broadly obovate–elliptic, the same colour or slightly more heavily veined than the standards with a round dark brown or blackish signal patch and a 2–2.5 cm wide band of long yellowish or brownish-purple hairs; style arms the same colour as the standards with conspicuous reflexed lobes. **Capsule** 7.5 × 2.5 cm. **Flowering** April–June.

Observations
The largest-flowered of the Turkish species requiring drier and warmer conditions than those from farther north. Like the large species from farther south it seems to be especially susceptible to virus infection in cultivation. The plants introduced from near Siirt by S.D. Albury, M.J. Cheese and J.M. Watson (ACW 1230) varied in general colour from pale greyish to pale brownish.

I. haynei Baker

Synonym
I. biggeri Dinsm.

Distribution
Jordan and Israel, on Mount Gilboa in NE Samaria.

Description
Rhizome stout, forming clumps. **Leaves** 5–8, 1–1.5 cm wide. **Stem** 40–50 cm tall. **Flowers** fragrant, taller than broad, concolorous; standards 9–10 cm long,

6–7 cm wide, incurved, obovate, purple-veined and -spotted on a pale ground; falls 7–8 cm long, 4–6 cm wide, recurved, oblong–ovate, dark purple to brownish-purple, densely dotted and veined; signal patch blackish to dark purple, beard variable in colour, the hairs dark purple, white or dark-tipped yellow. **Capsule** not recorded.

Observations
Found growing on the edges of fields in terra rossa and on rocky hillsides. A semi-albino with a golden-yellow flower and a dark red signal patch is recorded.

I. hermona Dinsmore

Distribution
Israel and S Syria, on the central and northern Golan and on Mount Hermon. Found in heavy basalt soil and dark brown limestone soil, often on the edges of oak scrub.

Description
Rhizome stout, compact. **Leaves** 9, erect, to 30 cm long, 1.8 cm wide. **Stem** to 50 cm. **Flowers** 18 cm diameter, standards paler than the falls; standards 6.4–8.5 cm long, 5.5–7.5 cm wide, suborbicular with fine purple veins on a white to pale lilac ground; falls 6.5–8.5 cm long, 4.5–6.5 cm wide, obovate, strongly recurved, with purple veins and oblong purple-brown spots on a creamy ground; signal patch orbicular, almost black, 1.2 cm long, 1.5 cm wide; beard sparse, of purple brown hairs. **Capsule** not recorded. **Flowering** April–May.

Observations
Reported to be locally abundant in places. Bastow (1968) describes this species as a non-stoloniferous *I. bismarkiana*.

I. heylandiana Boiss. & Reut. ex Boiss.

Distribution
The type of this name originated in northern Iraq, just to the south of Mosul, but it may occur also in north-east Syria.

Description
Rhizome not known. **Leaves** 5–7, 1–1.2 cm wide, the outer falcate. Stem 15–40 cm tall. **Flowers** 8–9 cm diameter or larger; standards 5.5–7.5 cm long, 4.5–5 cm wide, broader than falls, white with fine veins and dots; falls 5.5–7.5 cm long, 3.5–4 cm wide, veined and spotted brownish-violet on a whitish ground; beard of rather sparse hairs. **Capsule** not recorded. **Flowering** April–June.

Section Oncocyclus

Observations
A very little-known species. The measurements given here for the standards and falls are taken from Chaudhary et al. (1975). Boissier (1882) confused this with *I. gatesii*. It has also been confused with *I. nectarifera* and both species are recorded from the Derbassieh area of Syria, possibly because what S.A. Chaudhary was describing from there as *I. heylandiana* is the same as the species later described as *I. nectarifera* from adjacent Turkey. True *I. heylandiana* needs to be collected again in the type area and then closely compared with *I. nectarifera* from Turkey.

I. iberica Hoffm.

Description
Rhizome compact, up to 1.5 cm in diameter. **Leaves** 4–6, to 0.6 cm wide, often falcate. **Stem** 13–30 cm tall. **Flowers** 10–15 cm diameter; standards 4.5–10 cm long, 3.5–7.4 cm long, more or less orbicular, variable in colour; falls 3.5–6.5 cm long, 2.7–6.5 cm wide, heavily veined deep purple or maroon on a white ground with a large black signal patch and purplish-brown hairs on the claw, orbicular, often emarginate. **Capsule** 7–9 cm long, 2–2.5 cm wide. **Seeds** about 0.5 cm diameter.
Three subspecies are recognised.

subsp. *iberica*

Distribution
Georgia, from the north-western environs of Tblisi, south-eastwards along the north side of the Kura river as far as Lake Adzhinour. Also around the sides of the Tblisi valley on the south side of the Kura.

Description
Stems tall, 20–30 cm (exceptionally 45–60 cm). **Flower**: standards white with pinkish or bluish shading, equal to or smaller than the falls; style branches not deflexed so that a pale area is left between the large signal patch and the style. Exceptionally variable: Gavrilenko, (1962) described the variation in *I. iberica* and recognised 19 colour forms.

Observations
A plant of steppe and semi-desert, particularly on clay or sandy, bare, eroded hills and in unstable soil. Formerly abundant, but now rare because of collecting for export in the 1930s.

subsp. *elegantissima*

Distribution
Southern Armenia, south of Erevan; in NE Turkey, from the Tortum area east to Dogubayazit and north to the Armenian border, south to the area of Lake Van. There are one or two records also from the extreme north-west of Iran, 60 km SW of Maku.

Description
Stems rather short, usually less than 20 cm. **Flower**: standards generally pure white or creamy-white, veined at the base but sometimes veined throughout, rarely pale blue; falls tightly reflexed, veined with reddish maroon; style arms black to speckled above, speckled at the apex below, usually strongly deflexed onto the falls. **Flowering** May–June.

Observations
A plant of bare hills with *Artemisia* and *Astragalus*, or on grassy steppes in rather dry areas at 1100–2250 m. Soil can vary from the heaviest clay to light volcanic. Rain may, or may not, fall before the snow comes in October, but autumn is generally cold and dry. Spring can be either dry and dusty after the snow melts, with an occasional shower, or wet with torrential rain.

This subspecies is common between Horasan and Dogubayazit, along the main road, growing on the uncultivated edges of cornfields and on the pass south of Iğdir across the lower slope of Mount Ararat. I have observed pollination by a large black bee (probably *Xylocopa vulga* Gerstaecker, which is common in eastern Turkey), which crashes into the signal patch and climbs up under the style. I have seen populations both in S Armenia and in N Turkey in which the standards are veined, sometimes quite heavily, with purple; these are quite distinct from subsp. *lycotis*. Rodionenko (1967) mentions forms with pale blue, sulphur-yellow and bright yellow standards from near Erevan. A form with pale blue standards has recently been collected by Nick and Inge Baker, near Muradiye in Turkey on the north-eastern corner of Lake Van.

subsp. *lycotis* (Woron.) Takht.

Distribution
Soviet Armenia, W Iran, from the border south to west of Esfahan and Deh Bid, and Turkey only on the edge of the Yukşekova plain.

Description
Stem rather short, up to 30 cm, usually less in the wild. **Flower**: standards and falls both heavily veined purplish on a white or purple ground, or sometimes pale and finely veined; falls not strongly deflexed, usually held at an angle to the stem; style branches horizontal or slightly deflexed, not overlapping the edge of the signal patch. **Flowering** April–June.

Observations
This species is very rare in Turkey, but Wendelbo (1977) records that it can be seen flowering in great quantities in some parts of the Zagros mountain. The photograph in Mathew (1989) shows particularly fine clumps. The populations studied by Grossheim on the Karagut mountains north of Julfa seem to have been exceptionally variable. I have found a collection from Turkey difficult to flower and temperamental to grow, being susceptible to root rot in humid weather.

I. kirkwoodii Chaudhary

Distribution
South Turkey, in the districts of Maraş, Gaziantep and Hatay. Northwest Syria, in the districts of Bismishly (var. *kirkwoodii*) and El Bara (var. *macropetala*). It grows in stony places on limestone, 750–1700 m.

Description
Rhizome stout, compact. **Leaves** 6–7, to 30 cm long, 0.5–1.5 cm wide, falcate. **Stem** 50–75 cm tall. **Flowers** 13–18 cm diameter, the standards paler than the falls; standards 7–10.5 cm long, 6–8.2 cm wide, spotted and veined deep purple on a whitish or pale blue ground, orbicular; falls 6.5–8 cm long, 4.5–6 cm wide, coarsely spotted and veined deep purple on a whitish or greenish ground, obovate with a small, round signal patch; beard sparse, 1.5–2.5 cm broad, of purplish-black, red-brown or yellow hairs; style arms with erect or reflexed lobes. **Capsule** about 9 cm long. **Flowering** April–May.

Var. *macropetala* Chaudhary *et al.* differs in its 8–9 longer leaves, to 50 cm long, 1 cm wide, and ovate falls about 10 cm long with a central signal patch.

Observations
This species is close in size to *I. gatesii*, but is easily recognised by its more heavily veined, dark purple flowers. Of the Turkish populations, those from the Maraş area are said to be the largest-flowered.

subsp. *calcarea* (Dinsm.) Chaudhary *et al.*

Distribution
North-west Syria; Deir Semaan district, from the ruins of Qala.

Description
Differs in its 7–9 broader leaves, 1.5 cm wide, and shorter falls, 8 cm long or less, obovate–orbiculate, with the signal patch nearer the apex.

Observations
Very rare and threatened.

I. lortetii W. Barbey

var. *lortetii*

Distribution
South Lebanon, upper Galilee and Mount Gilboa.

Description
Rhizome stout, short. **Leaves** around 8, 1–2 cm wide, erect. **Stem** 30–50 cm tall. **Flowers** large, the standards paler than the falls; standards 9–11 cm long, 7–8 cm wide, orbicular, with pale lilac veins on a nearly white ground; falls 5–8 cm long, 4–4.5 cm wide, oblong–obovate, recurved, with lilac to pink dots and veins on a pale cream or yellowish ground; signal patch rather small, purple-brown or reddish; beard of sparse brown hairs. **Flowering** May.

var. *samariae*

Synonym
I. samariae Dinsm.

Distribution
Samaria around Nablus (Schekem); about 800 m.

Description
Differs in its darker standards, with brownish to purple veins on a cream-coloured ground. **Flowering** April.

Observations
Usually found in stony terra rossa, or on rocky limestone slopes.

I. mairiae W. Barbey

Distribution
S Palestine and NE Egypt, in Sinai and the western and northern Negev, especially in Wadi-ul-Abyad and ul-Azrah to ush-Shawahim.

Description
Rhizome stoloniferous. **Leaves** 7–8, 0.4–0.6 cm wide, glaucous, strongly falcate. **Stem** 20–30 cm tall. **Flowers** lilac to violet, the standards larger and paler than

the falls; standards 6–6.5 cm long, 4 cm wide, erect–incurved, purplish or pinkish with deeper veins; falls 5 cm long, 2.5–3 cm wide, oblong-ovate, recurved; signal patch dark velvety; beard of nearly black hairs; stye arms with long upright lobes. **Capsule** not recorded. **Flowering** February–March.

Observations
Found growing in loose sand with *Artemisia monosperma*; fairly common in some areas.

I. meda Stapf

Synonym
I. fibrosa Freyn.

Distribution
Iran, in the southern foothills of the Elburz between Karadj and Tehran, especially around Mianeh, and southwards from Arak to Hamadan and Sanandaj.

Description
Rhizome thin, tufted, to 1 cm across. **Leaves** usually upright, not recurved, narrow, 0.15–0.4 cm wide. **Stem** 10–25 cm, usually about 15 cm tall. **Flowers** rather small, 5–7 cm diameter with ground colour white, yellowish or pale lilac with greenish or reddish-brown veins; standards broadly 6–7 cm long, about 3.5 cm wide, oblanceolate, falls about 6 cm long, often curled under with a purplish signal patch; beard narrow, of long yellow hairs; style arms greenish yellow, veined near apex. **Capsule** not recorded. **Flowering** April–May.

Observations
Found on dry hills, often in sandy or gravelly soils at 1400–2400 m. Plants have narrow, upright leaves and flowers with widely spaced veins rather similar to, but slenderer, than those of the Turkish *I. sari*. This species has not done well in cultivation in England, although one collection from near Mianeh survived for more than 20 years. It probably requires drier conditions than most of the northern species and particularly dry summers.

I. nectarifera Güner

Distribution
Turkey: Mardin, near Kiziltepe, on steppe in terra rossa at about 500 m. Also recorded in Brian Mathew's *The Iris* (1981) from NE Syria, near Derbassieh, and probably from N Iraq in the Sinjar and Tel Afar regions.

Description
Rhizome stout, with long stolons. **Leaves** 6–8, 0.8–.13 cm wide, falcate. **Flowers** 13–16 cm diameter, veined and flushed purple on a white or yellowish ground, standards paler than the fall; standards slightly veined, 7–8.5 cm long, 3.7–4.2 cm wide, obovate; falls heavily veined with a deep purple signal patch and a narrow yellow beard, 6–7.5 cm long, 2.3–2.5 cm wide, lanceolate or narrowly elliptic; nectary on each side of the base of the falls; style arms with erect to recurved lobes. **Capsule** 4.5–6.5 cm long. **Flowering** April.

var. *nectarifera* has the leaves 0.9–1.3 cm wide; the perianth tube 0.35–0.40 cm long.

var. *mardinensis* Güner has the leaves 0.8–0.9 cm wide; the perianth tube 2.0–2.5 cm long.

Observations
This species was discovered by A.T. Güner in 1979; I have not seen living specimens. In many ways the flowers are similar to those of *I. sari*, but the stoloniferous roots of *I. nectarifera* are characteristic. It is also close to the little-known *I. heylandiana* from N Iraq. Its requirements in cultivation should be similar to those of *I. gatesii*, which is found nearby.

I. nigricans Dinsmore

Distribution
Jordan, east of the Dead Sea, near and south of Amman; Es Salt, and Moab, in Madaba, Karak and Um-el-Ammud at 750–900 m.

Description
Rhizome smallish, compact. **Leaves** 6–13, 0.8–0.9 cm wide, more or less falcate. **Stem** 30–40 cm tall. **Flowers** dark brown-purple, concolorous; standards 9–10 cm long, 6–7 cm wide, larger and slightly paler than the falls with purple veins and dots on a paler ground; falls 6–7 cm long, 3–4.5 cm wide, recurved; signal patch black; beard with purple hairs on a cream to pale yellow ground; style arms not recorded. **Capsule** not recorded. **Flowering** April.

Observations
Grows in fallow fields and steppe.

I. paradoxa Steven

Distribution
Armenia, NW Iran and SE Turkey.

Description
Rhizome tufted, without elongated stolons, slender, usually less that 1 cm in diameter. **Leaves** narrow, 2–5 cm wide, often falcate–recurved. **Stem** 10–25 cm long. **Flower**: standards variable in colour form white to dark purplish-black, 5–9 cm long, 3.3–6 cm wide, broadly obovate; falls much reduced, 2.5–4 cm long, 1–1.5 cm wide, covered with blackish-purple or violet hairs, with a pale V-shaped mark in the centre. **Capsule** not recorded. **Flowering** April–June in the wild.

Habitat
Gravelly steppes and loose earthy slopes, on the edges of cultivated fields and on mountain plateaux at 1750–2800 m.

Several different colour forms of this species have been described.

f. *atrata* Stev.

Distribution
Southern Armenia.

Description
Standards blackish, less wavy than f. *paradoxa*.

f. *choschab* (Hoog) Mathew & Wendelbo

Synonym
I. medwedewii Fomin.

Distribution
S Armenia, SW Azerbaijan, in the Talysh and in Turkey; Iran between Khoi and the Turkish border.

Description
Standards white or pale lilac, lightly veined with blue; beard black, covering the upper half of the falls; style arms cream, speckled reddish-brown.

Observations
In Turkey this form is restricted to the mountains between Lake Van and the town of Başkale. It is common around the village of Hoşap, familiar to travellers in the area for its spectacular castle. Choschab, meaning 'good water', was an old spelling for the castle which gave this form of the iris its name. The village is now often given the Turkish name of Güzelsu. The climate in this area is rather similar to, or a little drier than, that found east of Erzurum, described under *I. iberica* subsp. *elegantissima*.

Section Oncocyclus

I. paradoxa f. *choschab* has often been introduced into cultivation and has usually done well for a time before dwindling away.

f. *mirabilis* Gavrilenko

Distribution
Azerbaijan, near Akstafa; NW Iran, near Julfa.

Description
Standards pale blue or pure white; falls with a golden-yellow beard and dark apex.

Observations
This form was found in Soviet Azerbaijan by Gavrilenko and described as having pale blue standards. Paul and Polly Furse collected a similar form near the border in Iran in 1966 which had white standards. This was introduced into cultivation (PF 7035), but is probably now lost.

f. *paradoxa* (syn. f. *vulgaris*)

Distribution
Southern Armenia.

Description
Standards bluish-purple with wavy edges; beard black or yellow covering the upper half of the falls; style arms yellowish-green suffused or speckled with purple.

Observations
Through the kindness of Prof Gabrielian, I received some rhizomes from the Botanical Garden in Erevan; they grew well in similar conditions to f. *choschab*.

I. petrana Dinsmore

Distribution
Jordan and Israel: mainly to the east and south of the Dead Sea and in the central Negev, mainly between Yesoham and Dimona, and in Jordan from Shaubak southwards; generally at 1200–1500 m.

Description
Rhizome thin, stoloniferous. **Leaves** 6–10, 0.3–0. 6 cm wide, strongly falcate, glaucous, much shorter than the stem. **Stem** to 20–25cm tall. **Flowers** dark brown to dark purple, usually concolorous, rarely with standards paler than the fall; stan-

dards 5.5–6 cm long, 3.5–4 cm wide, incurved, nearly uniform in colour or with darker veins; falls recurved, uniform in colour; signal patch black; beard of dark-tipped yellow hairs on a creamy ground; style arms brownish lilac. **Capsule** not recorded. **Flowering** April.

Observations
Found growing in semi-desert, in gravelly sand with *Anabasis articulata* and *Retama raetum* and on low chalky hills with *Artemisia herba-alba* and *Anabasis syriaca*.

I. sari Schott ex Baker

Distribution
Turkey, scattered across central and SE Turkey from Çankiri, Amasya and Ankara, east to Bayburt, Erzurum and the mountains south of Lake Van; 900–2700 m.

Description
Rhizome compact, up to 2 cm diameter. **Leaves** 5–7, 0.3–0.15 cm wide, often falcate in the small form. **Stem** 6–30 cm tall. **Flowers** 12–15 cm diameter with a yellowish ground colour variably veined with reddish-brown, reddish-purple or black; standards the same colour as the falls, obovate or suborbicular, 6–8.5 cm long, 3.5–5.8 cm wide, crenate and undulate, often darker than the falls; falls 5.5–8 cm long, 2.8–4.5 cm wide, elliptic, obtuse or rounded, usually with the sides bent back and often with the apex curled under, with a deep maroon signal patch and a golden yellow beard; style arms pale yellowish, finely streaked with brown. **Capsule** 4.5–6 cm long. **Flowering** April–June.

Observations
There are two forms of this species. In the west the plants have narrow, usually falcate leaves, short stems (up to 9 cm) and small flowers, usually yellowish with red-brown veins; these are found on bare, open steppes. One of these small forms was called *I. manissadjianii* Freyn, described from Karaman dağ, near Amasya.

In the south and east the plants have broader, upright leaves, taller stems (to 30 cm) and larger flowers often very heavily veined with black; these are found on richer mountain slopes, often with tall umbellifers, paeonies and vetches or among oak scrub. I have had both forms in cultivation for 25 years and the size and colour differences have been retained. Plants from the Gaziantep area are large-flowered with purple veins. Neither form is difficult to grow, although the small form tends to be more free-flowering.

I. sofarana Foster

Distribution
Lebanon, in the Falougha district between Beirut and Damascus, especially in the Zehleh pass. Found growing on rocky and stony hillsides, 1400–1500 m.

Description
Rhizome stout, compact. **Leaves** 8–9, 25 cm long, 1.2–2.5 cm wide, sometimes slightly falcate. **Stem** to 40 cm. **Flowers** 15–18 cm diameter, standards same colour as the falls; standards lightly marked with blue-purple to maroon-purple veins and dots on a white to inky-blue ground, 9.5–10.5 cm long, 7–8 cm wide, orbiculate; falls similar in colour, 8–8.5 cm long, 5–6.5 cm wide, obovate, orbiculate, pointing downwards with the signal patch orbiculate, wider than long, closer to the apex; beard thin, dark purple; style arms curved to form half a tube, dark purple. **Capsule** 10.5 cm long, inflated and 6-lobed. **Flowering** April.

Observations
A similar plant is reported from Turkey in the Belen district of Hatay in the supplement to the *Flora of Turkey* (Mathew 1984). Peter Werckmeister (1957) reports that this species is pollinated, in Lebanon, by the blue bee *Xylocopa violacea*. A yellow to white form, f. *franjieh* Chaud. *et al.*, is known from the Falougha area.

subsp. *kasruana* (Dinsm.) Chaud.

Synonym
I. kasruana Dinsm.

Distribution
Central Lebanon: Naba-al-Asal and Laqlouq. Found in rocky places in the mountains at about 1300 m.

Description
Differs from subsp. *sofarana* in having up to 10 leaves; standards paler than the falls, 8–11 cm long, 6–8 cm wide, obovate; falls 8–10 cm long, 6–7.5 cm wide, ovate, with tear-shaped signal patch, 1.5–2.5 cm long, 0.6–1.5 cm wide; styles curved almost into a tube.

I. sprengeri Siehe

Synonyms
I. elizabethae Siehe; *I. ewbankiana* Foster var. *elizabethae* (Siehe) Dykes.

Distribution
Turkey, in central Anatolia in the provinces of Nigde and Konya, to the southeast of Tuz golu.

Description
Rhizomes slender, with long creeping stolons, forming patches up to 1 m across. **Leaves** falcate, 0.3–0.5 cm broad. **Stem** 6–15 cm tall. **Flowers** veined with reddish or purplish-brown on a white ground, the veins rather coarse; standards elliptic–oblanceolate, acute, 5.7–5.8 cm long, 2.3–2.7 cm wide, with a brown signal patch, curled under at the apex; beard narrow with yellow or cream-coloured hairs. **Flowering** April–May.

Observations
From rocky steppes and unstable pumice slopes; 1000–1980 m. This species is similar in flower to *I. meda*, but differs in its long stolons. *I. nectarifera* from SE Turkey also has long stolons, but differs in having broader leaves and stout rhizomes.

I. swensoniana Chaudhary, Kirkwood & Weymouth

Synonym
I. barnumae var. *zenobiae* Mouterde.

Distribution
S Syria, Tell Chehan.

Description
Rhizome small, compact, with dense sprouts. **Leaves** up to 8, 20 cm long, 1 cm wide. **Stem** to 40 cm tall. **Flowers** scented, with standards slightly paler than falls; standards 6–8.5 cm long, 3.5–5 cm wide, purple or dark maroon; falls 6–7 cm long, 3–3.5 cm wide, recurved and often folded back, dark purple to almost black with darker veins; signal patch to 2 cm wide, notched, wider than long, velvety, almost black; beard dense, of bright yellow hairs 0.5 cm long, purple-tipped; style arms keeled, orange, strongly streaked with purple. **Capsule** 8–10 cm long, about 2.5 cm wide. **Flowering** April.

I. westii Dinsmore

Synonym
I. sofarana forma *westii* (Dinsm.) Mouterde.

Distribution
S Lebanon: endemic to the heights of the Mesghara–Jezzine area at 1800 m.

Description
Rhizome small or medium in thickness, compact. **Leaves** 6–8, 0.6–1 cm wide, slightly falcate. **Stem** 30 cm. **Flowers** 12.5–15 cm diameter, standards much paler than the falls; standards 6–9 cm wide, 5–6 cm wide, obovate, orbiculate, with lilac-blue veins and minute dots on a pale lilac ground; falls 5–8 cm long, 5–5.5 cm wide, elliptic–obovate, veins and spots embossed, brown-purple to purple, dense; signal patch round, in the middle of the fall, 1.5 cm across; beard of long, sparse purple hairs, very wide, extending to the margin of the falls; style arms horizontal to oblique. **Capsule** not recorded. **Flowering** April–May.

Observations
This species is reported to be very rare, or possibly extinct in the wild. It was recorded from loose terra rossa in rock crevices, with *Spartium junceum*. It is similar to *I. hermona*.

I. yebrudii Chaudhary

Distribution
Central Syria, Yebrud area between Damascus and Homs.

Description
Rhizome small, compact. **Leaves** 5–8, 21 cm long, to 1 cm wide, often strongly falcate. **Stem** 15–18 cm tall, to 30 cm in cultivation, usually hidden by the leaves. **Flowers** 13 cm diameter, standards slightly paler than the falls; standards 7.5 cm long, 6.5 cm wide, orbicular, pale yellow with fine purple veins and dots; falls 7 cm long, 5 cm wide, oval to obovate, with dark brown-purple embossed veins, except below the signal patch, where it has only spots, on a pale yellow ground; signal patch 1 cm diameter, dark purple; beard of long purple hairs reaching below the spot on either side. **Flowering** May.

subsp. *edgecombii* Chaudhary

Distribution
Central Syria, in the Kastel area.

Description
Differs in its larger flowers and more reddish-purple overall colour. Standards 10 cm long; falls 9 cm long. Flowering April.

Observations
Reported by Chaudhary to be very rare and in danger of extinction.

Other species

I. susiana L.: the Mourning Iris or Lady in Mourning.

This was the first oncocyclus iris to be cultivated in northern Europe. It was brought to Clusius in Vienna by Ogier Ghiselain de Busbecq, the ambassador from the Holy Roman Emperor, in 1573. Busbecq found hyacinths, tulips and irises being cultivated by the Turks. The name Susan is the Arabic for iris. *I. susiana* has been cultivated in Europe ever since; Brian Mathew is of the opinion that the plants cultivated now could be the original clone introduced by Busbecq. The plant cultivated as *I. susiana* at present is closest to *I. sofarana* and *I. basaltica*. It has a large, very rounded flower with bluish standards with fine dark grey lines and stipples, and falls with heavier brownish lines.

The following are considered hybrids; all come from Transcaucasia.

I. camillae Grossheim

Distribution
Transcaucasia; Georgia, Lake Kazan Gel, and on hills of the Kura.

Description
Rhizome medium in thickness. **Leaves** narrow, strongly falcate. **Stem** 20–40 cm tall. **Flowers** 6–8 cm diameter, standards larger than the falls; standards broader than, and the same colour as, the falls, violet, pale blue or pale yellow; no measurements for standards or falls recorded; signal patch dark purplish; beard of yellow hairs. **Capsule** not recorded. **Flowering** April.

Observations
Found in stony places. Possibly of ancient hybrid origin between *Ii. iberica, paradoxa* and *acutiloba*.

I. annae Grossheim
A hybrid found only a few times in the wild.

I. grossheimii Woronow
Probably *I. lineolata* × *I. lycotis*.

I. kazachensis Grossheim
Probably *I. lineolata* × *I. paradoxa*.

I. koenigi Sosn.
I. paradoxa × *I. iberica*♀. The reciprocal cross produced *I. sosnowskii* Matv. and *I. ketzhoweli* Matv.

I. schelkownikowii Fomin
: Close to *I. acutiloba*, possibly of hybrid origin and with some characters of *I. camillae*. Flowers scented; standards larger than falls; beard yellowish; style branches brown, well developed and conspicuous forming a tube. From between Baku and Tblisi on the left bank of the Kura River; Karadshi-dag and Boz-dag.

I. sinistra Sosnowsky

I. tatianae Grossheim

I. zuvandicus Grossheim
: Probably a hybrid between *I. paradoxa* f. *choschab* and *I. acutiloba* susbp. *lineolata*.

Section Regelia   Lynch

Map 4, Figure 4.

This section has been considerably reduced over the past two decades as putative members were transferred to other groups; new members have been added, however.

In general these are plants from high, open, mountainous sites in a relatively small area of north-eastern Iran, northern Afghanistan and Russian central Asia. These areas are extremely cold in winter, are adequately watered in spring and very late autumn, and have hot, arid summers. So success in cultivation requires a completely dry and warm dormancy period. This is most easily achieved by using a plant frame, because the plants tend to be too vigorous for pot culture. In zones with a continental or mediterranean climate, plants can be grown out of doors with drastic drainage and protection from winter rain, but only if the subsoil also permits rapid clearance of water. Although in many ways they are easier to grow than the oncocyclus irises, to which they are fairly closely related, it is as well to discourage them from coming into growth too early in the autumn. In the wild, late rain shortly before the winter snowfalls encourages the preliminary root growth, which is essential; the snow then discourages further activity until the spring thaws, after which growth is rapid. It is possible, indeed it can be worthwhile, to lift the plants after they have died down, store them in sand in a warm, dry place and re-plant as late as possible in the autumn. A low-nitrogen, balanced inorganic fertiliser should be added to the potting compost and, if necessary, liquid feed when in vigorous growth.

They are distinguished from the oncocyclus irises by usually having more than one flower and that with a vertical orientation rather than globular. There are also similarities with Psammiris, Pseudoregelia and Hexapogon groups, not least in that we would like to know a lot more about all of them. They are best known for having thin beards on the standards as well as thick ones on the falls: if the falls were pendant like the standards, all beards would be on the same, upper surface; as it is, the standards are more or less erect and so the beards are on the inner surface. Many of these plants may be short-lived in the wild; it is probable that more attention should be given to seed setting in cultivation and the raising of new stock in spite of the difficulties with germination.

I. afghanica Wendelbo

Distribution
Afghanistan, Hindu Kush north of Kabul.

Description
Rhizome brown, making clumps up to 20 cm across; fairly compact. **Leaves** to 30 cm tall, 0.2–0.6 cm wide, outer usually falcate, inner erect, green, membranous margin, sheathing stem at base. **Stem** to 23 cm tall, 1–2 cauline leaves; bracts 6–8 cm long, inner slightly shorter, acuminate, green, almost membranous. **Flower** solitary; tube 2–2.5 cm long; falls 6–9 cm long, 2–3 cm wide, cream to yellow with heavy purple veining and conspicuous deep purple signal surrounding central beard of long variably coloured hairs, elliptic–oblong to oblanceolate, acute; standards 6–7.5 cm long, 1.8–2.7 cm wide, cream-yellow, erect, obovate or elliptic–obovate, subacute, occasionally notched at apex, central beard with greenish hairs on lower $1/2$–$1/3$ of inner surface; styles 3–3.5 cm long, 1.6–2 cm wide, arched, obovate, bilobed, pale brown with darker spots; lobes 0.3–0.5 cm long, rounded, slightly crenate; stigma entire; filaments 1.2–1.4 cm long, greenish; anthers 1.2–1.8 cm long, pale yellow; pollen cream; ovary 1.5–2 cm long, beaked. **Capsule** and seeds unknown. **Flowering** May–June. $2n = ?$

Cultivation
From sheltered slopes with limestone outcrops and weathered granite grits, around 2000 m. Standard alpine house or plant frame composts as above. New growth usually starting in October.

Observations
This plant was extensively cultivated after its first introduction in 1966 by Paul and Polly Furse, but is in short supply at present. It should be noted that *I. afghanica* seems to be an exception to the Regelia rule of 2 flowers although oddly enough Paul Furse, in describing the original finds in the 1968 *Iris Year Book*, says 'with two flowers'. The size of the plants in the wild tends to be related to altitude: the higher, the smaller. The colouring, too, is variable from dark to pale shades.

Section Regelia

I. darwasica Regel

Synonyms
I. lineata Foster, *I. suworowii* Regel.

Distribution
Russian central Asia, Bokhara, and Darwas near Afghanistan, Tadjikistan.

Description
Rhizome short, slender, stoloniferous; persistent fibrous remains of old leaf bases; sometimes making large clumps. **Leaves** 15–40 cm long, 0.4–0.8 cm wide, of tufted habit, basal falcate, others linear. **Stem** to 30 cm; 2-flowered; spathes keeled. **Flowers** 5–6 cm diameter, greenish-yellow or greenish-buff heavily veined brown-red or purple; falls to 7 cm long, about 2.7 cm wide, haft cuneate to same width as blade, beard sparse, appearing dark; standards similar in size to falls, dark beard in caniculate haft, but hairs not always present; style pale yellowish-green, narrow; crests small, triangular; stigma entire. **Flowering** June. $2n = ?$

Cultivation
Probably never in cultivation. Try a plant frame.

Observations
A rare and inadequately known species from rocky slopes at an altitude around 2000 m. Descriptions are unsatisfactory and it is sometimes difficult to believe that the writers were referring to the same plant. It appears closely related to *I. lineata* and *I. korolkowii,* possibly being a subsp. of the latter and distinguished from the former by rounded ends to the longer falls and standards, also rather wider leaves.

I. heweri Grey-Wilson & Mathew

Distribution
Afghanistan, Baghlan Province.

Description
Rhizome 0.5 cm diameter, slender with stolons 1–5 cm long, forming loose clumps; fibrous remains of old leaf bases. **Leaves** 4–7, shorter than stem at flowering, to 14 cm long, 0.3–0.5 cm wide at fruiting, falcate, green, with membranous margins. **Stem** 3.5–9(–30) cm tall; spathes about 3.7 cm. **Flowers** 1–2, 5 cm diameter; tube about 1.5 cm long; falls 4 cm long, 1.25–1.5 cm wide, deep purple-blue, elliptic, oblanceolate, blade sharply reflexed, dense central beard of white hairs tipped purple or lilac, haft whitish veined purple; standards 3.5 cm long, 1–1.25

cm wide, deep purple-blue, erect, narrowly obovate, haft cuneate with sparse beard extending onto blade; style 1.4 cm long, pale purple, trilobed, elliptic; lobes 0.9 cm long, triangular, margins slightly wavy. Filament 0.5 cm long, white; anthers 1.1–1.3 cm long; pollen creamy-white. **Capsule** 3.5 cm long, 1.5 cm wide, short pedicel, remains of tube persist at apex. **Seeds** 0.4 cm long, 0.25 cm wide, dark brown, rough, aril white. **Flowering** May. $2n = ?$

Cultivation
From hilly slopes, field banks and alpine meadow, 1400–1900 m. Alpine house or bulb frame.

Observations
The original material was collected by Prof Hewer, for whom the plant is named, in 1969 but, although further collections have been made, it is not widely cultivated at present. It is similar in many ways to *I. falcifolia,* but differs in the much looser rhizome system and in the leaves. The flower colour appears to be consistent.

I. hoogiana Dykes

Distribution
Turkestan, Pamir Alai mountains and Varzob Valley.

Description
Rhizome stout, spreading rapidly by stolons several centimetres long, very similar to that of *I. stolonifera* when dormant. **Leaves** 38–46 cm long, 2 cm wide, slightly glaucous, green, often tinged purple at base. **Stem** to 65 cm tall; spathes 45–89 cm long, up to 2 cm wide, sharply keeled, membranous in upper 1/3, faintly tinged purple; ovary about 2.5 cm long; tube under 2.5 cm; 2–3-flowered. **Flowers** uniform pale lavender in all parts; falls 7.5 cm long, 3 cm wide, pendant, beard dense brilliant orange, broad on haft but narrowing to a point 1/3 along blade; standards about 7.5 cm long, widening gradually to from haft to rounded apex, strongly bearded; style about 2.5 cm long, crests triangular, erect; stigma entire; anthers long; pollen cream. Pedicel short. **Capsule** long, narrow, tapering to apex, dehiscing laterally. A pure white form, also a good grower, exists and comes true from segregated seed. **Flowering** May–June. $2n = 24$

Cultivation
From well-drained areas and mountain slopes up to 1800 m. If in doubt employ the procedures outlined at the beginning of this section, but this plant is semi-hardy in very well-drained beds and a warm site. It grew at the foot of a south-facing wall at Kew for some years. It is possible to provide overhead shelter from

summer rain; seedlings may differ in their degree of hardiness. The purity of colour makes both forms the loveliest of this Section.

Observations
The species was named for the Hoog brothers, who propagated and distributed it. It is closely related to *I. korolkowii* and *I. stolonifera*, but is a rare exception among irises: colour is a main determinant of specific status and in both forms the absence of veining distinguishes it.

I. korolkowii Regel

Distribution
NE Afghanistan, Russian central Asia, Tien Shan and Pamir Mountains.

Description
Rhizome thick, short-branched; fibrous remains of old leaf bases. **Leaves** 3–5, 35–45 cm long, 0.5–0.9 cm wide, basal, more or less straight, greyish-green, membranous margins. **Stem** 20–30 cm tall, 2-flowered, 1–2 sheathing leaves; spathes 9–10 cm long although that of the higher flower much less, greenish, submembranous; tube 2.5–3 cm long. **Flowers** 6–8 cm diameter; falls 7–11 cm long, haft 3–4 cm long, around 1.5 cm wide, pale with heavy brown-purple veining, central beard of fairly long dark hairs, brown-purple signal near base, elliptic oblong–cuneate, blade sharply pendant; standards 7–11 cm long, 2.5–3 cm wide, pale with lighter veining than falls, beardless, varying from narrowly elliptic–obovate to obovate; styles 4.5 cm long, 1.3 cm wide, elliptic–obovate; lobes 1.3 cm long, 0.7–0.9 cm wide with irregular margin; filament 0.8 cm long; anthers 1.8–1.9 cm long, green. **Capsule** about 6 cm long, ellipsoid–ovoid, beaked. **Seeds** 0.7 cm long, with irregular surface, brown, arillate. **Flowering** May. $2n = 22, 33, 44$.

Cultivation
A plant from dry, open slopes or open forest areas and too vigorous for pot culture, but good in a frame and can be planted out in suitable sites.

Observations
A popular species because of the relative ease of cultivation. There are a number of colour variants and a range of chromosome numbers, but the elegantly balanced, distinctive flower shape makes it easily recognisable. Note the absence of beard from the standards.

The following forms are sometimes found listed as varieties:

concolor: violet veining; spathes strongly keeled, purple marking

leichtliniana: creamy-white with almost black veining and purple-black signal

Section Regelia

venosa: dark-veined almost to the point obscuring the base colour

violacea: heavy violet veining and less compact appearance

Hybrids are known with other Regelias and with members of the Oncocyclus, Psammiris and Pogon groups.

I. kuschkensis Grey-Wilson & Mathew

Distribution
NW Afghanistan, very limited area N of Herat on the road to Kushk.

Description
Rhizome 2.5 cm diameter, stout; densely covered in persistent fibrous leaf bases. **Leaves** 7, tufted, 24–50 cm long, 0.8 cm wide, erect, lineate, apex cucullate, margins narrowly membranous. **Stem** 2–3 cm long; spathes 3, 5–5.5 cm long, 1–1.4 cm wide, subacute, papery, green, margins hyaline. **Flowers** 2, about 6 cm diameter, purple-bronze, heavily veined purple; tube 2.5 cm long; falls 5 cm long, 1.2 cm wide, elliptic, apex obtuse and slightly dentate, abruptly recurved, beard dense, light purple; standards 4–4.4 cm long, 1.4 cm wide, erect, oblanceolate, apex toothed, with sparse beard on inner haft; style 2.5–2.7 cm long, pale purple; lobes 0.5 cm long, triangular, margins ruffled. **Capsule** 5.5 cm long. **Seeds** undescribed. **Flowering** April. $2n = ?$

Cultivation
A plant of grassy hill slopes and sandy gullies at about 1600 m, it is so rare that very little is known. In the wild it consorts with *I. fosteriana* and *I. drepanophylla*, so similar conditions should be effective in a bulb frame.

Observations
First introduced in 1971, there are similarities with *I. darwasica* and *I. stolonifera*, but there are a number of distinguishing features of which the most conspicuous is the very thick, close-packed rhizome structure with the conspicuous, tufted remains of the old leaf bases.

I. lineata Foster ex Regel

Synonym
I. karategina B. Fedtsch.

Distribution
NE Afghanistan, Kataghan, upper Farkhar Valley.

Description
Rhizome short, generally stoloniferous; persistent fibrous remains of old leaves. **Leaves** (15–)25–40 cm long, 0.3–0.6 cm wide, straight. **Stem**: (6–)17–32 cm tall; spathes about 5.5 cm long, subequal, upper parts flushed purple, with pale margins, membranous; tube 2.5 cm long; 2-flowered. **Flower** to 6 cm diameter, brownish; falls about 4.5 cm long, to 1.3 cm wide, elliptic–oblanceolate to narrowly obovate, beard blue, long, on central mid-rib; standards 4.5 cm long or more, elliptic oblanceolate to narowly obovate, with thin beard on lower part of inner side; style 2.5–3 cm long, pale blue, lobes purple brown; anthers 1.4–1.7 cm long, filaments 1–1.2 cm long, pollen yellow. **Capsule** and seeds ? **Flowering** April. $2n$ = ?

Cultivation
A plant of stony granite slopes around 2400 m. It was last introduced by Paul Furse around 1967, but seems to have disappeared since then, probably because of virus infection.

Observations
If the photograph in the 1968 *Iris Year Book* (P.F. 8026, between p. 72 and 74) is any guide, this plant can be spectacularly coloured, but is very narrow in all its flowering parts. There is some doubt about its precise relationship to *I. darwasica* and descriptions made from dried herbarium material are not really helpful.

I. stolonifera Maxim

Synonyms
I. leichtlinii Regel, *I. vaga* Foster.

Distribution
Tadzikhistan, Pamir-Alai Mountains.

Description
Rhizome slender, red-skinned, stolons to 20 cm. **Leaves** 30–40 cm long, to 2 cm wide. **Stem** 30–60 cm tall, spathes 6.5 cm long, sharply keeled, with slightly scarious edges, possibly flushed purple; tube about 2.5 cm, funnel-shaped; 2–3 flowers. **Flower** 7–8 cm diameter, colour very variable but usually with brownish margins; falls 6–8 cm long, 2.5–3.5 cm wide, haft broad, beard yellow or cream; standards 6.5 cm long, 3 cm wide, beard to about 1/3 of length; styles around 3 cm long, colour variable, crests narrow, triangular, stigma conspicuous; filaments short, anthers twice as long, pollen cream or blue. **Capsule** about 7.5 cm, long, sharply trigonal, tapering at both ends, with beaked apex. **Seeds** light brown, white aril. **Flowering** April–June. $2n$ = 44.

Cultivation
A plant of rocky hillsides at 800–2400 m. The *Flora of the USSR* (Komarov 1935) suggests that it is found near streams or in wet meadow land. It is one of the easiest of this group, but is still best in a bulb frame in temperate areas: its wandering habit makes it unsuitable for pots. Outside, it needs an extremely free-draining soil and a warm, sunny position where it can be protected from rain in the dormant period.

Observations
A 'once seen, never forgotten' flower. A number of colour forms with cultivar names are in circulation.

Section Hexapogon (Bunge) Baker

Map 5, Figure 5.

A pair of curious irises, closely related to the Regelias, and found over a surprisingly wide geographical area. The outstanding characteristic is fully developed beards on standards as well as falls.

I. falcifolia Bunge

Synonym
I. filifolia Bunge.

Distribution
Near-deserts of Central Asia with plants widespread into south Afghanistan, Baluchistan, north-eastern Iran, Kara Kum and Kyzl Kum in Kazakhstan. About 1200 m.

Description
Rhizome compact, gnarled, with thick roots, covered with fibrous remains of old leaves, forming small dense tufts. **Leaves** to 25 cm long, 0.2–0.4 cm wide, greyish-green, falcate, generally with a minutely hairy covering. **Stem:** 10–35 cm long, 1–2 sheathing leaves fairly erect from base; slender; spathes 3–4, 2.5–5 cm long, greenish flushed purple, submembranous, margin narrowly hyaline, usually 2–3-flowered, sometimes 5; pedicel very short; tube about 3 cm; ovary close to stem and hidden by leaves. **Flower** 3–4 cm diameter, tube about 3 cm long, lilac-violet; falls to 4 cm long, 0.6–0.9 cm wide, narrowly oblong, lanceolate, with whitish beard to middle of blade; standards to 4 cm long, narrowly lanceolate, caniculate haft, sparsely bearded near base; styles to 2.7 cm, keeled, narrow; lobes 0.8 cm long,

narrowly triangular; anthers slightly shorter than filaments. **Capsule** 3–3.3 cm long, nearly oval, narrowing to points at both ends, dehiscing below apex. **Seeds** about 0.5 cm long, pyriform; aril at smaller end, ring-shaped. **Flowering** March–April. $2n = ?$

Cultivation
This plant has only been in limited cultivation and it seems pointless trying to grow it in either temperate or continental climates. It comes naturally from areas with hot, dry summers and very cold, nearly dry winters. Comparable zones can be found in N America and possibly in parts of Australia where it might flourish.

Observations
This very widespread species is geographically coincident with the Regelias, but at much lower altitudes, and is virtually xerophytic.

I. longiscapa **Ledebour**

Distribution
Apparently confined to the Kara Kum and Kyzl Kum deserts.

Description
This plant seems only to differ from *I. falcifolia* in its leaves, which are 0.05–0.15 cm wide (much narrower) and straighter and the flower may be slightly larger, up to 4 cm diameter.

Cultivation
Probably as for *I. falcifolia,* but even less is known about *I. longiscapa.*

Observations
The only serious differences between these species appear to lie in the size of the leaves and flowers, which naturally affects the appearance of the plants. They could be different forms of the same species in spite of the differences in distribution, but the difficulty of collecting them, let alone growing them, makes detailed analysis out of the question at present.

Section Pseudoregelia   Dykes

Map 6, Figure 6.

An interesting little group of plants found in the high mountain valleys of the Himalayas. Possibly one of the reasons why it is not seen often in cultivation is that although it requires winter protection in temperate climates and so qualifies for the plant frame, it needs plenty of water in the growing season, not usually the case with other plants kept under those conditions. Also, it has become clear that it is not easy to distinguish one species from another by way of herbarium specimens as the recent arrival on the scene of *I. dolicosiphon* shows. An additional difficulty is that, as with the other, nearly related, arillate species, seeds can be very slow and erratic in germinating. Dykes makes the point that these plants should only be lifted when dormant, or nearly so, since the death rate in the growing season is high.

I. cuniculiformis Noltie & K.Y. Guan sp. nov.

Synonym
I. goniocarpa Baker var. *grossa* Y.T. Zhao.

Distribution
China, Yunnan, ?Sichuan.

Description
Rhizomes to 1.5 cm long, 0.7 cm in diameter, upright, forming dense clumps with fibrous roots. **Leaves** 13.5–30 cm long, 0.2–0.9 cm wide, sharply pointed at apex, matt on both sides, veins inconspicuous. **Stem** 14–29 cm long, brown markings, conspicuous in upper part; 2, or more, basal bracts; spathes 2, 3–5 cm long, pointed, subequal, outer closely overlapping the inner, with a purple flush at the base; tube 1–2 cm long, thick; pedicel very short; single flower. **Flower** varying shades of lilac, 6–7 cm in diameter; falls 4.25–5.5 cm long, 1.9–2.3 cm wide at blade, beard to mid-point, hairs yellow or grey, white signal around beard with vivid violet markings, slightly flared; standards paler than falls, 3–4 cm long, 1–1.2 cm wide at blade, haft very narrow, with inturned margins, widening sharply to blade, with notched edges and squarish apex; style to 2.8 cm long, 1.6 cm wide, arching, lobes about 0.5 cm, reflexed, triangular, similar colour to petals, but paler margins; filament to 1.5 cm long, anther to 1.4 cm long, pollen cream. **Capsule:** unknown. $2n = ?$

Cultivation
Found in conjunction with *I. bulleyana* in open, grassy areas, this plant has only recently been introduced to Britain and so no definite advice can be given. So far it has been grown successfully in the Royal Botanic Garden, Edinburgh, in a loam based on John Innes no.2 compost with additional grit. The container was sunk in a soil filled frame in light shade. Since the flowers persist for several days this could prove to be a popular plant.

Observations
The outstanding characteristic of this plant is the upright standards; hence 'rabbits ears'. The sheer size of the flowers distinguishes it from *I. goniocarpa* as well as its

more limited distribution range. It appears that there is another large-flowered species much more closely resembling *I. goniocarpa* and it will be important to distinguish the two and to keep them separate in cultivation.

I. dolicosiphon Noltie

Distribution
Bhutan and possibly further east through the Himalayas into SW China.

Description
Rhizome about 1 cm diameter, short, forming dense clumps. **Leaves** short at flowering, later to 54 cm long, with rounded tip at which the cartilaginous margin forms a slight point, linear, tapering to an acute point, dark green with 'waxy' surface, forming basal flowerless tufts. **Stem** very short; spathes 3–4, outer leaf-like, inner spathe-like, overlapping at bases, membranous, dying down after flowering. **Flowers** up to 8.5 cm diameter, single; tube up to 14 cm long, widening to 1 cm diameter, brownish-violet, glossy, covered by bracts for 2/3 height; falls with blade to 3.3 cm long, 1.8 cm wide, violet, darker at centre, spathulate, apex with wide shallow notches, haft about 1 cm long, 0.8 cm wide, oblong; beard dense, of club-shaped white hairs with orange tips reaching junction of haft and blade; undersurface of falls greenish marked white in centre; standards 3 cm long, 1.5 cm wide, deflexed, dark violet, broad-ended, narrowing abruptly to haft, haft 0.6 cm long, margins curled under, brownish, glossy; style branches to 1.8 cm long, 1.5 cm wide, generally elliptic, dark violet, with paler edges; lobes about 1 cm long, 0.6 cm wide, sharply acute, slightly dentate, reflexed; stigma about 0.8 cm wide, shallow, with slight division at centre; anthers 1 cm long, 0.2 cm wide, very pale violet; filaments 1.2 cm long, blue at top, cream below; pollen whitish. **Capsule** to 5 cm long, ellipsoid, slender, acute at apex, dehiscing below apex. **Seeds** to 0.35 cm long, aril large. **Flowering** June. $2n = 22$.

Cultivation
From steep, sunny hillsides at about 3,500 m, seed was collected in 1984 by David Long and Alan Sinclair of the Royal Botanic Garden, Edinburgh. It germinated, and plants have been grown successfully in the rock garden there in a well-drained, sunny site. It flowers more freely than either *I. kemaonensis* or *I. hookeriana*, but may be self-incompatible since there has been difficulty in obtaining any seed; propagation has had to be by division.

Observations
It seems clear that this plant has been collected several times in the past without being recognised as a distinct species: many herbarium specimens of this section bear a stronger resemblance to *I. dolicosiphon* than to their alleged species. Much more work is needed to establish its exact distribution.

It differs from the other Pseudoregelias in that the newly opened flowers do not appear to have the characteristic blotches, although these are evident if the plant is held up to the light and appear clearly as the flowers wither.

I. dolicosiphon subsp. *orientalis* Noltie subsp. nov.

Distribution
China, NW Yunnan, SW Sichuan; India, Arunachal Pradesh State; Burma, Seinghku Wang.

Description
Very similar to ssp. *dolicosiphon*, but smaller in all parts and with more strongly marked blotching on the falls.

Cultivation
Probably similar to the type.

Observations
Although there are intermediate forms between these subspecies, their geographical dispersion seems to justify the division at present.

I. goniocarpa Baker

Synonym
I. gracilis Maxim.

Distribution
Sikkim, Nepal through to W & C China.

Description
Rhizome slender, very compact, growing points close together. **Leaves** 15–25 cm long, 0.2–0.3 cm wide, erect. **Stem** 10–30 cm long, with sheathing leaf at base; 2 spathes, to 4 cm long, 0.8 cm wide, membranous, reddish; pedicel very short; 1-flowered. **Flowers** 2.5–3 cm diameter, lilac or blue-purple with darker blotches; falls to 3 cm long, 1.0 cm wide, obovate, cuneate, beard short, of dense whitish hairs with orange tips, 2 small swellings at each base; standards about 2 cm long, 0.5 cm wide, erect, apex retuse; styles about 1.75 cm long, with narrow triangular crests; stamens 1.5 cm long, anthers yellow; tube very short; ovary to 1.5 cm long. **Capsule** to 4 cm long, 1.75 cm diameter, cylindrical, with 3 ridges, pointed at apex, yellowish-red. **Seeds** ? **Flowering** May–June. $2n = $?

Cultivation
A plant of mountain valleys, hillsides and open forest areas, 2700–5500 m. Not a plant that seems to stay in cultivation. W.R. Dykes kept it in a sunny garden in

'light, vegetable soil'; probably quantities of leaf mould were added and the drainage was good. Col. D.G. Lowndes writes of finding it in birch wood as well as other sites.

Observations
R.J. Farrer did find a white form of this plant in the wild.

var. *grossa* Y.T. Zhao

Distribution
China, Sichuan Province, Yunnan; and Tibet.

Description
Similar to the type, but larger in many parts. **Leaves** to 30 cm long, 0.3–0.5 cm wide. **Flowers** 6–7 cm diameter.

var. *tenella* Y.T. Zhao

Distribution
China, Quinghai province.

Description
Similar to type, but apparently smaller. **Leaves** to 20 cm long, 0.2 cm wide. **Flowers** about 2.5 cm diameter.

I. hookeriana Foster

Synonyms
I. gilgitensis Baker, *I. kumaonensis caulescens* Baker.

Distribution
W Himalayas, particularly Kashmir; nearby areas of India and W to Chitral.

Description
Rhizome slender, fairly loose, knobbly; remains of old leaf bases persist; roots slender, fleshy. **Leaves** to 20 cm long, 1.5 cm wide at flowering, later to 30 cm or more long, 2 cm wide, yellowish-green, herbaceous. **Stem** 5–12 cm long, hidden in short sheathing leaves; spathes 7.5 cm long, remaining green after flowering, sometimes tinged purple; pedicel very short; ovary around 1.25 cm long, trigonal–cylindrical; 2-flowered. **Flowers** blue-purple with darker blotches, fragrant; tube 1.25–2 cm long, purple-striped on green ground; falls 5–6.5 cm long, 2 cm wide, haft cuneate, blade rounded oblong, beard white with coloured tips; standards 5 cm long, 2 cm wide, haft caniculate, blade oblong, often bluer than falls;

style sharply keeled, crests triangular, recurving, stigma entire, edge serrated; filaments blue, same length as anthers; anthers cream; pollen cream. **Capsule** 5 cm long, narrowing to conspicuous beak, which retains dried flower parts, dehiscing laterally. **Seeds** pyriform, browny-red, wrinkled, aril yellowish. **Flowering** June–July. $2n = ?$

Cultivation
From subalpine meadows between 2600 and 4400 m. It needs a rich, well-drained soil and sunny situation. Ample water in the growing season, but protection from rain in winter.

Observations
The plant is sufficiently similar in appearance to *I. kemaonensis* for it to have been in circulation in Britain under that name for some time about 20 years ago. It can be distinguished by the slender rhizomes, fairly long stem and short perianth tube.

I. kemaonensis Wallich ex D. Don

Synonyms
I. duthieii Foster, *I. kingiana* Foster, *I. tigrina* Jacquemont.

Distribution
Himalayas: Kashmir and eastwards across the mountain states to Sichuan and Yunnan.

Description
Rhizome thick, gnarled; fleshy roots; few fibrous remains of old leaves. **Leaves** to 20 cm long at flowering, later to 45 cm, 0.2–1 cm wide, light green, glaucous, linear, rounded at apex. **Stem** usually very short, but sometimes to 7.5 cm tall; spathes about 5 cm long, 1.75 cm wide, keeled, acuminate, scarious at tip, green, sheathing base of tube; pedicel to 1.5 cm long, ovary subcircular in section; tube up to 6 cm long; bracts 2–3, up to 6 cm long, up to 2 cm wide, pointed; 1–2 flowers. **Flower** 4–5 cm diameter, lilac-purple with darker blotches; falls 4.5 cm long, 2.5 cm wide, haft cuneate, blade ovate; beard dense, of white hairs with yellow or orange tips; white with broken purple veining; standards 4 cm long, 1.5 cm wide, obovate, upright; style about 3 cm long, 0.5 cm wide, dark at centre, paler at edges; crests small, triangular, crenate; stigma entire, crenate; filaments blue; anthers lavender; pollen white. **Capsule** up to 2.5 cm long, tapering to a pointed apex, dehiscing laterally. **Seeds** pyriform, brown-red, small, aril cream. **Flowering** May–June. $2n = ?$

Cultivation
The plant is from alpine meadows, 3000–4500 m. Probably the easiest of this section in a temperate climate, with a preference for a neutral to slightly acid soil,

equally happy in sun or part shade, with good drainage. The foliage tends to persist well into the winter; this makes it easier to locate plants.

Observations
This plant, variously and commonly known as *I. kamaonensis* or *I. kumaonensis*, has now fallen prey to the priority rule and should be known as *I. kemaonensis,* since Brian Mathew's research has shown that J.F. Royle's 1839 description alone fits botanical requirements. The variations are simply due to different transliterations of the local pronunciation.

I. leptophylla Lingelsheim

Distribution
China: Sichuan, Gansu.

Description
Rhizome tuber-like, fleshy, irregular in shape, brownish, rooting at base of stem; roots soft, hairy; few fibrous remains of old leaves persist. **Leaves** 20–25 cm long, about 0.25 cm wide, with prominent mid-rib, tapering gradually to an acute apex. **Stem**: about 17(–35) cm long, slender, with 1 stem leaf on lower half, to 9 cm long, acuminate; 3 spathes about 4 cm long, 1.0 cm wide, with distinct mid-rib, lanceolate, acuminate, green, membranous; pedicel very short; tube about 3.75 cm long, to 1.5 cm diameter, flaring at top; 2-flowered. **Flower:** 5–6 cm diameter, violet; falls 5 cm long, 2 cm wide, blade obovate, with bearded mid-rib; narrow haft; standards about 3.5 cm long, 0.5 cm wide, oblanceolate; style possibly 3 cm long, with whitish markings on blue ground; stamens about 1 cm long, anthers white. **Capsule** to 2.5 cm long, 2 cm diameter, rounded, 6-ribbed, with persistent dry remains of tube. **Seeds** roundish, dark brown; reddish-brown aril. **Flowering** April–May. $2n=$?

Cultivation
No information is currently avilable. This is a plant of shaded areas in and on the edge of forests.

Observations
Material of this plant has only recently become available. Preliminary results indicate that it should be included in this section rather than in subgenus *Nepalensis*.

Mathew notes that the flowers open at midday and when dying twist into a spiral, as do those of the *Nepalenses,* but studies of the leaves by Qi-Gen Wu and David Cutler suggest that it is closer to the Pseudoregelias. Tony Hall of Kew noted that the capsules differed from those of *I. decora,* which it has been thought to resemble.

I. narcissiflora Diels

Distribution
Sichuan, fairly widely distributed.

Description
Rhizome slender, stoloniferous. **Leaves** 20–30 cm long, to 0.3 cm wide, linear. **Stem** to 30 cm long, with 2–3 very narrow sheathing leaves, all carrying gland-like protuberances; 2 spathes about 3 cm long, 1.0 cm wide, acuminate, brownish, papery, tips everted tube to 7 cm long; ovary to 1.5 cm long, spindle-shaped; no pedicel; 1-flowered. **Flower** to 6 cm diameter, yellow. Falls 3 cm long, 2 cm wide, apex rounded, haft cuneate, with thin beard on midrib; standards similar in size, slightly narrower, held horizontally; style to 1.5 cm long, 0.8 cm wide, narrower at both ends, crests irregular; stamens to 1.3 cm long, filaments longer than anthers. **Capsule** unknown. **Seeds** 0.5 cm long, pyriform? **Flowering** April–August, depending on location. $2n = ?$

Cultivation
This is a plant of open areas such as hillsides, forest edges and clearings. It is coming into cultivation; treatment as for the rest of the Section is probably the best way to start.

Observations
The plant was first found in 1922 and described in 1924, but it is only over the past few years that it has been observed on any scale and seeds have been made available from China. Zhao (1985) placed this plant with Section Hexapogon because of its beard, but while its exact status is still unclear it does seem to fit better here.

I. pandurata Maxim

Distribution
China: Gansu and Qinghai, Yellow River Valley.

Description
Rhizome short, tuberous, with rather fleshy roots; covered in fine, persistent fibres from old leaves. **Leaves** to 25 cm long, 0.4 cm wide, with parallel veins, narrowing gradually to a point. **Stem** to 12 cm tall, with small, membranous basal leaves; spathes 2–3, to 6 cm long, 1.5 cm wide, membranous, pointed; pedicel absent or very short; ovary to 1.5 cm; tube to 3 cm long, widening to perianth; 2 flowers. **Flower**: red-purple; falls to 4.5 cm long, 1.4 cm wide, narrow, obovate, flaring; standards to 3.5 cm long, 0.8 cm wide, erect, oblanceolate; stigma about 2 cm long, crests triangular, erect, divided; stamens to 2.5 cm long; anthers purple.

Capsule to 3.5 cm long, 1.5 cm diameter, roundish oval, with 6 conspicuous ribs, pointed at apex. **Seeds** 0.4 cm long, 0.2 cm diameter, roundish, rough, red-brown, with no aril. **Flowering** May. $2n = ?$

Cultivation
Only recently available. Very little is known about its requirements, but they are possibly similar to those of *I. tigridia*.

Observations
Recent work in China suggests that this plant can be distinguished from *I. tigridia* by the possession of roots with a constant diameter down their length.

I. sichuanensis Y.T. Zhao

Distribution
China: Gansu and Sichuan.

Description
Rhizome to 1.5 cm diameter, tuberous, brownish; thin short roots; some fine fibres left from old leaf bases. **Leaves** to 35 cm long, 1 cm wide, with conspicuous midrib, narrowing gradually to apex. **Stem**: to 20 cm long; 1–2 small, sheathing leaves; spathes 3–4, to 8 cm long, 1.8 cm wide, narrowing sharply to apex, membranous; pedicel very short; ovary to 3 cm long, slender, spindle-shaped; tube to 5 cm long, diameter very narrow at base widening greatly to top; 2–3 flowers. **Flowers**: violet, to 6 cm diameter; falls to 5.5 cm long, to 2 cm wide, haft narrowly cuneate; beard on mid-rib conspicuous, yellowish, hairs club-shaped; standards to 4 cm long, 1 cm wide, erect, lanceolate; style to 4.5 cm long, crests triangular, erect, divided; stamens very long. **Capsule** to 4 cm long, 1.3 cm diameter, pointed at apex; dehiscing from middle. **Flowering** April. $2n = ?$

Habitat
Grassy areas on hillsides and roadsides.

Observations
Originating from the same area as *I. leptophylla*. There has been some doubt as to whether it is a separate species, but there do seem to be considerable structural differences. B. Mathew finds the plant self-fertile.

I. sikkimensis Dykes

Distribution
Unknown.

Section Pseudoregelia

Description
Rhizome slender, gnarled, with fibrous remains of old leaves. **Leaves** 10–20 cm at flowering, later 30–45 cm long, 1.25–2 cm wide, pale green, ensiform. **Stem** 10–15 cm high, with sheathing leaf at base, surrounded by 4 reduced leaves; spathes 5–7.5 cm long, lanceolate, green, outer keeled, both scarious at top and along edges at flowering; pedicel 1.25–2 cm long; ovary 2 cm long, trigonal, green with faint purple markings. **Flowers** 2–3; tube 3.5–5 cm long, trigonal, deep purple; falls 6.5 cm long, 2.5 cm wide, dark purple-lilac, with darker mottling particularly round beard, haft cuneate, white with purple blotching; beard white with orange tips; standards 5 cm long, 1.75 cm wide, mauve-lilac, held diagonally, blade oblong, with wide margin, narrowing sharply to caniculate haft; styles 2.4 cm long, narrow, purplish-blue at centre, paler at edges; crests 1.25 cm long, triangular, revolute; stigma obscurely bilobed, edge irregularly indented; filaments pale violet, equal to anthers; anthers creamy white; pollen cream. **Capsule** and seeds unknown. **Flowering** time unknown. $2n = ?$

Cultivation
Dykes grew this plant successfully for four years, but without obtaining any seed and presumably under the same conditions as his other plants from this section. He, very properly, described it, but it has never been collected again and Dykes himself, observing several resemblances to both *I. hookeriana* and *I. kemaonensis*, thought it might be a natural hybrid. It certainly appeared to be self-sterile. Only further collection, or possibly hybridisation, will settle the matter.

I. tigridia Bunge ex Ledebour

Distribution
Altai Mts, Manchuria and Mongolia.

Description
Rhizome small, compact, with fibrous remains of old leaves; roots numerous and fleshy. **Leaves** to 12 cm long, 0.4 cm wide, erect, grassy. **Stem** to 4 cm, subterranean; spathes 2, small, narrow, yellowish; ovary to 1.2 cm long, slender, green; tube to 2 cm long, broadening at top; single-flowered. **Flowers** violet, to 4 cm diameter; falls 3.5 cm long, 1 cm wide, blotched purplish brown to white, haft cuneate, beard of blue-white hairs with yellow tips on mid-rib; standards to 3 cm long, 0.5 cm wide, darker than falls, erect; styles to 2.5 cm long, crests narrowly triangular; pollen blue. **Capsule** to 4 cm long, narrowing sharply to apex where dried perianth persists; texture corky. **Seeds** roundish, with creamy aril. **Flowering** May. $2n = 38$.

Cultivation
From sandy areas or stony, barren mountain slopes with cold, dry winters. An awkward candidate for the bulb frame in temperate areas. It must have a dry winter; however, as it grows all summer it needs some water, but drainage must be efficient. It will not tolerate standing water around leaf bases. Nor is it easy to propagate vegetatively, but Dr Boussard was given 'small, and more or less dry, plants' in one September, which grew on well in the following spring. This confirms Dykes's view that they should only be moved when dormant.

Observations
There are still lingering doubts about the precise classification of this plant, which does strongly resemble the other Pseudoregelias, but which is geographically well out of their range.

var. *fortis* Y.T. Zhao

Distribution
China, Jilin province, Shanxi and Inner Mongolia.

Description
Similar to *I. tigridia*. **Leaves** to 20 cm long, 0.6 cm wide. **Stem** to 20 cm tall, spathes to 5cm long, 1 cm wide. **Flower:** all parts marginally larger.

Observations
This form has only recently been identified and is not in general cultivation.

Subgenus Limniris: the Beardless Irises

Section Lophiris (Tausch) Tausch (the Evansia Irises)

CHRISTOPHER BREARLEY AND J.R. ELLIS

Map 7, Figure 7.

This section consists of a small number of rhizomatous and sometimes stoloniferous irises whose common distinguishing feature is the usually prominent cockscomb-like crest on the falls of the flower. Hence they are frequently called 'Crested Irises'. In most species the crest is a single linear structure, but in two species (*Ii. cristata* and *lacustris*) the crests are triple. Usually the margin of the crest is entire, but in some species the margin is dissected either coarsely (*I. tectorum*) or finely (*I. milesii*). In *I. tenuis* the crest has the form of a slightly elevated rounded central ridge or mid-rib, rather than a prominent cockscomb-like structure. Apart from the crest, the characteristics of its members are diverse, ranging in size from the dwarf *I. lacustris* to the comparatively vast *I. wattii*. An obvious relationship is often not apparent. Most species are essentially either thin- or soft-leaved plants associated with moist woodland.

Eleven species are recognised at present: *I. confusa* Sealy, *I. cristata* Solander, *I. formosana* Ohwi, *I. gracilipes* Gray, *I. japonica* Thunberg, *I. lacustris* Nuttall, *I. latistyla* Y.T. Zhao, *I. milesii* Foster, *I. tectorum* Maximowicz, *I. tenuis* Watson and *I. wattii* Baker.

Other than the North American species *Ii. cristata*, *lacustris* and *tenuis*, all the members of the section are native to East Asia. Several of the Asiatic species are easily mistaken for orchids. R.A. Salisbury, *Transactions of the Horticultural Society*, (1812) considered *I. japonica* so distinct from other known irises that he formed a subgenus, *Evansia*. This was in honour of Thomas Evans of the India House, who introduced *I. japonica* to England in 1794. W.R. Dykes (1913) considered *Evansia* a section and Lawrence, *Gentes Herbarum* (1953), a subsection of the genus *Iris*. Rodionenko (1961) classified them at subgeneric level in the subgenus *Crossiris* Spach (1846), which was divided into three sections, as follows.

Section 1 Crossiris:	(a) Series Japonicae: *Ii. japonica*, *speculatrix* and *wattii*	
	(b) Series Tectores: *Ii. milesii* and *tectorum*	
Section 2 Lophiris	Series Cristatae: *Ii. cristata* and *lacustris*	
Section 3 Monospatha	*I. gracilipes*	

Although this system is a useful classification it does not include all the known Evansia species and it is not as homogeneous as indicated. Wu & Cutler (1985) after an anatomical survey suggested that *I. speculatrix* had more features in common with members of the Series Chinenses of Section Limniris and should accordingly be transferred to this section. They also proposed that *I. rossii* should be transferred from Series Chinenses to the Evansia group (Lophiris). In addition, the authors also considered that the inclusion of *I. gracilipes* in the Evansias was perhaps spurious and while they did not propose any taxonomic change they noted that Rodionenko (1961) had earlier transferred this species from the Evansia group to form a monotypic section. *Ii. confusa, formosana* and *tenuis* were omitted by Rodionenko; although the first two species obviously belong to Series Japonicae, the classification of the last is uncertain. The absence of a series name in Section Monospatha appears to be an inconsistency.

I. confusa Sealy

Synonym
I. wattii Baker.

Distribution
This species grows in western China in the provinces of Guangxi, Sichuan and Yunnan. It inhabits forest fringes, moist valleys and grass slopes.

Description
A very vigorous, stoloniferous, clump-forming plant that produces erect, 30–80 cm bamboo-like aerial rhizomes in summer, with fans of evergreen leaves at their apex. The leaves may attain a length of 40 cm and a width of about 5 cm. They are lighter in colour than those of *I. japonica*. From these fans arise widely branching inflorescence stalks that bear many small, 4–5 cm across, white flowers. The falls have a deep orange-yellow crest and central patch with the surrounding blade having orange-yellow (and sometimes pale mauve) spots. Flowers are usually produced during April–May and although individually short-lived, their abundance ensures a continuous display over several weeks. All plants are self-incompatible, but readily form cylindrical capsules containing numerous flat, mostly D-shaped seeds following compatible pollinations. $2n = 30$ (Simonet 1934; Snoad 1952).

Cultivation
In hardiness, it is intermediate between *I. japonica* and *I. wattii* and though it can be grown successfully in a sheltered position outdoors, a cool greenhouse is preferable during winter and spring. A 30–35 cm pot or tub, which is well drained, and John Innes No.2 potting compost will give satisfactory results. Because of the shallow root growth, a mulch of peat, well-decayed leaf mould, or similar material, will be beneficial to those growing in beds. Dead leaves and flowers

should be removed. After flowering the old stems should be cut out as near to ground level as possible so as to enable new shoots to develop freely.

Propagation is easily achieved by division after flowering. In addition, the old stems can frequently be rooted in water: the shoots that develop are detached and potted when roots appear.

Observations
This is widely represented in cultivation outside Asia by plants derived from those raised by Dykes. He obtained seeds in 1911 from Père Ducloux, a French missionary, who collected them in the Yunnan. Dykes (1915b) concluded, after comparison with specimens in the Kew Herbarium, that they were identical to *I. wattii*. In his monograph (1913a, p. 101) Dykes had reduced *I. wattii* to *I. milesii*, but now had no hesitation in re-instating it as a distinct species.

Dr O. Stapf of Kew in the years 1924 and 1926 had the Ducloux iris figured for the *Botanical Magazine*. Upon examination he came to the conclusion, unlike Dykes, that it was nearer to *I. japonica* than *I. wattii*. Unfortunately, he had not completed his investigations at the time of his death in 1933, although his notes indicate that he considered it to be a new species.

In 1931, Major Lawrence Johnson introduced a plant which he had collected near Tengyueh, Yunnan. This, following investigation, was identified at Kew as undoubtedly representing the true *I. wattii* of Baker. Dykes was obviously wrong in his identification of the Ducloux species.

Because of the confusion which arose some clarification was necessary. Hence, in 1917, J.R. Sealy described the Ducloux plant as a new species, *I. confusa*. The name is indicative of the fact that it was confused with *I. wattii*.

I. cristata Solander

Synonym
Neubeckia cristata Alef.

Distribution
Widespread in the south-eastern and central states of the USA, particularly around Washington, DC, and Ohio, south and west to eastern Oklahoma, also in the Allegheny, Appalachian and Ozark mountains. It grows on rocky hillsides, in ravines and in moist woods on neutral or slightly acid soils, and may form extensive colonies.

Description
From a tangled mat of slender greenish rhizomes arise fans of yellowish-green, short (15 cm), narrow (1–3 cm), deciduous, pointed leaves. These may elongate after flowering. The spathes bear 1, or occasionally 2, flowers during April to May.

They are about 5 cm across; typically pale lavender with a central white patch on the falls, edged purple with an orange crest comprising 3 crimped ridges. The flowers are almost stemless; it is the 4–8 cm perianth tube that is responsible for raising the flowers above ground level. Capsules ovoid, 1–2 cm long, triangular in cross section. Seeds ellipsoid with a small viscid spiral appendage. $2n = 24$ (Longley 1928), 32 (Simonet 1934).

Several clones including various shades of blue, violet, pink and white are known and some, mostly found as wild plants, have been named.

Cultivation
Culturally, it favours a cool semi-shaded position and a gritty well-drained soil which is rich in humus. Hence the incorporation of peat and sharp sand when preparing the planting position is recommended. Each rhizome annually produces several new ones, which may be severed and re-planted. Ideally this should be done in their period of active growth. Frequent propagation is desirable so as to maintain vigour and a compact clump. It is necessary after re-planting to ensure that the plants are kept moist until re-established. Slugs and snails are often a problem.

Observations
This species is extremely variable in habit and probably some of the smaller, poorer and paler forms are mistakenly grown as *I. lacustris*. It is figured in the *Botanical Magazine* t.412, 1798.

I. formosana Ohwi

Distribution
Formosa. The *Flora of Taiwan* (Lin & Ling 1978) states that it is uncommon and endemic at medium altitudes of 500–1000 m in the central part of the island.

Description
A vigorous plant that spreads by means of slender stolons. In appearance it is very similar to *I. japonica* except that the leaf fans are borne on short stems from which arise branching inflorescences, and the flowers are larger, 7–9 cm across. They are pale lilac-blue with yellow-orange markings surrounded by pale mauve spots. The flowers are produced mainly in April–May and are self-incompatible. Capsules oblong–ovoid, 3–4 cm long with remains of the perianth tube at the apex. $2n = 28$ (Yasui), 35 (J.R. Ellis & Y. Lim, unpublished data).

Cultivation
Requirements are identical to *I. japonica* except that it is less tolerant of low temperatures. It is rarely encountered in cultivation. Propagation is by seed or division.

Observations
This is a plant surrounded by some confusion as to whether it is a true species, a form of *I. japonica* or a hybrid. Ohwi considered it distinct from *I. japonica* mainly because of its larger flowers and its ability to set good seed in the wild. Therefore he accepted it as a new species.

It is represented in cultivation primarily by plants collected by Chow Cheng in Formosa. These were sent by a correspondent in Japan to Dr Maurice Boussard of Verdun and appeared to be intermediate between *Ii. japonica* and *confusa* (*B.I.S. Year Book* 1970, illustration only, p.80). The chromosome count $2n = 35$ refers to this plant and suggests that it may not be representative of the true species.

I. gracilipes A. Gray

Distribution
Occurs in the Japanese mountains of Hokkaido (SW district), Honshu (Kinki district and eastwards), and Kyushu, and also in China. It is a woodland species that appreciates a rich, slightly acid, moisture-retentive soil.

Description
From short branching rhizomes arise 20–30 cm long, narrow (0.5–1 cm), deciduous, gracefully arching, grass-like leaves. The flowers are about 4 cm across and borne on branching stems 10–15 cm tall. They are lilac or pinkish-mauve with a large white patch, veined lilac in the centre of the falls. The crest is predominantly white, but is orange at the tip. There is also a white-flowered form in cultivation with yellow markings around the crest. Distinctive features are that the spathes consist of a single valve and the capsules are almost globose. The seeds are obovate, about 0.3 cm long with a conspicuous cream-coloured raphe. $2n = 36$.

Cultivation
The cultural requirements are identical to those of *I. cristata*.

Propagation
By seed or division in spring while the leaves are still in the bud stages, or immediately after flowering. Any disturbance should be kept to a minimum; transplanting in autumn and winter should preferably be avoided.

Observations
Inter-specific crosses between *Ii. gracilipes* and *tectorum* and *Ii. gracilipes* and *cristata* are known. *I. gracilipes* is figured in the *Botanical Magazine*, t.7926, 1903.

I. japonica Thunberg

Synonyms
I. chinensis, I. fimbriata, I. repanda E. Berg, *I. squalens* Thunb., *Isis fimbriata, Evansia chinensis* Salisb., *E. fimbriata* Decaisne.

Distribution
Native to central China and Japan on shady moist hill slopes, in wood fringes or in sparse undergrowth.

Description
I. japonica is a very vigorous stoloniferous plant that spreads rapidly. It produces fans of dark, shiny evergreen leaves, which are lighter and duller on the undersides. They are up to 60 cm long and about 2.5 cm wide, tapering to a point. The 30–60 cm multi-branched stems carry many flowers, thus ensuring continuity although individually they are short-lived. Normally one could expect 3–4 flowers to be open on each stalk at any one time although occasionally there can be more. Each is 4–5 cm across and frilled. Pale lavender predominates; towards the centre they are patterned white, deeper lavender and orange. They possess a conspicuous orange crest. Flowering usually occurs April–May. Most forms are self-incompatible, but capsules with seeds develop after compatible pollinations. Capsules ovoid, 2.5–3 cm long, pointed at the apex. Seeds dark brown and irregular in shape. Various chromosome counts are reported: $2n = 34, 54$ (Ledger's form) (Kazao 1928); $2n = 36$ (Simonet 1932b); $2n = 31, 32, 33, 35$ & 55 (J.R. Ellis & Y. Lim, unpublished data).

Cultivation
This species can be grown successfully in a sheltered position outdoors. However, cultivation as for *I. confusa* in a cool greenhouse protects the delicate flowers from damage by adverse weather.

Propagation
Easily achieved by division following flowering.

Observations
The different chromosome numbers observed by Ellis and Lim were found in types collected as *I. japonica* and some may represent natural hybrids.

There are several clones of this plant in cultivation, such as 'Uwodu', introduced by S. Berry of California in 1928, and 'Ledger's Variety'. The latter was brought to England from the garden of the British Legation in Tokyo by Sir Frederick Lugard in 1912 and introduced in 1925. It is reputed to be more vigorous and hardier than the type.

Forms with variegated leaves are also known although they frequently lack vigour and flower irregularly and infrequently. Sometimes they revert to plain green leaves.

Hybridisation has occurred between *Ii. japonica* and *confusa*: it should be noted that the hybrid *I. japo-watt* is *Ii japonica* × *confusa*, not *Ii. japonica* × *wattii*. The name arose because *I. confusa* was originally classified as *I. wattii*.

I. japonica is figured in the *Botanical Magazine* t.373, 1797 under the synonym *I. chinensis*.

f. *pallescens* P.L. Chiu & Y.T. Zhao

Distribution
Zhekiang.

Observations
Differs from the type in having yellow-green leaves and spathes; flowers without the blue-purple markings on the falls. Not yet in general cultivation.

I. lacustris Nuttall

Distribution
Confined to the Great Lakes region in the northern USA (Wisconsin, Michigan, Ohio States) and Ontario Province, Canada. It occurs frequently in sandy coniferous woods and calcareous rocks and gravel.

Description
Although similar to *I. cristata* it possesses several positive differences. It has a distinctive chromosome complement and morphologically it is significantly smaller in all its parts and is less variable. An unmistakeable difference is the extremely short (0.2 cm or less) perianth tube. The slate-blue flowers with a whitish patch and triple yellow crest occur during May and sometimes again in autumn. There is also a white-flowered form. Capsules ovoid, about 1 cm long. Seeds oval with a coiled appendage. $2n = 42$.

Cultivation
Identical to *I. cristata*. It is frequently suggested that these species fail to succeed in close proximity because of their differences in relation to acidity and alkalinity. However, experience indicates that both species are tolerant of a range of soils.

Observations
Popularly known as the Dwarf Lake Iris, it was previously considered to be a variety of *I. cristata*, but because of the distinguishing features previously mentioned it has achieved specific status.

I. latistyla Y.T. Zhao

Distribution
Tibet, along margins of forests and paddy fields at altitudes of 3000–4000 m.

Description
Small inconspicuous rhizomes bear thick fleshy roots and brown fibrous remains of old leaf bases. Leaves: pale green, ribbed, 15–25 cm long, 0.2–0.3 cm wide, tapering to a point; leaf bases clasping. Flowering stems 6–14 cm long, unbranched or with a single side-branch. Spathes 3, grass-like (2.5–4.5 cm long, 0.6–0.8 cm wide) subtending 2 blue-purple flowers 5 cm across. These are produced May–June. Falls obovate, 3.5–4 cm long, 1.5 cm wide, claw gutter-shaped with a whitish-blue lined pattern. Crest irregularly serrated. Standards of similar dimensions to the falls; style branches flat and broad, 4 cm long, 1.5 cm wide with the broad stylar crests arching over towards the centre of the flower. $2n = ?$

I. milesii Foster

Distribution
I. milesii is geographically isolated from other species in the section, its natural distribution occurring over a wide area of the Himalayas.

Description
The habit of *I. milesii* is similar to that of a vigorous form of *I. tectorum*. The rhizomes are short, thick and fleshy, greenish above and conspicuously marked with scars of the previous season's growth. The rhizome apex is enlarged and bears fleshy roots, a terminal leafy flowering stem and two lateral non-flowering leafy shoots. The rhizomes formed on the lateral shoots become the active apices in the next season's growth so that the overall growth pattern of old and new rhizomes forms a series of repeated dichotomies.

Basal leaves thin, heavily ribbed, yellowish-green and slightly glaucous, about 60 cm long, 3–7 cm wide. The 2–4-branched flowering stems are 30–90 cm tall. They bear several reduced leaves at their base and about 8 foliage leaves, which are 20–45 cm long and 1.5–2.5 cm wide. These increase in size from the base up and then decrease towards the apex, where they are succeeded by a small bract or bracts. $2n = 26$.

Dykes (1913, p.102) says 'From a decorative point of view, this is a distinctly disappointing iris'. This was because he considered the flowers relatively small and fugitive and their colour not altogether pleasing. However, to consider the plant insignificant is somewhat unjust. Each flower is about 10 cm across and pinkish-violet with darker mottlings. They have a fringed yellow crest, which is so finely and deeply laciniate that it resembles the beard of the bearded irises. The flowers are individually very short-lived and are self-fertile. Their abundance ensures continuity for several weeks from June onwards. Capsule ovoid–cylindrical. Seeds pyriform and black.

Cultivation
Similar to *I. tectorum*, although the necessity to divide is less frequent. Although hardy, damage to flower buds may occur if late spring frosts are encountered.

Observations
I. milesii was named by Prof R.C. Foster to honour Frank Miles, who raised it about 1880 from seeds collected by his cousin in the Kulu and Parbutta valleys of the NW Himalayas. Some of these were also received by Max Leichtlin and it was from his plants that Foster described the species in the *Gardener's Chronicle*. It is figured in the *Botanical Magazine* t.6889, 1886.

I. tectorum Maxim.

Synonyms
I. cristata Miquel, *I. fimbriata* Klatt, *I. rosthornii* Diels., *I. tomiolopha* Hance.

Distribution
This species is commonly referred to as the Roof Iris of Japan, indicative of the fact that it is said to frequently grow there on the roofs of straw-thatched houses. However, although long cultivated in Japan it was probably introduced there from central and south-western China. It is also recorded from northern Burma.

Description
I. tectorum produces stout, pale, buff-coloured rhizomes from which arise long (30–40 cm), broad (2.5–5 cm wide), ribbed, pale green leaves that taper to a point. The sturdy branching flower stems, bearing reduced leaves, attain 30 cm or more and carry several large, flattish purple-blue flowers, about 10cm across; the falls are mottled with deeper colour and have a conspicuous, coarsely and irregularly laciniate white crest with brownish-purple flecks. The flowers are self-fertile. Capsules oblong-ovoid, 4–6 cm long, 2–2.5 cm across with six ribs; on ripening it splits in three from the top. Seeds numerous, pyriform with a small cream-coloured aril. The seed and flower are illustrated in Dykes, *The Genus Iris* (1913). $2n = 28$.

Several clones with flowers in varying shades of purple to blue are cultivated. There is also a yellow-crested white form, 'Alba', which, though less vigorous, is a plant of outstanding beauty. Less spectacular is a purple-flowered variegated form, which has leaves with pale-coloured stripes that vary in intensity depending upon the time of year.

Cultivation
The cultural requirements pose few problems providing that it is given a well-drained soil and a sunny position, thus facilitating the necessary ripening of the rhizomes in late summer. It is also important to note that the plant's rapid growth and short roots quickly exhaust the soil in which it grows. Therefore division and re-planting are recommended at least every other year just after flowering when root growth is active. Alternatively, an annual top dressing of well-rotted compost may be incorporated amongst the clumps during spring or autumn. Failure to observe their nutritional requirements inevitably results in a rapid decline in vigour and few, if any, flowers.

I. tectorum is also an exceptionally fine plant for the cold greenhouse where it can be grown in a 15 cm diameter pot. Annual division and re-potting are necessary to maintain vigour.

A virus disease that causes serious leaf discoloration is common; infected plants should be carefully discarded to avoid further transmission. Fortunately, clean plants can be raised from seed, which germinates readily. They usually flower in their second year. Seeds from 'Alba' produce all white-flowered plants provided there has been no cross-pollination with the blue form.

I. tectorum is figured in the *Botanical Magazine*, t.6118, 1874.

Observations
It was first introduced into Europe by Philipp Franz von Siebold, who sent plants to St Petersburg sometime during the early nineteenth century.

Inter-specific hybrids are known with *Ii. pallida, cengialti, cristata* and *wattii*.

I. tenuis S. Watson

Distribution
Confined to Clackamas County, Oregon, USA, where it inhabits cool shady places in moist leafy soil. It can be found in the Douglas fir forests or in dense undergrowth.

Description
I. tenuis produces slender, widely spreading rhizomes with fans of leaves at their tips. Hence it does not form dense clumps. The thin, pale green, deciduous leaves are about 30 cm long and 1–1.5 cm wide. In appearance it resembles *I. cristata* except that it is taller and that the 3–3.5 cm flower stems are deeply forked. These bear lilac flowers, 3–4 cm across, from May onwards. The inconspicuous crest consists of a low, yellowish, undissected ridge and is uncharacteristic of the section. Capsules ovoid. Seeds D-shaped, pitted, pale brown with a whitish raphe. $2n = 28$.

Cultivation

Unfortunately this plant is rarely encountered in cultivation, perhaps because it is difficult to establish. For success one needs to simulate its natural environment of slightly acid, well-drained, humus-rich soil and a shady position. It is frequently reported that this species is not hardy, but such statements are questionable.

Observations

For many years this species was included in the Series Californicae. However, because of its distinctive features Q.D. Clarkson (1958) re-classified it as monotypical of Subsection Oregonae, then newly established by him. Subsequently (1959) he reduced the category from subsection to series to be taxonomically correct. Lenz (1959) transferred it to Subsection Evansia.

I. wattii Baker

Synonym
I. milesii Dykes.

Distribution
Manipur, Assam and China (Yunnan).

Description

This is the largest member of the section. It has erect bamboo-like stems, similar to those of *I. confusa*, although they are considerably taller and more robust. Generally, they attain a height of about 120 cm but are often considerably taller, depending on environmental conditions. The evergreen, sword-shaped, heavily ribbed leaves may attain a length of 70 cm and a width of 6.5 cm. From the apical leaf fans arise branched flower stalks bearing numerous flowers, each about 6 cm across. They are predominantly lavender-blue with a whitish area in the centre of the falls spotted with orange and deep lavender. The crest is a conspicuous whitish ridge with orange spots. An obvious characteristic is that the falls are strongly reflexed whereas in *Ii. confusa* and *japonica* they are near horizontal. Some clones are self-incompatible and never set seed when grown in isolation. $2n = 30$.

Cultivation

As for *I. confusa*, the fact that *I. wattii* is not hardy must be emphasised. It is also advisable to stake the stems, for they have an irresistible urge to lie in an untidy manner upon the ground.

Observations

Baker described this species from herbarium specimen No. 6337 collected by George Watt from the summit of Khongui Hill, Manipur, Eastern India, in 1882. In his observations he failed to observe the crest and so referred the species to

Section Pardanthopsis.

Further confusion was created by Dykes, who identified *I. wattii* with *I. milesii* and later with what we now call *I. confusa*.

Plants collected by Major Lawrence Johnson in 1931 corresponded closely to the original description by Baker and, after careful investigation, were considered to be the true *I. wattii*. This clone received an Award of Merit from the Royal Horticultural Society in 1938. Recently, considerable doubt has been expressed as to whether this clone is a possible hybrid rather than a true species (Ellis 1979).

A second clone of *I. wattii* received from the Kunming Botanical Garden, Yunnan, occurs in cultivation and differs from the original type introduced by Johnson. In this more recent form the falls are rounder and more horizontal, bluer in colour with purple-blue veining on a white ground, and with a yellow area near the apex of the crest. The flowers of this clone are self-fertile.

Hybrids are known with *Ii. tectorum* and *confusa*.

Author's additional comments

The present text lists eleven described species, but it should be noted that collected representative specimens do not always conform with either the species description or the concept of valid biological species. The recently introduced *I. formosana* has a chromosome count of $2n = 35$ and presumably would not be true-breeding for chromosome number, which is a fundamental criterion for any species. Likewise the uneven chromosome numbered cytotypes of *I. japonica* ($2n = 31, 33, 35$) would not be true-breeding. In addition the 'Ledgers Variety' of *I. japonica*, which has an even chromosome number of $2n = 54$, does not breed true for chromosome number. Progeny raised from it (in crosses with *I. confusa*) originated from gametes having chromosome numbers that ranged from $n = 24$ to $n = 28$ (Chimphamba 1971). Kazao (1928) considered that Ledger's Variety was a triploid, and the variation in the gametic chromosome numbers would support this suggestion: it therefore does not show the characteristics of a 'good' species.

The original plant of *I. wattii* introduced by Johnson was reported by Ellis (1979) to show cytological features of hybridity; similar characteristics have also been observed by Y. Lim (unpublished data) in the recently introduced second clone of *I. wattii*. Because hybridity is often associated with greater vigour, it is possible that such hybrid forms are the ones selected for introduction into horticulture.

A cytogenetic survey of some Asiatic collected specimens (J.R. Ellis and Y. Lim, unpublished data) has revealed the occurrence of two potential new species. One, with a chromosome number of $2n = 30$, was received as a form of *I. confusa*. However, it had a growth habit similar to that of *I. wattii* and had large white self-fertile flowers, thus making it distinct from *I. confusa* and probably also from *I. wattii*. Unfortunately the supplier of this new iris had no details of its origin; and this lack of information precludes any attempt to justify describing it as a new species. For recording purposes, it has been referred to as *I. 'nova'* (Ellis 1980). In its growth form and chromosome number it shows close affinity to both *Ii. confusa* and *wattii* and will readily hybridise with both species.

The second potential new species was a specimen collected from Sichuan and was received as a possible form of *I. confusa* via Mrs J. Witt of Seattle. This plant had a chromosome number of $2n = 30$ and a growth habit like a small form of *I. confusa* but with glossy dark evergreen leaves. Its slender, 25–30 cm, branched inflorescences produce deep lavender flowers with a prominent purple zone surrounding a white area adjacent to a deep yellow crest. The flowers are self-incompatible and in size and coloration are very distinctive in the bamboo-type of evansia irises. In growth habit and chromosome number it shows close affinity to *Ii. confusa, wattii* and '*nova*'; vigorous hybrids with all three have been obtained. These were studied by Y. Lim (University College, London) for an assessment of relative cytological affinities. He concluded that the new type was distinctive and that it should be given full specific status and referred to as *I. wittii*.

Section Limniris

SERIES CHINENSES (DIELS) LAWRENCE

Map 8, Figure 8.

There have been considerable changes in the irises attributed to this series in recent years. Probably there will be more to come, but it would help if more of these forms could be cultivated. In the meantime it is convenient to give them separate status in spite of a strong resemblance to some of the Lophiris species. There is one curious characteristic: nodules are sometimes found on the roots, just as on those of clover plants.

I. henryi Baker

Synonym
I. gracilipes Pampan.

Distribution
Central China in Hupeh Province and possibly eastern Szechuan.

Description
Rhizome wiry, stoloniferous, very thin. **Leaves** 12–40 cm long, 0.1–0.5 cm wide, possibly with pale central stripe, prominently ribbed, taller than flowering stems. **Stem**: to 25 cm, with 1–2 leaves; spathes 2–3 (to 4 cm) long, green, narrow, acuminate; pedicel to 4 cm long; ovary to 0.7 cm long, spindle-shaped; tube about 0.3 cm; 2-flowered. **Flower**: blue (or white), about 3 cm diameter; falls to 2 cm long, 0.7 cm wide, rounded, narrowing to haft, with yellow markings; standards shorter,

narrower; styles about 1 cm long, crests narrowly triangular; stamens to 1 cm. **Capsule** and seeds? **Flowering** May. $2n = ?$

Cultivation
No information available.

Observations
The extremely narrow leaves are indicative of this species; Mathew compares them to those of crocuses.

I. koreana Nakai

Distribution
Central and southern Korea.

Description
Rhizome slender, with creeping stolons. **Leaves** to 35 cm long, 1.3 cm wide, with a number of prominent veins, possibly red at bases. **Stem**: shorter than leaves; tube much longer than spathes, 2-flowered; **Flower**: yellow; falls obovate, held horizontally, hafts with brownish markings and raised, pleated edges; standards paler, nearly erect, with a notch at the end and a narrow haft, rather shorter than falls; styles same colour as standards; crests narrow, acuminate. $2n = ?$

Cultivation
No information available; it does not seem to have been in cultivation for a very long time.

Observations
It seems likely that the plant described by Dykes was a form of *I. henryi*. *I. koreana* is now considered to be a larger form of *I. minutoaurea*.

I. minutoaurea Makino

Synonyms
I. koreana auct. non Nakai; *I. minuta* Franch. & Sav.

Distribution
Probably China or Korea.

Description
Rhizome wiry, very thin, with short stolons. **Leaves** about 8–10 cm long, 0.2–0.7 cm wide, but growing after flowering, strong, ribbed. **Stem** very short; spathes

lanceolate, keeled, to 4 cm, 1-flowered; pedicel 1.25–2 cm long; ovary sharply trigonal, about 2 cm; tube 2–2.5 cm. **Flower**: falls yellow, obovate, spoon-shaped, with brown markings particularly on the haft; standards paler, held diagonally, shorter than falls, hafts caniculate, marked brown; styles rather flat, oblong, pale yellow; crests lanceolate, overlapping; stigma triangular; filaments colourless, anthers brown; pollen yellow. **Capsule** trigonal, subglobose, with short beak. **Seeds** globose, yellow-brown with white raphe round half of the circumference. $2n = 22$.

Cultivation
All that can really be said is that this species is surprisingly difficult to grow, at least in Britain. Something as small as this needs good drainage and probably access to good food reserves since it will rapidly exhaust its immediate area.

Observations
In autumn, the roots develop tiny nodules which require further investigation. It is not known in the wild state, having originated from Yedo, in Japan, probably from cultivation.

I. odaesanensis Y. Lee

Distribution
Korea.

Description
Rhizomes long, thin, stoloniferous. **Leaves** 1 cm wide, lengthening to 12–35 cm after flowering, narrowing gradually to apex, glaucous, ribbed. **Stem**: to 20 cm, pedicels long, tube 0.5 cm or less, 2-flowered. **Flowers** 2; white, 3–5 cm diameter; pedicels long; falls and standards in same horizontal plane; falls obovate rounded, yellow signal in middle, haft ridged and wrinkled?; standards narrower, less rounded; styles small, lobes acute. **Capsules** 1–1.3 cm wide, 2–3 cm long, strongly trigonal, hanging clear of spathes from long pedicels. $2n = ?$

Cultivation
No information available. This plant appears to come from deciduous forest on low, level slopes with a western aspect, so probably needs good drainage and possibly light shade.

Observations
So far this plant is not in general cultivation and detailed information about it is lacking. In some ways it resembles *I. henryi*, but is clearly differentiated by the wide leaves, very short tube and having the leaves so much shorter than the stem at flowering time. This is another plant that suggests affinity with the Lophiris group.

I. proantha Diels

Synonym
I. pseudorossii Chien.

Distribution
Eastern coastal region of China, also Shanghai area on hills near Nanking and Chinkiang.

Description
A small plant of stoloniferous habit. **Rhizomes** spreading, nodose, branching, covered with numerous rigid fibres. **Leaves** 5–15 cm long, 0.1–0.25 cm wide, linear, acuminate, rigid, glaucous or grey-green; margins white and scarious near base. **Stem** to 7 cm tall, with 1–2 sheathing leaflets on lower part; 2 spathes, 3.5–5.5 cm, 0.5 cm wide, green; pedicel to 1 cm long; ovary to 0.5 cm long; tube about 3 cm long; single-flowered. **Flowers** to 4 cm diameter, variable in colour, blue, purple-blue, white or white suffused blue or pink, with a zone of darker colour around the apex of the distinct yellow crest on the falls; falls up to 2.5 cm long, 1 cm broad, blade obovate, haft oblong and shorter; standards about 2 cm long, about 0.6 cm broad, erect; style pale-coloured, about 1.8 cm long, 0.4 cm wide; crests triangular, slightly toothed; stamens about 1 cm long, anther and filaments white. **Capsule** globular, to 1.5 cm diameter, with conspicuous apex. **Seeds** ? **Flowering** March–April. $2n = ?$

Cultivation
No information available.

Observations
From open hillsides, this species was described in 1931 as *I. pseudorossii* by the Chinese botanist Chong-Shu Chien. However, comparison between this and the original specimen of *I. proantha* Diels reveals that there are no significant differences. Despite some inaccuracies in Diels' description, Zhao (1985) considers it to be valid. Therefore the name *proantha* would take priority. It should not be confused with *I. rossii* Baker. The crest on the falls is reminiscent of those on plants of the Lophiris group and Prof Zhao has classified it with them. More study is required.

var. *valida* (Chien) Y.T. Zhao

Synonym
I. pseudorossii var. *valida* Chien.

Distribution
Western Chekiang, eastern China.

Description
Differs from *I. proantha* in that it is larger and of a more robust habit. Up to 28 cm tall. Leaves over 28 cm long, to 0.7 cm wide, flaccid. Falls to 2.6 cm long; standards 2–2.2 cm long.

Observations
I. proantha and *I. p.* var. *valida* appear to be rare and have not apparently been introduced into cultivation.

I. rossii Baker

Distribution
Northern China, Korea; Honshu, Shikoku and Kyushu in Japan.

Description
Rhizome slender, tough, with old leaf bases persisting around new growth. **Leaves** 0.2–0.5 cm wide, 7.5 cm at flowering, later to 30 cm, grassy-looking, ribbed, slightly glaucous. **Stem**: 4–10 cm, bearing reduced leaves; spathes 3.5–6 cm long, acuminate, green, tapering to apex, similar to leaves, 1-flowered; pedicel to 1.25 cm; ovary 0.5 cm; tube 4–6 cm or more, slender. **Flower:** 3–4 cm diameter, falls and standards verge on horizontal. Falls: blade obovate narrowing sharply to a short haft, colour variable from purple through lilac to white, although yellow is doubtful; standards oblanceolate, unguiculate. **Capsules** rounded, stemless, produced in July. $2n = ?$
A good white form described as *alba* has been found.

Cultivation
No information available since this plant never seems to reach the West. Its natural habitat is scrubland and well-drained, grassy banks. Dr S. Hirao found it fairly easy to grow given well-drained acid soil in full sun, flowering from seed in two years, but the flowers surviving only one day. Clumps were very floriferous.

Observations
Another confusing species which could have affinities with the Lophiris species. It is also possible that doubtful forms attributed to this species may belong to *I. pseudorossii* instead.

f. *alba* Y. Lee

Distribution
Uncertain.

Description
As for type, but the flowers are white with yellow markings in the middle.

Found in deciduous forest clearings.

I. speculatrix Hance

Synonym
I. grisjii Maxim.

Distribution
Hong Kong and the adjacent mainland where it grows on windy slopes among grass, scrub or wood fringes.

Description
Rhizomes slender, short, creeping. **Leaves** 30–90 cm or more long, 0.5–1 cm wide, shiny, cross-veined, evergreen. **Stems,** 20–30 cm tall, carry several reduced leaves and each is 2-flowered. **Flowers**: about 5 cm diameter, basically lavender-coloured; falls obovate, conspicuously marked towards the centre with a dark purple zone surrounding a central pale area with purple veining lateral to a low yellow undissected ridge; standards, about 1.5 cm, wide are narrowed to a short claw at the base and spread obliquely; style crests 1.5 cm long, 0.5 cm wide; self-fertile. **Capsule** 2.5 cm or more long, narrowly ovate, ridged with a prominent beak. **Seeds** dark brown with a conspicuous succulent white attachment, which later withers. **Flowering** April or May. $2n = 44$ (Snoad 1952; Lenz 1959).

Cultivation
This plant has been considered difficult to keep. Dykes recommended a temperate greenhouse; a frost-free environment is considered essential. Mrs Judd in New Zealand grows it successfully on the top of a hill near Wellington where it is exposed to wind and rain in winter, but gets all the light available. Dr Boussard, of Verdun in France, finds that it will survive frost. It clearly requires free-draining soil, plenty of light, probably a generous root run, and regular re-potting if grown under cover. The present need is obviously for seedlings.

Observations
Until recently this species has been included in Section Lophiris, but research by Prof Zhao shows that it is definitely better placed in Series Chinenses.

SERIES VERNAE (DIELS) LAWRENCE

Map 9, Figure 9.

I. verna Linn.

var. *verna*

Synonyms
I. nana (*I. verna* Pallas is probably *I. ruthenica* Ker-Gawl.).

Distribution
USA, Appalachian Mountains southward from Pennsylvania through Kentucky to Georgia over sand-dune areas; again in the Ozark Mountains in Oklahoma and Missouri.

Description
Rhizome slender, side shoots 5–15 cm apart, forming a dense clump. **Leaves** 3–8 cm wide. **Stem** very short, hidden in spathes which overlap perianth tube. Spathes acuminate, green, divergent 4–5 cm long; single-flowered. **Flower** 3–5 cm in diameter, clear bright blue- lilac. Falls obovate–cuneate, 1 cm wide, 3 cm long, a vivid orange, slightly pubescent band stretches out from the haft, edged with white veined lilac-brown, and continues halfway down the blade. Standards obovate, narrowing gradually to a long haft. Styles joined for about 0.75 cm above base, keeled; crests triangular, large, paler in colour. Anthers pale blue, filaments longer than anthers, pale lilac or colourless; pollen cream. Ovary trigonal, up to 1 cm. No pedicel. **Capsule** up to 1.8 cm long, trigonal, grooved, rounded at base, tapering to point at apex. **Seeds** pyriform, pale yellowish-brown, with white raphe. $2n = 42$.

var. *smalliana* Fernaldii

Description
Very similar to var. *verna*. **Rhizomes** short, plump, side shoots 1.3 cm apart. **Leaves** 0.6–1.3 cm wide. **Capsules** 1.7–2.3 cm long.

Cultivation
For such a widespread plant, *I. verna* has been surprisingly difficult to keep in cultivation. It seems to do best on acid soil, with light shade, which is not allowed to dry out in summer. Lynch (1923) recommends a topping of live sphagnum moss as a guide for this purpose and suggests drier conditions in winter. Another suggestion is to plant near a small rock garden cascade where local splashes and humidity help to keep it sufficiently moist in summer while, since the machinery

is probably turned off in winter, the drainage will be good. It is possible that, like many small irises kept in pans or equivalent shallow soil conditions, this iris requires far more in the way of nutrients than is usually realised, and more frequent re-planting.

Observations
Out of the flowering season, the plants are curiously like either a very small pogon or *I. cristata*, from which it may be distinguished, in both forms, by the much thicker rhizomes. It has a habit of apparently growing the flower spike separately from any leaves. Lynch (1923) remarks that what appears, from his description, to be var. *smalliana* grew in the Cambridge Botanical Garden, where the leaves persisted through the winter. Several colour forms have been selected and propagated, but, except for the white form, do not seem to be in cultivation now.

A hybrid is known with *I. pallida*.

SERIES RUTHENICAE (DIELS) LAWRENCE

Map 10, Figure 10.

This is either one species, or a group, which requires considerable further study: the taxonomic resemblances between forms that have been distinguished as species until now are such that B. Mathew considers that there is probably one species with a number of distinct variants. Considering the immense geographical area covered by these plants, a good deal of variation is to be expected. Until, however, someone comes up with a reasoned survey it will be convenient for iris growers to keep the forms under species names.

I. ruthenica Ker-Gawl

Synonyms
I. alpina Pallas ex Roem. & Schult., *I. nana* Maxim., *Ioniris ruthenica* Klatt, *Isis ruthenica*, *Xiphion ruthenicum* Alefeld.

Distribution
From Transylvania in the west, through the Altai zone into northern China and the Korean peninsula.

Description
Rhizome much branched, slender, thickly coated with hairy remains of leaf bases. **Leaves** about 15 cm long at flowering time to 30 cm, 0.25–0.5 cm wide, of grassy appearance. **Stem** 2.5–20 cm long, usually with 2 reduced leaves at base and 1 further up; spathes 2.5–4 cm long, lanceolate, green sometimes with pinkish edging, inflated; pedicel 1–5 cm long; ovary about 0.75 cm long, sharply trigonal; 1 or 2

flowers. **Flower** 5 cm diameter, deep violet colour; falls 4 cm long, 1 cm wide, blade broadly ovate, creamy-white ground with bright blue-purple veins and dots, which are very conspicuous on lower part, with slight central ridge tipped violet; standards 3.5 cm long, 0.5 cm wide, erect, conspicuous deep purple or blue-violet, lanceolate; styles 3.5 cm long, usually red-purple, held well clear of falls; crests triangular, sometimes serrated; stigma prominent, triangular; filaments mauve, attached unusually high on falls; anthers pale mauve; pollen cream; tube 1.2–2.5 cm long. **Capsule** short, rounded, barely ribbed, dehisces sharply when ripe and ends curl back distinctively. **Seeds** globose, with conspicuous raphe, which seems to disappear on drying. **Flowering** May–July depending on form and conditions. $2n = 84$.

Cultivation
Light, fairly rich soil, sun or light shade. Well-established plants are quite drought-resistant, but resent being moved. Transplanting should be restricted to large clumps about the size of a clenched fist and should be done while the plant is in active growth and before cold weather sets in. It comes easily from seed. Dykes thought the degree of floriferousness related to leaf width.

Observations
Because this species is difficult to transplant unless re-planted promptly, it has not been easy to obtain plants with a guaranteed provenance. Nor has this been helped by its habit of broadcasting all the seeds from a pod. The raphe is curious because although it drys out after seed dispersal it reconstitutes in damp weather, becomes sticky and adheres to the feet of birds or animals, so ensuring wide distribution. Many forms are scented more or less pleasantly.

var. *brevituba* Maxim.

Distribution
From the mountains of Kirgizia and Xinjiang.
Distinguished by a very short perianth tube around 0.75 cm.

f. *leucantha* Y.T. Zhao

Distribution
Xinjiang Province, China.
It has white flowers, but is otherwise similar to the type.

var. *nana* Maxim.

Distribution
Found over much of China and in Tibet.
Very much smaller than the type in all respects; colour varies from light blue to violet. **Flowering** April–May.

I. caespitosa Pallas ex Link

Distribution
Possibly far-eastern Siberia.
  With large quantities of very narrow leaves only up to 15 cm long and single, deep violet flowers in May. Possibly best in peaty conditions and full sun. Its status is uncertain.

I. uniflora Pall. ex Link

Distribution
Eastern Siberia, inner Mongolia, NE China and Korea.

Description
Rhizome slender, fibres of old leaves persist, nodes enlarged; creeping. **Leaves** to 20 cm long, to 1 cm wide, lengthening after flowering, linear, acuminate, sheathing at base. **Stem** about 15 cm long, slender, with 1 leaflet on lower half; 2 bracts about 3 cm long, 1 cm wide, yellow-green, reddened margins, ends obtuse; pedicel very short; ovary to 0.5 cm long, spindle-shaped; tube to 1.5 cm long, narrow, expanding to at top; 1-flowered. **Flower** violet, to 4.5 cm wide; falls to 3 cm long, 0.8 cm wide, ovate–orbicular; standards to 3 cm long, 0.3 cm wide, upright; style about 3 cm long, 0.3 cm wide, flattish; crests semi-circular, slightly toothed; stamens to 1.5 cm long. **Capsule** to 1 cm diameter, 6-ribbed, rounded, bracts persistent. **Seeds** ? **Flowering** May–June. $2n = ?$

Observations
Given specific rank in the *Flora of the USSR* by B. Fedtschenko, 1968, and clearly supported by Chinese botanists. B. Mathew, 1989, finds it smaller than the type and similar to *I. caespitosa*.

var. *alba*
  A good white variety, rather smaller than the type overall.

  Distribution
  Siberia; Inner Mongolia, Heilongjiang, Jilin, Liaoning; Korea.
  Flowering May–June.

var. *caricina* **Kitagawa**
  Has exceptionally narrow leaves.

  Distribution
  Heilongjiang, Jilin, Liaoning, China; Inner Mongolia.
  Flowering May.

SERIES TRIPETALAE (DIELS) LAWRENCE

Map 11, Figure 11.

This series is exceptional in that it encircles the globe in the northern hemisphere and also in the way that the natural distribution of *I. tridentata* in the wild is so sharply demarcated from other forms in N America. It has been suggested that ice caps formed in the last ice age may be responsible.

I. setosa Pall. ex Link

Synonyms
I. brachycuspis Fischer ex Sims, *I. brevicuspis* Fischer ex Sims, *I. douglasiana pygmaea*, *I. hookeri* Penny, *I. yedoensis* Franch. & Savat., *Xiphion brachycuspis* Alef., *Xyridion setosum* Klatt.

Distribution
Northern Asia, Kamchatka Peninsula; China, Jilin; Japan, Labrador and the east coast of N. America into New England.

Description
Rhizome: thick, covered with fibrous bases of old leaves. **Leaves** ensiform, 1.5–2.5 cm wide, 30–60 cm long, slightly glaucous, green and usually purple at base. **Stem** solid, several branches with the first bearing flowers on a level with those at the apex; a reduced leaf at each node; spathes narrow, acuminate, unequal, 3.5–5 cm long, the outer shortest, green sometimes flushed or margined purple, sometimes slightly scarious, 2–3-flowered; pedicel 2.5–4 cm long; tube 0.5–0.75 cm long, shorter than ovary with very little constriction between them; ovary acutely trigonal, green or flushed purple. **Flower**: falls orbicular with blade narrowing sharply to short haft, which is veined purple on yellowish-white ground which continues onto the fall for a short way before becoming a uniform purple on which darker veins can be discerned, the colour varies from blue-purple to red-purple; standards: are variable in shape, but characteristically are very small, up to 2 cm long, broader at the base and often narrowing to a point; styles about 2.5 cm long, oblong, whitish, with purple keel; crests overlapping, subquadrate, with coarsely serrate edges; stigma rounded–triangular; filaments purple or yellow stained purple anthers purple, pollen cream. **Capsule** trigonal, with grooved sides, much inflated so that it is half as broad as long; seeds become loose quite early. **Seeds** with obvious raphe down one side, light, glossy brown. **Flowering** June–August. $2n = 38$.

There is also a white-flowered form.

Cultivation
In general, a very tolerant species although it does not care for heavy clay soils which dry out, nor for calcareous ground (although a very slight presence of lime

will be tolerated by most forms). Plenty of water is needed in the growing season. Plants come easily from seed; transplanting is best carried out in late summer or early autumn. Since these plants are very fertile, different species should be kept well separated to avoid cross-pollination. There is some indication that individual plants are not very long-lived.

Observations
There are a large number of named forms and varieties of this species. Any plant that does not correspond to a description, or breed true, should be regarded as a garden form. Dykes noted a plant in Maine growing where the salty spray from the sea could could reach it; this salt-tolerance could be useful in some situations. A form with variegated leaves has recently appeared as a seedling in an American nursery and is being propagated. Other variants have currently been narrowed down to those listed below.

- f. *alpina* Komarov, from Siberia, has a very short stem.
- var. *arctica* from Alaska, a dwarf form with white variegated purple flowers.
- subsp. *canadensis* (syn *I. hookeri*) from eastern Quebec, Newfoundland and Nova Scotia. A variable plant: stem nearly leafless, unbranched; usually only 1 lavender flower. A white form, known as *pallidiflora*, exists and also forma *zonalis*, which has yellowish or white transverse bands on the leaves.
- subsp. *hondoensis* Honda, a large, robust form, stem up to 75 cm, leaves about 110 cm long, 5.5 cm wide, large purple flowers. Mr Michio Cozuca comments that this plant is sterile.
- subsp. *interior* (Anders) from Alaska: narrow leaves, spathes violet-coloured, papery, shorter than those of the type.
- var. *nasuensis* Hara from Japan is exceptionally large: very wide leaves, stem up to 100 cm and flowers comparable in size to those of *I. laevigata*, but with reduced standards.
- f. *platyrhyncha* Hulten from Alaska has solitary flowers and rather larger standards than usual, occasionally nearly equal to falls.
- f. *serotina* Komarov from Siberia: to 60 cm, free-flowering, large clumps.

Hybrids: are known with *I. ensata, I. laevigata, I. sanguinea* and *I. sibirica*.

I. tridentata Pursh.

Synonym
I. tripetala Walt.

Distribution
USA: Florida, Tennessee and Carolinas.

Description
Rhizome slender, almost stoloniferous. **Leaves** 0.75–2 cm wide, 30–40 cm long, linear, finely ribbed, green, sometimes edged red. **Stem** up to 70 cm, sometimes with 1 or 2 branches and with reduced leaves; spathes narrow, pointed, stiff, outer usually half as long as inner, with 1–2 flowers; pedicel about 2.5 cm; tube funnel-shaped, less than 2.5 cm long; ovary trigonal, grooved at each angle. **Flower:** falls suborbicular, blade as long as haft, 2.5–3.5 cm diameter, violet or bluish-purple with darker veining and sometime paler mottling, signal whitish with yellow area at centre, whitish haft with netted brown veining, 5 cm long; standards violet, erect, to 1.5 cm long, sometimes slightly tridentate with centre point longest; styles narrow, relatively small, crests large, subquadrate; stigma entire, rounded; filaments very short, anthers to match. **Capsule** 2–2.75 cm long, obscurely angled so that there is little concavity at sides. **Seeds** thick, compressed, semi-circular, with a single line in each locule, dark red-brown.

Cultivation
This plant apparently requires semi-tropical conditions and is doubtfully hardy in Britain. Sir Michael Foster flowered it in a greenhouse at Shelford. Adequate water in the growing season is needed and probably also good drainage in a doubtful climate during the winter. Nevertheless, B. Mathew has grown it in somewhat sandy soil in a sunny border where conditions were not particularly wet; it spread well and flowered freely.

Observations
For some time this plant was classified apart from the rest of the Tripetalae, but resemblances are so close and the distribution of other forms so extreme that they have been united. In its natural habitat, it is a plant of damp meadows and ditches.

SERIES SIBERICAE (DIELS) LAWRENCE   CHRISTOPHER GREY-WILSON

Map 12, Figure 12.

The members of the Sibericae are amongst the most popular in cultivation, the species being widely grown, easy plants for the open garden, thriving best in moist, sunny or semi-shaded positions on a variety of soils from mildly acid to alkaline, although preferring the heavy loams to lighter soils.

    Two species in particular, *I. sibirica* itself and the closely related *I. sanguinea* (syn. *I. orientalis*), have been cultivated for several centuries and are classed among the finest irises for the herbaceous border, for streamsides, or the marginal areas of ponds and lakes. There are a variety of flower colours from white to various shades of blue, mauve, violet and purple among the named cultivars of these two species. All are easy to grow and from a small plant will produce a sizeable clump in a relatively short period. These, as with other species in the series, can be simply divided by breaking up the parent clump into smaller sections with a spade; this can be done in the autumn, or the spring before growth commences. Both species are confined in the wild to the northern temperate zone: *I. sibirica* to western and southern Europe, eastward to Asia Minor and the central former USSR, extending to the west of Lake Baikal, while *I. sanguinea* has a more easterly distribution from east of Lake Baikal to N China, Korea and Japan. These two species have a chromosome count of $2n = 28$ and hybridise freely to produce fertile hybrids; indeed most of the named cultivars in gardens today, often referred to as '*I. orientalis*', are such hybrids. These, together with cultivars of *I. sibirica*, have largely replaced the true wild species in cultivation.

The other species in the series have a distinct distribution in the Himalayas and SW China (mainly Yunnan and Sichuan provinces). Familiar in cultivation are *I. chrysographes*, *I. delavayi* and *I. forrestii*, but there are others less familiar. These species are slightly more exacting in cultivation: generally disliking the herbaceous border environment with its drier regime, but preferring damper and more boggy ground that is not subjected to prolonged winter flooding. In the wild these species inhabit seepage zones and moist pastures, but do not grow actually in shallow water. All these species have a chromosome count of $2n = 40$ and hybridise readily in cultivation, and, incidentally, also in the wild when two species grow in close proximity; in contrast these species do not hybridise readily with either *I. sibirica* or *I. sanguinea* and the resultant hybrids are nearly always sterile. In cultivation this has meant that seed-raised stock is often unreliable, as the results often prove to be hybrids. As a result there has been much confusion over the species' identities in gardens. Fortunately in recent years a number of these species have been re-collected in the wild and the 'true species' are now re-established in gardens. Having learnt of their tendency to promiscuity in cultivation, these species should, unless carefully isolated, be increased by division rather than by seed.

Wild-collected seed germinates very freely and, autumn-sown, the seedlings can be pricked out the following summer. Most will flower in the three years from seed, some sooner.

Strangely, very little controlled hybridisation has been carried out on this group, but the potential horticulturally is great when one considers the range of possible colours and plant heights. All the species flower during May and July in cultivation, rather later in the wild.

Plants are herbaceous perennials with stoutish brown rhizomes with many roots and root-fibres; these form dense tufts in time with the densely packed rhizomes branching freely. **Leaves** narrow, linear, not flushed with pink at the base, turning brown and dying down completely in the autumn. **Flower stems** hollow (except *I. clarkei*) bearing a simple or few-branched inflorescence. **Flowers** blue, mauve, purple or yellow (white in some cultivars) with a short perianth-tube; falls beardless; style split to the base, with a small triangular, tongue-like, stigmatic flap; perianth not persistent in fruit. **Fruit** capsule trigonous, sometimes nearly rounded in section. **Seeds** D-shaped or cubical.

Series Sibericae

Key to Species

1. Stems solid, often branched in the lower half; standard petals held horizontally — *I. clarkei*

   Stems hollow; standard petals held obliquely or erect — 2
2. Flowers yellow — 3

   Flowers blue, violet or purple, or reddish violet — 4
3. Standards erect; pedicels short in fruit, not more than 2.5 cm long; leaves shorter than the stems — *I. forrestii*

   Standards inclined obliquely; pedicels long, at least 5 cm, in fruit; leaves as long as the stems — *I. wilsonii*
4. Spathes scarious at flowering time — 5

   Spathes herbaceous at flowering time — 6
5. Pedicels long, up to 10 cm in fruit, markedly unequal; spathe pairs bearing 2–5 flowers — *I. sibirica*

   Pedicels short, not more than 5 cm long in fruit, more or less equal; spathe pairs bearing 2 flowers only — *I. phragmitetorum*
6. Spathe as long as or exceeding the flowers — *I. bulleyana*

   Spathe valves shorter than the flowers — 7
7. Leaves ensheathing the stem for the greater part — *I. dykesii*

   Leaves only ensheathing the stem towards its base — 8
8. Leaves twisting, as long as the stems or longer — *I. typhifolia*

   Leaves not twisting, generally shorter than the stems — 9
9. Flowers with erect standards and a suborbicular blade to the falls; falls with well-marked veins on a paler ground — *I. sanguinea*

   Flowers with oblique standards and an oblong or oval blade to the falls; falls with rather obscure veins on a dark ground — 10
10. Robust plant to 150 cm tall; leaves much shorter than the stems; flowers mid- to deep bluish-purple with a 'broken' white signal patch in the centre — *I. delavayi*

    Plant less robust, not more than 45 cm tall; leaves equalling the stems; flowers intense reddish violet or blackish violet with a gold signal patch in the centre of the falls; occasionally without signal patch — *I. chrysographes*

I. bulleyana Dykes

Distribution
SW China (NW Yunnan & S Sichuan, SE Xizang); precise distribution uncertain. Growing on alpine meadows, moist mountainsides and river flats in similar habitats to *I. forrestii* and *I. wilsonii*; 2300–4000 m.

Description
Plant small, 45–60 cm tall, forming rather dense tufts with the stems and leaves about the same length. **Leaves** glossy green above but grey-green beneath, 4–10 cm wide. **Stems** unbranched; spathes foliaceous, with a scarious tip at flowering time, the outer long and sometimes exceeding the flower, each pair bearing two flowers. **Flowers** 5.5–6.5 cm diameter, pale to mid-violet blue, or bluish-purple; the fall rather pale or whitish with darker veins and suffused with colour in the upper 1/3; standards narrow–oblanceolate, more or less erect; falls 4.5–5.5 cm long, with an oval blade and a narrowed, rather short, haft. **Capsule** oblong–cylindrical, 4–5.5 cm, nearly 4 times as long as broad, borne on short pedicel less than 5 cm long. **Flowering** June–July. $2n = 40$.

Observations
There has been a good deal of confusion over this species. It was described in 1910 from introductions made by George Forrest. The original plants from which Dykes prepared his description were raised by A.K. Bulley and were believed to have come with an early batch of *I. forrestii* collected in 1908. However, George Forrest stated shortly after that he had no recollection of seeing any plant like *I. bulleyana* in the wild. It was then suggested that the plant had arisen as a hybrid in Bulley's garden, but this simply was not possible. The possible parents were either *I. forrestii* or *I. wilsonii* and *I. chrysographes*; *I. wilsonii* was only collected in 1907 and the other two the following year and none flowered until 1909. This clearly left no time for *I. bulleyana* to have risen and flowered by 1910. The likelihood therefore was that hybrid seed was introduced by Forrest from the wild, assuming the plant to be a hybrid, that is! The original plant, and indeed most plants under the name '*bulleyana*' in gardens today, have a flower with a pale straw-coloured or yellowish background, suffused and lined, especially on the fall, with violet or bluish-purple. Similar and rather variable plants can be produced artificially by crossing *I. forrestii* and *I. chrysographes* in the garden. Thus it came to be supposed that *I. bulleyana* was in fact a natural hybrid between these two named species and so the situation remained until recent expeditions started returning to China. While I (C.G.-W.) was in NW Yunnan I came across extensive colonies of a plant on the Yangtse–Mekong Divide which I have come to recognise as the true *I. bulleyana*. This plant often is very uniform in the wild and indeed seedlings are also very similar. The flowers do not have a yellowish undertone and are not variable as in cultivated '*I. bulleyana*'. In NW

Yunnan it is sometimes found in association with *I. chrysographes*. In such instances hybrid swarms result.

What remains to be sorted out is whether or not the original plant described by Dykes was a hybrid. If this were so then the true species would require a new name; this clearly requires further research. Another factor that now needs to be borne in mind is that it is likely that if Dykes's *I. bulleyana* is of hybrid origin then the likely putative parents are *I. forrestii* and the true *I. bulleyana*, for both grow in close proximity in the wild.

The description above is based on wild authenticated *I. bulleyana* and not cultivated material.

f. *alba* Y.T. Zhao (1980)

Distribution
China: NW Yunnan.

Description
Flowers milky white.

I. chrysographes Dykes

Distribution
SW China (Guizhou, Sichuan, Yunnan & E Xizang) and N Burma growing in marshy places and alpine meadows, sometimes in open scrub, 2250–3700m.

Description
Plant densely tufted, 35–45 cm tall, the leaves and stems about the same length. **Leaves** grey-green on both surfaces, 0.7–1.5 cm wide. **Stems** unbranched; spathes foliaceous at flowering time, each bearing 2 flowers. **Flowers** deep bluish-purple with, or without, a 'broken' gold signal patch in the centre of the falls; falls 5.5–7 cm long, with an oblong or oval blade and a rather short, narrowed haft; standards narrow-oblanceolate, inclined at an oblique angle. **Capsule** 4–6 cm long, fusiform to oblong–cylindric, about 3 times as long as wide, borne on short unequal pedicels not more than 5 cm long. **Flowering** June–July. $2n = 40$.

Observations
This is one of the finest species in the series and certainly a fine garden plant which is widely grown, especially in its intensely dark-coloured forms. Var. *rubella* is a name given to a particular fine deep-coloured form seen in gardens. Although the true *I. chrysographes* is well established in cultivation, many paler-flowered plants in gardens under this name are in fact probably of hybrid origin: the species crosses readily, and often unaided, with most of the other members of the series.

I. clarkei Baker ex Hook.

Distribution
Himalaya from E Nepal to Sikkim, Bhutan, SE Xizang (Tibet), NE India and Upper Burma where it inhabits wet montane meadows and woodland margins and clearings between 2500 and 4300 m, often forming extensive colonies.

Description
Plant vigorous, to 60 cm, the leaves rather shorter than the stems. **Leaves** glossy green above, grey-green beneath, 1.3–2 cm wide. **Stems** solid, usually with 2–3 branches, often branching rather low down; spathes foliaceous at flowering time. **Flowers** 6.5–7.5 cm diameter, variable in colour from mid-blue to reddish-purple with a distinct white signal patch in the middle of the falls; standards oblanceolate, curved and inclined into a horizontal position; falls oval to suborbicular, narrowed into a relatively short haft. **Capsule** ellipsoidal, about 2.5 times as long as broad. **Flowering** June–July. $2n = 40$.

Observations
This interesting species was known as early as 1848 when Joseph Dalton Hooker sketched it in the Himalaya. However it was not described until 1892 and only came into cultivation in 1907. Since then it has proved a fine and relatively easy garden plant, thriving best in the boggy marginal ground around ponds and streams.

I. clarkei is an unusual member of the Sibericae series and certainly holds an isolated position within the series, some believing that it might perhaps best be placed in a separate series of its own. The possession of a solid stem, the low branching of the stems and the horizontal standard petals are all unique features not shared with any other member of the series.

Plants recently collected in central Nepal, which are superficially similar to *I. clarkei*, differ in details of petal shapes and positions and may represent another, undescribed species; this requires further investigation.

I. delavayi Micheli

Distribution
Bhutan and SW China (SE Xizang, W Yunnan & Sichuan) where it grows in wet or marshy mountain meadows and ditches; 2700–3100 m.

Description
Plant stout, to 150 cm tall, forming large tufts, the stems considerably longer than the leaves. **Leaves** grey-green on both surfaces, usually 0.6–1 cm wide. **Stems** generally with 2–3 short branches; spathes foliaceous with brown tips at flower-

ing time, each pair bearing 2 flowers. **Flowers** 7–9 cm diameter, deep violet with a conspicuous 'broken' white signal patch in the centre of the falls; standards oblanceolate, inclined at an oblique angle; falls 6–7.5 cm long, obovate. **Capsule** 5–7.5 cm long, oblong–cylindrical, about 3 times as long as wide. **Flowering** May–July. $2n = 40$.

Observations
This is by far the largest species in the series. It was named in honour of Abbé Delavay, the French missionary and plant collector, who discovered it in Yunnan in 1895, but it was introduced into cultivation later by George Forrest and others. It was rather rare in gardens until more recent times when several expeditions have introduced fresh wild-collected seed. It has proved to be an easy and vigorous species, but greatly dislikes becoming too dry during the summer. Despite its large stature this is an attractive plant with vividly coloured flowers.

I. dykesii Stapf

Distribution
Not known in the wild.

Description
Plant with the vigour of *I. chrysographes*, perhaps more robust, with the stems and leaves about the same length. **Leaves** grey-green on both surfaces, ensheathing the stem for most of its length (an unusual feature in the series), broad, to 1.8 cm wide. **Stems** unbranched. Spathes foliaceous at flowering time and rather long, each pair bearing 2 flowers. **Flowers** large, 6.5–8.5 cm diameter, vivid deep purple-violet, the falls with a 'broken' signal patch with white and pale yellow; standards oblanceolate, inclined at a slight angle; falls with an oblong to oval fall and an equally long narrowed haft. **Fruit** unknown. **Flowering** June? $2n = ?$

Observations
Little is known about this species. It grew among other 'sibiricas' in Dykes's collection, but did not flower while he was alive. However, it flowered (in 1926) after his death in the garden of Charles Musgrave in Godalming. Its general characteristics would suggest a plant of hybrid origin, perhaps between *I. chrysographes* and *I. delavayi*, although this is pure speculation. The original plant does not appear to be in cultivation today and plants sold under the name bear little resemblance to the plant originally figured in *Curtis's Botanical Magazine* (tb. 9282, 1932); most of these can be more clearly associated with *I. chrysographes*.

I. forrestii Dykes

Distribution
SW China (W Yunnan, S Sichuan, E Xizang) and N Burma where it inhabits alpine meadows at 2900–4300 m.

Description
Plant slender, rather small, 30–40 cm tall, though sometimes only 15 cm tall in the wild, with stems much longer than the leaves. **Leaves** usually narrow, 3–5 cm wide, glossy green or yellowish-green on one side, but grey-green on the other. **Stems** unbranched. Spathes foliaceous at flowering time, each pair bearing 2 flowers only, occasionally 1. **Flowers** scented, 5–6 cm diameter, clear yellow; the falls oval narrowed to a short haft with broken brownish-purple lines on the haft and the lower part of the blade; standards 5.5–6.5 cm long, small and erect, oblanceolate. **Capsule** 4–4.5 cm long, narrow–ovoid with a slender beak, about twice as long as broad, borne on short pedicels seldom more than 2 cm long. **Flowering** May–June. $2n = 40$.

Observations
This is the daintiest species in the series and reasonably easy in cultivation provided it is not swamped by more vigorous neighbours. The young shoots can be rather susceptible to slug damage. *I. forrestii* was first discovered in the Lijiang (Lichiang) mountains of north-western Yunnan in 1908 and first flowered in cultivation during 1910. Unfortunately, many of the plants under this name in cultivation today are not pure *I. forrestii* but hybrids between this species and another, most probably *I. chrysographes*; these can be identified by their more vigorous habit and by the varying amounts of blue that suffuse the yellow of the falls, standards and sometimes the style arms. Fortunately, however, recent expeditions to China have re-introduced fresh seed of this charming species.

I. forrestii and *I. wilsonii* represent the only two yellow-flowered species in the series. *I. wilsonii* is altogether more vigorous, with larger flowers, obliquely held standards and long pedicels. Apparently more vigorous forms of *I. forrestii* sometimes seen in gardens may well be hybrids between *I. forrestii* and *I. wilsonii*.

I. phragmitetorum Hand.-Mazz.

Distribution
China (NW Yunnan) growing in reed (*Phragmites*) swamps.

Description
Plant about 50 cm tall, with the stems overtopping the leaves. **Stems** unbranched. **Leaves** 1–1.5 cm wide. Spathes scarious at flowering time, each pair bearing 2

flowers. **Flowers** 6–8 cm diameter, dark blue with a 'broken' white signal patch in the centre of the falls; falls with an oblong blade and a slightly shorter haft; standards narrow–oblanceolate, erect. **Capsule** unknown. $2n = ?$

Observations
seed has recently been introduced from the wild and young plants are now available, but very little is known although it was collected in 1926. It appears to come closest to *I. sanguinea*, but there is a considerable difference in geographical range between the two.

I. sanguinea Hornem ex Donn

Synonyms
I. extremorientalis Koidzumi, *I. nertschinskia* Lodd., *I. orientalis* Thunb. non Miller.

Distribution
East of Lake Baikal in the former USSR to N China (Heilongjiang, Jilin, Waoning, Inner Mongolia), Korea and Japan. Damp meadows, river and lake margins.

Description
Plant variable in height from 30 to 70 cm, with leaves and stems about the same height. **Leaves** somewhat glaucous, usually 0.5–1.2 cm wide. **Stems** usually unbranched. Spathes herbaceous at flowering time, each pair bearing 2, sometimes 3 flowers. **Flowers** 6–8 cm diameter, rich bluish- purple, the falls 4.5–6.5 cm long, with darker veins and a yellow haft, sometimes pure white; falls with a broadly elliptic or suborbicular blade narrowed into a short haft; standards obovate, distinctly narrowed at base, erect. **Capsule** 3.5–5.5 cm long, oblong in outline, at least three times as long as broad, borne on more or less equal pedicels. **Flowering** May–June. $2n = 28$

var. *sanguinea* f. *albiflora* Makino

Distribution
Japan & NE China (Heilongjiang).

Description
flowers white.

var. *yixingensis* Y.T. Zhao

Distribution
China: Yixing & Jiangsu Provinces.

Description
Distinguished by its numerous sheath-like leaves only 0.2–0.4 cm wide; reddish-brown spathes and deep violet flowers. The fruits are 3-ridged (in var. *sanguinea* there are 6 ridges).

Observations
For additional notes see under *I. sibirica*. There are many cultivars, including 'Alba' with pure white flowers (many plants sold under this name are not true albinos and often reveal a trace of pale blue) and 'Violacea', a plant said to have originated in Korea, which has spathes and flowers of a rich violet-purple.

I. sibirica Linn.

Distribution
N Italy eastward through central and eastern Europe to NE Turkey and the former USSR, east to just west of Lake Baikal, at varying altitudes.

Description
Plant 50–120 cm tall, the stems distinctly longer than the leaves. **Leaves** green, up to 0.5–0.8 cm wide, occasionally more. **Stems** frequently branched above, often rather slender. Spathes scarious at flowering time, each pair bearing 2–5 flowers. **Flowers** rather small, 5–7 cm diameter, mid-blue to bluish violet, the falls 5.5–7 cm long, with darker veins and generally with a paler or whitish zone in the centre; falls oblong to obovate, narrowed somewhat in the haft; standards narrowly obovate to elliptic–obovate, erect. **Capsule** 2–3.5 cm long, short and rather dumpy, ellipsoidal, never more than two times longer than broad and carried on markedly unequal pedicels, the longest of which is up to 10 cm. **Flowering** May–July. $2n = 28$.

Observations
This species has been much confused in cultivation with *I. sanguinea* (syn. *I. orientalis*). However, although the two species are clearly very loosely related they differ in a number of important respects: the former has rather small flowers in usually branched inflorescences with brown scarious spathes and very unequal pedicels at the fruiting stage, whereas the latter species generally has an unbranched stem, large flowers and longer fruit pods borne on rather equal pedicels.

There are many cultivars of *I. sibirica* and of hybrids with *I. sanguinea,* which range in colour from white to the palest blues to deep blues, mauves and violet-purples.

I. typhifolia Kitagawa

Distribution
NE China (Jilin, Liaoning, Inner Mongolia) to Manchuria, where it apparently inhabits marshy ground and streamsides, often close to lakes.

Description
Plant slender, 60–110 cm tall, with leaves as long as, or longer than, the stems. **Leaves** deep green, slender and twisted, 0.15–0.22 cm wide. **Stems** generally unbranched; spathes foliaceous or partly membranous at flowering time, each pair usually bearing 2 flowers. **Flowers** 7–8 cm diameter, blue-violet, the falls with a reddish flush on the hafts; falls 5–5.5 cm long, oblanceolate, the rounded blade constricted into a rather narrow haft; standards oblanceolate, erect. **Capsule** up to 7 cm long, borne on markedly uneven pedicels. **Flowering** May–June. $2n = 28$.

Observations
This is a little-known species which has been recently brought into cultivation. Cultivated plants appear to differ from the original description of the species in their broader leaves (originally said to be very slender, only 0.15–0.22 cm wide) and this requires further investigation.

I. wilsonii C.H. Wright.

Distribution
W China (N Yunnan, Gansu, Shaanxi, Sichuan, W Hubei & Shensi) where it inhabits alpine meadows, streamsides and forest margins.

Description
Plant vigorous, 60–70 cm tall, the leaves and stems about the same length. **Leaves** grey-green on both surfaces, 0.3–0.8 cm wide, with a more or less distinct mid-rib. **Stems** unbranched; spathes foliaceous at flowering time, each pair bearing 2 flowers usually. **Flowers** fragrant, pale yellow, 6–8 cm diameter, the falls veined and dotted with pale brown or purple in the lower 2/3; falls obovate, narrowed rather gradually into the haft; haft with a pair of purple-brown auricles close to the base; standards oblanceolate, held at an oblique angle. **Capsule** 3–4 cm long, ellipsoidal, rather more than two times longer than wide, beaked, borne on long pedicels 7.5–10 cm long. **Flowering** May–June. $2n = 40$.

Observations
I. wilsonii was discovered and introduced into cultivation by Ernest Henry (Chinese) Wilson, who was sponsored by the firm of Veitch & Sons, in 1907. This species is

rather rare in cultivation today; this is regrettable, as it is an attractive species. As with *I. forrestii, I. wilsonii* hybridises readily in cultivation with other members of the series, notably *I. chrysographes,* and this has undoubtedly been its undoing as many plants under this name are not pure *I. wilsonii*; it is to be hoped that some of the recent introductions of seed from China may contain this species. Although both it and *I. forrestii* appear to overlap in general distribution, there is no indication that the two species grow together in the wild and therefore hybridisation in the wild is probably precluded. Reports that a purple-flowered form occur in the wild have never been substantiated; such plants, if they do occur, may be referable to *I. chrysographes.* Hybrids between these two species in cultivation have the stature of *I. wilsonii,* but the yellow flowers are variously flushed with blue or pinkish-purple.

SERIES CALIFORNICAE (DIELS) LAWRENCE V.A. DICKSON-COHEN

Map 13, Figure 13.

This series contains some of the most beautiful irises known years ago as the 'Californians' and later, through the revision of the series by Dr Lee W. Lenz, as the 'Pacific Coast Irises' because they do in fact extend outside California to Oregon and one species, *I. tenax,* is to be found also in the state of Washington. The Californicae are found growing west of the crests of the Sierra Nevada in California and the Cascades in Oregon and they extend through to the Pacific Ocean. They are generally plants of mountainous and forested areas, often growing in light shade. *I. douglasiana* is an exception: a coastal species of open headlands and grassy slopes, often down to the beaches, occasionally penetrating inland up the river valleys or roadsides. No irises are found within the Great Valley of California. In southern California one isolated subspecies is to be found in the San Bernardino and San Gabriel Mountains.

The Series Californicae is a grouping of beardless rhizomatous irises, generally small and compact with slender wiry rhizomes and narrow grass-like leaves. Two of the species produce leaves that are long, broad and sword-like. With one exception the stems are unbranched. The majority of the species bear two flowers to each stem, the second flower opening as the first dies away. *I. munzii* can have up to four flowers per stem and *I. hartwegii* subsp. *columbiana* has three flowers. *I. douglasiana* often produces three flowers from within the spathes and is unique in forming branched stems carrying up to nine flowers on a stem.

The flowering season extends from about mid-March until June with the peak around the end of April and the beginning of May. Californian species flower earlier than those in Oregon, but both latitude and altitude influence this. The first to flower are usually the first to fruit. Ripened seed is found from the end of May onwards, later species ripening their capsules in June and July.

For the size of the plants which bear them, the flowers of this series are quite large. The standards and falls are well developed, often distinctively veined and marked with rich colours. Most of the species possess style crests which are rounded or deltoid, often toothed on the upper edge, but two species form extremely long and narrow crests. With one exception (*I. purdyi*) the stigmas are triangular or tongue-shaped.

Colour is often a variable factor. Several species have quite a range of colours from lavender-blue through purple to pink, cream, white and yellow shades. Even where a species is known to have a restricted range of colour this can vary from plant to plant and between one colony and the next. It has even been noted that wild plants of *I. munzii* transferred to the garden have shown a more intense range of blue colours.

The chromosome count for all the species and subspecies of the Series Californicae is $2n = 40$ (somatic). They are all fertile and hybridisation within the series is always probable when different species are within range and in flower at the same time.

Hybrids are known with the series Sibericae, particularly *I. chrysographes* and *I. forrestii* which have the same chromosome count, and are usually sterile.

Members of the Californicae possess some distinct taxonomic features which in combination help in their identification. Eleven species and five subspecies are recognised in accordance with Dr Lenz's re-classification of the series.

It has been found possible to split the series into two main groups with three species remaining outside. They separate not only by perianth tube length, but also through the spathe formation. The short-tubed group all have open spathes: that is, the spathe valves diverge from each other at the top like the letter V and are separated at the base where they clasp the stem. The long-tubed group all have closed spathes (open only at the top where the flower emerges) with the spathe valves clasping the stem opposite each other.

Short tube with open spathes: *I. tenax; I. tenax* subsp. *klamathensis; I. hartwegii; I. hartwegii* subsp. *australis; I. hartwegii* subsp. *columbiana; I. hartwegii* subsp. *pinetorum; I. munzii*.

Long tube with closed spathes: *I. macrosiphon; I. fernaldii; I. chrysophylla; I. tenuissima; I. tenuissima* subsp. *purdyiformis; I. purdyi*.

The four remaining species, *I. douglasiana; I. bracteata, I. innominata* and *I. thompsonii*, do not fit into either of the preceding groups, but are quite distinct and not to be confused with the others.

Cultivation

One of the most important points is to avoid any lime in the soil, except for *I. douglasiana*, which will grow almost anywhere and tolerate most conditions. If the plants are growing in pots they should be carefully re-potted in mid-September because by then the roots will have found their way to the base and edge of the pan. A covering of new soil will protect from winter frosts. Also, by September, new roots are just emerging from leaf bases at the joint with the rhizome.

I. bracteata S. Watson

Distribution
Del Norte County, California; Josephine County, Oregon.

Description
Rhizome moderately stout to 0.9 cm diameter. **Leaves** up to 60 cm long, 1 cm wide, few and scattered in habit, thick, rigid, strongly ribbed, deep green with red staining on bases. **Stem** up to 30 cm usually bearing 2 flowers and often shorter than the leaves; several short cauline leaves, lower bracts strongly overlapping, upper-stem bracts longer and free towards the tips; all usually stained red up to, and including, the spathes; large spathe valves, average 7 cm long, 0.8 cm wide; spathes closed, unequal in length, lanceolate, acuminate; pedicels up to 5 cm. Ovary barrel-shaped, tapering sharply into the perianth tube, averaging 2 cm long; perianth tube averaging 0.9 cm long, thick, funnel-form in shape. **Flower** colour cream to rich buff-yellow veined with maroon or purple-brown lines; falls averaging 6.5 cm long, 2.5 cm wide, oblanceolate; standards averaging 5.5 cm long, 1.4 cm wide, oblanceolate; style crests averaging 1.2 cm long; stigma triangular. **Capsule** averaging 2.5 cm long, oblong, circular in cross section tapering at both ends. **Seeds** irregular and dark brown.

Observations
This is an attractive, large-flowered species from the pine forests of the Siskiyou Mountains in southern Oregon and northern California. Restricted to shades of yellow, the flower colour varies from cream to deep brownish yellow, but usually veined, or netted, over the blade of the falls with red or purple-brown lines and with a deep golden signal patch running along the centre. The form usually seen has large flowers with spreading falls and wavy standards held well aloft.

With its very distinct character of bract-like leaves clasping the flowering stem and its short tube emerging from closed spathes, this species must be set apart from the two main groups. It is well known in cultivation: Dykes grew it for many years and also wrote about some beautiful hybrids where *I. bracteata* was the seed parent. I have a plant, collected as a seedling near Waldo in Oregon, which is still growing and flowering in my garden in a dark shaded area.

I. chrysophylla Howell

Distribution
Del Norte County, California; Benton, Coos, Douglas, Jackson, Josephine, Klamath, Lane, Linn, Marion and Polk Counties, Oregon.

Series Californicae

Description
Rhizome slender, to 0.5 or 0.6 cm diameter. **Leaves** longer than stems, 0.3–0.5 cm wide, finely ribbed, light green and often glaucous, bases coloured pink or red. **Stem** up to 20 cm, but many plants are nearly stemless; up to 3 cauline leaves; large, broad spathes, 2-flowered, unequal, averaging 7 cm long, 0.8 cm wide. Pedicels usually short, averaging 0.8 cm at anthesis; ovary averaging 1.5 cm long, tapering into perianth tube and abruptly into the pedicel. Perianth tube very long and slightly swollen at the top, averaging 6 cm. **Flower** creamy-yellow to creamy-white usually veined in gold or lavender; falls averaging 5.5 cm long, 1.4 cm wide, narrowly oblanceolate; standards averaging 4.8 cm long, 0.8 cm wide, lanceolate; style branches averaging 2 cm, slender; style crests averaging 2 cm, very long, often exceeding the length of the branches; style triangular. **Capsule** averaging 2.5 cm long, oblong and sharply beaked. **Seeds** brown, irregular in shape.

Observations
This is really just an Oregon species since it technically crosses the state line with California in only one area for a short distance. It is found within, or near, the dry, open pine and fir forests of southern and mid-western Oregon. The flowers tend to look thin and fragile and are carried over spathes that appear too heavy for them. They are often short-stemmed, produced among a mass of leaves which hide them.

Hybrids between *I. chrysophylla* and *I. tenax* can be found in a number of areas; most of them are more attractive than *I. chrysophylla* itself. The same can be said for hybrids with *I. bracteata* found in S Oregon and N California.

I. douglasiana Herbert

Synonyms
I. beecheyana Herbert, *I. humilis* Beechey, *I. watsoniana* Purdy.

Distribution
Del Norte, Humboldt, Marin, Medicino, Monterey, San Francisco, San Luis Obispo, San Mateo, Santa Barbara, Santa Cruz and Sonoma Counties, California; Coos and Curry Counties, Oregon.

Description
Rhizome 0.8–0.9 cm diameter, base covered with remains of old leaves. **Leaves** up to 10 cm long, 2 cm wide, variable in size, shape and colour, but usually ribbed and deep green, bases often pink or red. **Stem** up to 27 or 28 cm, usually shorter than leaves, often branched, especially on mature plants; up to 4 side branches; spathes 2–3-flowered; spathe valves averaging 8 cm long, 0.9 cm wide, usually opposite, but sometimes separated and divergent, lanceolate-acuminate; pedicels

2–5 cm (averaging 3 cm); ovary averaging 3.8 cm long, elliptic–oval, triangular in cross section, tapering to both ends, upper end with nipple-like projection; perianth tube averaging 2.3 cm long, usually widening to a bowl shape at top. **Flower colour** variable, ranging from deep red-purple through lavender and grey-blue to shades of cream and white, veined and marked in gold, blue or purple; falls 5–9 cm long (averaging 6.5 cm), 2.4 cm wide, oblanceolate–obovate; standards averaging 6 cm long, 1.4 cm wide, oblanceolate; style branches averaging 2.7 cm long; style crests rounded to oblong, reflexed with toothed edges; stigmas triangular. **Capsule** 2.5–5 cm long, triangular in cross section, tapering at each end with nipple-like apex. **Seeds** dark brown and wrinkled.

Observations
Perhaps the best known species in this series although it is not the most attractive. It is unique in forming branched stems on mature plants and can carry eight or nine flowers on each main stem. Extremely varied in size, it can produce leaves up to 2 cm wide and as much as a metre in length. A large form was at one time described as *I. watsoniana*, but apart from large, stiff leaves spreading semi-horizontally from bases lacking the red staining, it does not differ greatly from more typical forms.

With a long, narrow distribution pattern confined to the coastal areas of S Oregon and California, *I. douglasiana* is found along a 700 mile stretch from Coos County in Oregon to Santa Barbara County in southern California. Tolerant of a wide range of conditions, it is most often found on grassy hillsides and cliff tops within sight of the Pacific. It often grows right down to the beaches. North of Eureka in Humboldt County, California, where the highway runs along the coast, this iris is aggressively abundant, covering adjacent pastures with enormous plants which have grown into each other in the manner of heathers on European moors. Where logging and roads have opened up the territory it has spread inland to hybridise with any other species encountered.

In cultivation, *I. douglasiana* is the easiest of all to grow under almost any conditions and will even tolerate lime in the soil; other species require lime-free soils.

Hybrids are known with *Ii. longipetala, missouriensis, sanguinea* and *I.* 'Caesar's Brother'.

I. fernaldii R.C. Foster

Distribution
Lake, Napa, San Mateo, Santa Clara, Solano and Sonoma Counties, California.

Description
Rhizome averaging 0.6 cm diameter. **Leaves** up to 40 cm long, 0.8 cm wide, slim, often grey-green and glaucous, with bright pink staining on lower parts

extending down to base. **Stem** 20–38 cm tall with 2 or more cauline leaves tight to the stem for half their length; spathes usually 2-flowered; spathe valves averaging 7.5 cm long, 0.8 cm wide, opposite, broadly lanceolate; pink staining often extending up stems and onto spathes; ovary averaging 2 cm long, elliptical with short pedicel; perianth tube averaging 5 cm long, upper portion funnelling out to form a wide throat. **Flower** pale creamy-yellow veined gold or grey; falls averaging 5.5 cm long, 1.8 cm wide, oblanceolate; for half their length the falls are held horizontally to their tips and not usually reflexed as in many other species; standards 5 cm long, 0.8 cm wide, narrow, oblanceolate; style branches averaging 2.6 cm long; style crests oblong to spear-shaped, reflexed, slightly toothed; stigma triangular. **Capsule** averaging 3 cm long, oblong, beaked. **Seeds** brown.

Observations
I. fernaldii is seen in its purest form in the vicinity of the Petrified Forest in Sonoma County. Here large colonies show all the typical characteristics of this species: dark grey-green leaves heavily stained with deep pink or beetroot-red at the base, this often extending up the stems and onto the spathes, which are quite broad. Some were bearing pale creamy-yellow flowers typical of the species when I was there in 1965, but already the process of hybridisation with other species was going on apace in many areas and this species in its purest form may now be in danger of extinction.

I. hartwegii Baker

Distribution
Amador, Butte, Calaveras, Eldorado, Fresno, Kent, Madera, Mariposa, Nevada, Placer, Plumas, Sierra, Tulare, Tuolumne and Yuba Counties, California.

Description
Rhizome 0.5–0.8 cm diameter, slender, bearing remains of old leaves. **Leaves** averaging 35 cm long, 0.3–0.6 cm wide, deciduous, pale green and slightly glaucous, bases usually green. **Stem** to 30 cm long, slender, 1–3 cauline leaves, free in upper half; 1–2 flowers; spathe valves linear–lanceolate, outer averaging 0.8 cm long, 0.5 cm wide, usually divergent and separated for 1.5 cm on average, herbaceous; pedicels averaging 4.5 cm; ovary averaging 1.5 cm long, cylindrical; perianth tube averaging 7.5 cm long. **Flower** usually pale yellow to cream, but lavender and deep yellow forms are found in some areas; falls averaging 5 cm long, 1.5 cm wide, oblanceolate; standards averaging 4.5 cm long, 0.6 cm wide, narrow, oblanceolate; style branch averages 2 cm long; style crest averaging 0.9 cm long, rounded, overlapping; stigma triangular. **Capsule** averaging 2.5 cm long, oblong and tapering sharply at either end. **Seeds** light brown and wrinkled.

Observations

I. hartwegii produces a thin, scattered, loose appearance with few leaves; its flowers are thin-looking and rather unattractive. Dykes dismissed it when he wrote 'the somewhat insignificant plant that goes by the name of *I. hartwegii* has pale, straw-coloured flowers of no great merit'. The perianth tube is short and about as wide as the ovary when the flower opens.

Hybrids are known with *I. prismatica*.

subsp. *australis* (Parish) Lenz

Synonyms
I. hartwegii var. *australis* Parish, *I. tenax* var. *australis* (Parish) R.C. Foster.

Distribution
Riverside and San Bernadino Counties, California.

Description
Rhizome 0.6–0.9 cm in diameter bearing remains of old leaves; open and rather loose habit; relatively few leaves in each clump. **Leaves** averaging 40 cm long, over 0.6 cm wide, slender, bases usually pink. **Stem** almost same length as leaves with 1–3 cauline leaves free in the upper half; spathe valves usually 2-flowered, divergent and separated, outer averaging 6.5 cm long, 0.8 cm wide; pedicels vary, averaging 3 cm long on the first flower with that for the second being longer; perianth tube short and stout, averaging 0.85 cm long. **Flower** colour purple to blue-violet and cobalt blue; falls averaging 6 cm long, 2.3 cm wide; standards averaging 5.5 cm long, 1 cm wide; stigma triangular. **Capsule** oval to oblong averaging 2.5 cm long and tapering sharply at both ends. **Seeds** light brown and wrinkled.

Observations

I. hartwegii subsp. *australis* is isolated geographically from all the other Pacific Coast irises. It is found in the San Bernardino Mountains and San Gabriel Mountains of southern California with the Mohave Desert separating it from other irises.

A good deal of confusion and misunderstanding has surrounded this iris from 1897 onwards; some writers have confused it with *I. tenax*. Although the flowers are similar, the respective habits and appearance of the leaves differ: the leaves on this subspecies look grey and hard. The ecological situation is also quite different with subsp. *australis* existing under alpine conditions, emerging from its snow cover at the end of April or even later. This attractive iris was grown at Kew some 25 years ago in the old species bed under the label of *I. hartwegii*.

subsp. *columbiana* Lenz

Distribution
Tuolumne County, California.

Description
Rhizome 0.7–0.9 cm diameter, bearing remains of old leaves. **Leaves** up to 90 cm long and 1 cm wide, sometimes glaucous and faintly pink at base. **Stem** up to 35 cm long, unbranched and shorter than the leaves, 2–3 cauline leaves free in the upper half; spathe valves separated and divergent, up to 12 cm long and averaging 0.7 cm wide, outer linear–lanceolate; usually 3 flowers produced from within the spathes; pedicel of first flower averages 2.5 cm long, those of second and third longer; ovary oblong, round in cross section, tapering abruptly into the perianth tube and more gradually below; perianth tube averages 0.8 cm long. **Flower** colour pale, but clear yellow with darker golden veining; falls averaging 5 cm long and 1.4 cm wide, oblanceolate; standards averaging 5 cm long, 1.4 cm wide, narrowly oblanceolate; stigma triangular. **Capsule** oval to oblong, up to 4 cm long, 1.8 cm wide, tapering sharply at both ends. **Seeds** brown and slightly wrinkled.

Observations
I. hartwegii subsp. *columbiana* is narrowly endemic to a small area in Tuolumne County, California. With mixed woodland and Yellow Pine forest as background, this subspecies is found growing on hillsides near the old historic mining town of Columbia. This is a tall iris in comparison with most others in this series. Indeed, I believe that *I. munzii* is the only one constantly taller, with the odd and occasional *I. douglasiana* topping them both. I remember seeing many plants in flower at the Rancho Santa Anna Botanical Garden, in southern California, and most of the stems were up to 75 cm in height.

subsp. *pinetorum* (Eastwood) Lenz

Distribution
Plumas County, California.

Description
Rhizome 0.5–0.6 cm diameter, bearing remains of old leaves. **Leaves** up to 40 cm long, 0.5 cm wide, pale green, finely ribbed, bases usually pink. **Stem** up to 28 cm tall, slender, bearing 2–4 cauline leaves on its lower section, free for about 1/3 of their length; spathe valves separated and divergent, outer one averaging 6 cm long and linear–lanceolate; 2-flowered and, uniquely, both flowers open at the same time; pedicel averaging 1 cm long; ovary 1.4 cm long, tapered at each end; perianth tube averaging 1.4 cm long, rather thick and, in some specimens, dilating to form a broad throat. **Flower** colour pale, creamy-yellow with deep gold veining on the

falls; falls averaging 6 cm long, 1.5 cm wide, narrowly oblanceolate; standards averaging 5 cm long, 1 cm wide; style branches averaging 2.6 cm; style crests averaging 1 cm long; stigma triangular. **Capsule** 2 cm long, broadly oblong. **Seeds** brown.

Observations
I. hartwegii subsp. *pinetorum* is confined to Plumas County, California. It is native to the Yellow Pine Forest from 1200–1500 m near the northern end of the Sierra Nevada.

A most unusual feature is that both flowers on each stem can be open at the same time (in all other irises of the Californicae where two or more flowers are produced in the spathes they open consecutively). This twin-flowering gives a most distinctive look to these irises.

I. innominata Henderson

Distribution
Coos, Curry, Douglas and Josephine Counties, Oregon.

Description
Rhizome averaging 0.4 cm diameter. **Leaves** abundant, up to 35 cm long, 0.2–0.4 cm wide, dark glossy green above and lighter greyish-green below; bases bright pink or purple-red. **Stems** may be shorter or longer than leaves, up to 20 cm; 2–4 cauline leaves, not inflated, free only in upper 1/3; spathe usually single-flowered; spathe valves averaging 4.6 cm long, 0.6 cm wide, opposite, subequal, broadly lanceolate to ovate; pedicels averaging 0.8 cm long; perianth tube averaging 2 cm long. **Flower** colour variable from dark golden-yellow through bright, or pale, yellow to pink, usually heavily veined with maroon-red, purple or golden-brown; falls averaging 5.4 cm long and 2.5 cm wide, broadly oblanceolate; standards averaging 4.7 cm long and 1.2 cm wide, oblanceolate with wavy edges; style branches averaging 2 cm long; style crests averaging 1 cm long, crests rounded, reflexed, edges toothed or minutely lobed; stigmas triangular. **Capsules** oblong to ovoid. **Seeds** brown, rounded and finely wrinkled.

Observations
I. innominata, probably the most attractive species in the Californicae, has achieved great popularity in a very short space of time. In its best-known form, from the Rogue River area of SW Oregon, it produces rich golden flowers beautifully veined and netted in Indian red or burnt sienna, but it has a much wider range than was at first assumed: from pale cream and apricot through beige and various shades of yellow to a rich deep orange. Nearly all are beautifully lined and marked with darker colours.

I. innominata has been deliberately used to a great extent in producing attractive garden hybrids, mostly with *I. douglasiana* as the other parent, but the true species has proved to be most accommodating in gardens: its compact tussocks of evergreen leaves make an attractive sight, and even more so when they are bearing their very beautiful flowers.

Series Californicae

I. macrosiphon Torrey

Synonyms
I. amabilis; I. californica.

Distribution
Butte, El Dorado, Glen, Lake, Marin, Mendocino, Napa, Nevada, Placer, Santa Clara, San Mateo, Santa Cruz, Sierra, Sonoma, Tehama, Trinity and Tuolumne Counties, California.

Description
Rhizome up to 0.8 cm diameter, bearing remains of old leaves. **Leaves** up to 40 cm long, 0.5 cm wide, often glaucous, exceeding stems; bases usually colourless. **Stem** up to 25 cm, sometimes only a few centimetres, tall with a few cauline leaves clinging through half the lower part; spathes usually 2-flowered; spathe valves averaging 7 cm long, 0.6 cm wide, opposite, lower valve lanceolate; pedicels averaging 0.9 cm long at anthesis; ovary averaging 2 cm long, ovoid; perianth tube up to 8.5 cm long, averaging 5.5 cm wide, slender with bowl-shaped enlargement at top. **Flower** colour range wide, from deep indigo-blue through purple and lavender to white, cream and yellow, usually veined, sometimes with a conspicuous white centre on the falls; some flowers in Lake County very fragrant; falls vary in shape and size up to 7 cm long, 2 cm wide, narrowly oblanceolate to broad ovate in shape; standards slightly shorter, up to 1.6 cm wide; style branches averaging 2.5 cm long; style crests broadly spear-shaped, roughly toothed and reflexed; stigmas triangular. **Capsule** up to 3 cm long, oblong to ovoid. **Seeds** dark brown, angular and wrinkled.

Observations
I. macrosiphon is an extremely variable species, to be found over a large part of northern and central California. I have usually found this species in open, sunny situations on the fringe of wooded slopes, roadside banks and grassy clearings. It is seldom seen in cultivation in Britain. Perhaps this is because the species is growing in areas free from frost and generally on the mild side. Maybe plants originating from higher altitudes in the northern counties of California would fare better.

I. munzii R.C. Foster

Distribution
Tulare County, California.

Description
Rhizome up to 1.2 cm diameter, rather thick, bearing remains of old leaf bases. **Leaves** up to 54 cm long, 2 cm wide, of evergreen habit, grey-green and quite

glaucous; quite green at bases. **Stems** up to 70 cm tall, rather stout; 1 or 2 cauline leaves free for half their upper length; spathes carrying up to 4 flowers, usually 3; spathe valves separated, sometimes by as much as 9 cm, usually about 6 cm and divergent; lower spathe averaging 9.5 cm long, 1.2 cm wide; length of pedicels variable from 2 cm in the first flower to a much greater length in the later flowers; ovary averaging 2 cm long, rounded in cross section, tapering gradually at base and sharply into the perianth tube; perianth tube averaging 0.8 cm long, short, stout and somewhat funnelled. **Flower** colour from pale powder-blue through lavender to violet, often frilled and fluted and finely veined in violet or darker blue; falls averaging 7 cm long, 2.8 cm wide, varying from oblong–ovate to broad oblanceolate; standards averaging 7 cm long, 1.6 cm wide, oblanceolate to spathulate; style branches averaging 3 cm long; crest subquadrate, reflexed, often roughly toothed; stigmas triangular. **Capsule** up to 5 cm long, oblong, sharply tapered at each end. **Seeds** D-shaped or irregular, brown and wrinkled.

Observations
I. munzii is a very handsome species, bearing some of the largest flowers to be seen in this series, and is only found in Tulare County, California. I chose to observe and record it at Coffee Creek Camp, an attractive locality in the foothills of the southern Sierra Nevada, where Foster collected the type. Approaching Springville, a few miles west of the area, the scent of the orange blossom from nearby citrus groves gave an indication of the mild climate of the area and underlines the fact that *I. munzii* is less tolerant of cold conditions than other species of the Californicae.

These beautiful flowers are to be found only in shades of blue or purple. Lee W. Lenz has discovered that under cultivation *I. munzii* can produce some of the bluest shades of colouring in rhizomatous irises; he has also raised some richly coloured turquoise-blue and sky-blue hybrids.

I. purdyi Eastwood

Distribution
Humboldt, Mendocino, Sonoma and Trinity Counties, California.

Description
Rhizome 0.4–0.6 cm in diameter, dark reddish-brown and bearing remains of old leaves. **Leaves** averaging 30 cm long, 0.8–12.5 cm wide, dark green above, greygreen and rather glaucous below; bases stained brilliant pink or red. **Stems** averaging 12.7 cm, stained red upwards from base, usually covered by several short, overlapping and inflated bracts, also red-flushed and free only towards the tips; large spathes usually 2-flowered, slightly inflated; spathe valves averaging 6.8 cm long and 1 cm wide, outer one broadly lanceolate–ovate; pedicel averaging 1.5 cm,

shortish; ovary averaging 1.8 cm long, narrowly ovate; perianth tube averaging 4 cm long often dilated toward the top in funnel shape. **Flowers** large, spreading, with very broad flattened standards and falls coloured pale creamy yellow or white flushed with pale lavender, strongly veined with dotted lines of purple or cerise-pink on the falls; falls averaging 6.5 cm long, 2 cm wide, oblanceolate; standards averaging 5.5 cm long, 1.5 cm wide, lanceolate, not upright, but wide-spreading; style crest narrowly triangular, reflexed, toothed slightly at apex; stigmas unique, never triangular as in rest of series, but rounded or truncate, or even bilobed and often edged with minute teeth. **Capsule** averaging 2.5 cm long, oblong ovoid, sharply tapering to a beak. **Seeds** irregular or D-shaped, light brown, or wrinkled.

Observations
When Carl Purdy found this iris it was common in the woods and Redwood region of northern California. Today it is very rare in its pure form and is in great danger of extinction. Disturbance of land, mainly by logging and creation of highways, opened the way for other species, mainly *I. douglasiana* and *I. macrosiphon*, to move in and hybridise with *I. purdyi*. In places these hybrids are most abundant.

At no time has *I. purdyi* been common in cultivation. It was known as the Redwood Iris and perhaps those very same unique conditions are needed for it to survive. I grew it for some fifteen years under shaded conditions where it never saw the sun, but it was finally killed, as were hybrids raised with *I. purdyi* as pollen parent, after adverse winter conditions although one could hardly expect it to prosper in London!

I. tenax Douglas ex Lindley

Synonyms
I. gormanii Piper 1924 (a yellow-flowered form); *I. tenax* var. *gormanii* (Piper) R.C. Foster 1937.

Distribution
Benton, Clackamas, Clatsap, Columbia, Coos, Douglas, Lane, Marion, Multnomah, Washington and Yamhill Counties, Oregon; Clark, Cowlitz, Grays Harbor, Lewis, Pacific, Skamania, Thurston and Wahkiakum Counties, Washington.

Description
Rhizome slender. **Leaves** up to 50 cm long, 0.3–0.5 cm wide, deciduous, narrow, light green, rather grassy and lax and usually longer than the flower-stem; bases often pink- or red-stained. **Stem** 15–35 cm long, slender, unbranched; spathes usually 1-flowered, sometimes 2; spathe valves linear–lanceolate, outer averaging 6.5 cm long, 0.4 cm wide, herbaceous; ovary averaging 2 cm long, tapering gradually below, but sharply above; perianth tube very short, averaging 0.6 cm

long, stout and slightly funnel-form upwards. **Flower** colour ranges from deepest purple-blue through pinks and lavender to white, cream and yellow; falls averaging 5.8 cm long, 1.6 cm wide, ovate to oblanceolate; standards averaging 5.8 cm long and 1.6 cm wide, very upright; style crests averaging 1 cm long, rounded, reflexed with toothed edges; stigmas triangular. **Capsule** averaging 3 cm long, oblong. **Seeds** D-shaped to irregular, brown, wrinkled.

Observations
I. tenax was first noted by David Douglas nearly 170 years ago and recorded in his journal as he explored through what then was mainly Indian territory. It is distributed through the south-western part of Washington and is widespread in most of Western Oregon. Seldom found in shaded areas, it is most abundant on the open hillsides of the Williamette and Umpqua Valleys. With quite a wide range of colour it will usually vary by locality although I have been to an area where on one hillside lavender, purple-red, orchid-pink and pure white blossoms can be seen together in riotous profusion. In the SW area of Oregon it was usually deep blue-purple and NW of Portland I could find only pale lavender forms. West of Portland, following Scoggin Creek up towards the crests of the Coast Mountains, one can find a yellow form known in the past as *I. gormanii,* which Gabrielson (1932) thought 'the most dainty and appealing of all native irises'.

Usually single-flowered, *I. tenax* is elegantly poised over the narrow, open spathes and, sitting on its smooth, barrel-shaped ovary, the large upright standards and ample curving falls give it the most perfect proportions.

subsp. *klamathensis* Lenz 1958

Distribution
Humboldt County, California.

Description
Rhizome 0.4 cm diameter, slender. **Leaves** up to 40 cm long, 0.4 cm wide, slender, usually taller than flower-stem; bases stained a brilliant pink or red. **Stem** slender, usually single-flowered, with several clasping cauline leaves free for 1/3 of upper length; spathe valves narrowly lanceolate, divergent and separated, outer valve averaging 5.8 cm long, 0.4 cm wide; ovary averaging 2 cm long; perianth tube averaging 1.8 cm long, rather stout and funnel-shaped. **Flower** colour pale buff-yellow, falls usually distinctly veined with maroon or purple-brown veins; falls averaging 6 cm long, 2 cm wide; standards averaging 1 cm wide, more widespread than species; style crests rounded to deltoid, reflexed or toothed; stigmas triangular. **Capsule** 3 cm long, oblong. **Seeds** brown and wrinkled.

Observations
I. tenax subsp. *klamathensis* is endemic to a small area around the village of Orleans in Humboldt County, California. The one feature that connects it with *I. tenax* is

the formation of the spathe valves: divergent and separated in both. Otherwise this little iris is unique and, to my mind, deserving of specific rank. The very slim, polished, dark green leaves are reminiscent of *I. innominata* and not in the least like those of *I. tenax*. Most of the plants I studied had flowers of a very pale shade of creamy-apricot; some were a pale cream, but all had a patch of golden-yellow running into the throat and they were all beautifully lined on the falls with red-brown veins.

I. tenuissima Dykes

Distribution
Butt, Glen, Humboldt, Shasta, Siskiyou, Telehama and Trinity Counties, California.

Description
Rhizome slender with very few roots. **Leaves** up to 40 cm long and averaging 0.6 cm wide, grey-green and slightly glaucous; bases sometimes pink or red. **Stem** with up to 3 cauline leaves; broad spathes usually 2-flowered; spathe valves averaging 6.8 cm long and 0.75 cm wide, opposite, lanceolate; ovary averaging 1.5 cm long, tapered at each end; perianth tube averaging 4.5 cm long, abruptly dilated in upper section. **Flower** colour pale cream, strongly veined in purple or red-brown; falls averaging 6 cm long, 1.5 cm wide, lanceolate; standards averaging 5.3 cm long, 0.8 cm wide; style crests very long and narrow, strongly reflexed; stigmas triangular. **Capsules** oblong, tapered below and forming a beak above. **Seeds** brown.

Observations
Dykes described *I. tenuissima* from dried herbarium material, so that he probably did not notice the unusual and most distinctive character of the perianth tube, which is abruptly dilated in the upper portion in the manner of an upturned bottle. He did, however, draw attention to the elongated and narrow, reflexed style crests. California's *I. tenuissima* is somewhat like Oregon's *I. chrysophylla*: both tend to produce short-stemmed flowers more often than long-stemmed ones.

subsp. *purdyiformis* (R.C. Foster) Lenz

Distribution
Plumas and Sierra Counties, California.

Description
Rhizome slender. **Leaves** up to 40 cm long, 0.5 cm wide, bases stained pink. **Stems** up to 35 cm tall, slender; covered with 3 or 4 closely clasping, but not

overlapping, stem bracts slightly inflated, but free at the tips, often flushed with pink or red staining; spathes carrying 2 flowers; spathe valves averaging 4.5 cm long, 0.8 cm wide, opposite, broadly lanceolate, flushed pink; ovary averaging 1 cm long, small; perianth tube averaging 3.5 cm long, slender with dilated upper part nearly half the length of the tube. **Flower** colour cream or pale yellow, sometimes showing a few dark veins; falls averaging 4.8 cm long, 1.4 cm wide, narrowly oblanceolate; standards averaging 4 cm long, 0.9 cm wide; style crests narrow, rather long and reflexed; stigmas broadly triangular to rounded. **Capsule** up to 2 cm long, oblong–ovate. **Seeds** brown.

Observations
I. tenuissima subsp. *purdyiformis* shows many features found in *I. tenuissima* and *I. purdyi* and it has been suggested that this subspecies originated as a hybrid between those two species, but *I. purdyi* has never been found in this part of California and in places where the two species do meet the resulting hybrids are rather different from this little subspecies.

I. thompsonii R.C. Foster

As the result of studies carried out at Portland State University, Oregon, USA, in the Departments of Biology and Chemistry, as well as research in the field in northern California and SW Oregon, *I. thompsonii* is now given full specific status. (Revision authorised 8 November 1989.)

Previously considered by Lenz to be the blue or purple form of *I. innominata* found near High Divide in Northern Del Norte County, California, we should now regard this taxon as *I. thompsonii*. It differs from *I. innominata* mainly in its slightly larger perianth tube (0.3 cm longer); also in colour, having mainly blue to purple flowers. Petal length, bract, sepal and stigma lengths are all slightly longer in *I. thompsonii*, but leaf width is 0.1 cm less. Details are not available for seeds or capsule.

All the study populations of *I. thompsonii* were located from south of Powers in Coos County, Oregon, southward into Northern Del Norte County, California. Other species studied were *I. innominata* in Oregon and *I. douglasiana* in N. California and SW Oregon as well as numerous hybrids of all three species.

SERIES LONGIPETALAE (DIELS) LAWRENCE

Map 14, Figure 14.

This is a gloriously confused group of irises from the botanical point of view and nothing is improved by the fact that certain forms have been introduced into various countries under specific names. Originally *I. missouriensis* and *I. longipetala* were recognised as separate species, but the latter has now been subsumed under the former.

I. missouriensis Nutt.

Synonyms
I. longipetala Herbert, *I. montana* Nutt., *I. pelogonus* Godding, *I. tolmeiana* Herbert.

Distribution
Both sides of the Rocky Mountains range from the border with Canada south to Mexico and from sea level to around 3000 m.

Description
Rhizome short, creeping, wide-spreading with remains of old leaf bases persisting. **Leaves** 1–2 cm wide, 45–60 cm long, glaucous, dark grey-green, semi-evergreen. **Stem** about 60 cm, with 2–3 reduced leaves and 1–2 branches, valves 7.5–10.5 cm long, green, lanceolate, outer usually 3–7.5 cm below inner, all 3–6-flowered; pedicels 2.5–7.5 cm long depending on flowering state; ovary oblong, trigonal, 1.5–2.5 cm long, uneven surface, ridged on each side; tube about 0.75 cm, funnel-shaped, green with purple lining in continuation of line of standards. **Flower:** falls obovate, unguiculate, flared at hafts, but dependent from the mid-point, central ridges flanked yellow with small purple dots, blade white, veined violet with purple dots, haft similar, blade to 3.5 cm, 6–7.5 cm overall; standards slightly divergent, oblong–unguiculate, 1.5 cm wide, 6–7.5 cm long, violet veining less distinctive than on falls; styles 3.75 cm long, narrow at base, but widening outwards, pale violet; crests almost quadrate with irregular toothing; stigma obscurely bilobed. Filaments equal to anthers, white, with pale violet mottling; anthers purple. Pollen creamy-white. **Capsule** 1.75 cm wide, 4–5 cm long, tapering to both ends, cross section almost circular in spite of 6 equidistant ribs. **Seeds** large, nearly globular with slightly wrinkled skins. $2n = 86, 88$.

Cultivation
Essentially, these plants should only be moved in the growing season. There seems to be a preference for fairly heavy, loamy, slightly alkaline soils. Although they are not fully evergreen, there is a tendency to produce new leaves during the winter, which may indicate that they are unsuited to very cold climates. In general, too, they require plenty of water in the growing season and good drainage against a wet winter.

Observations
This series does not have a consistent distribution over its known area and local forms have achieved distinctive appearance with variations in flower colour, stem length and behaviour which are easily recognised by gardeners. In fact, it is a classic example of the need to retain a flexible approach to what constitutes a species. In addition, yellow-flowered and albino forms have been recorded. Unfortunately, the various forms hybridise indiscriminately. It would be helpful if

a control garden could be established where they could be grown for direct comparison in a climate compatible with the range from which the forms are collected. Recent research shows that they still grow as far south as Monterrey in Mexico.

Allegedly named forms are in circulation, but should be regarded as unreliable without a detailed history.

Hybrids
these are known with *I. douglasiana*, *I. forrestii* and *I. sibirica*.

SERIES LAEVIGATAE (DIELS) LAWRENCE   SIDNEY LINNEGAR

Map 15, Figure 15.

This series of aquatic irises is found over most of the northern hemisphere and consists of five species. Two, *Ii. versicolor* and *virginica*, are found in N America; two, *Ii. laevigata* and *ensata*, are confined to Asia; the fifth, *I. pseudacorus*, is distributed throughout Europe and into Asia. All the species are likely to develop black 'water marks', or superficial lines, running between the veins on the leaves; on *I. ensata* these are caused by a thrip, *Frankliniella iridis* (syn. *Bregmatothrips iridis*), which can also affect other members of the series, but does not show so early. Plants all make large, compact clumps.

I. ensata **Thunb.**

Synonyms
I. caespitosa Pall. ex Link., *I. doniana* Spach., *I. kaempferi* Sieb. ex Lam., *I. longispatha* Fisch., *I. laevigata* var. *kaempferi* (Sieb.) Maxim., *Ioniris doniana* Klatt, *I. longispatha* Klatt, *Xiphion donianum* Alef.

Distribution
Japan, N China, eastern Siberia in the Sakhalin, Amus and Ussuri regions.

Description
Rootstock a stout short rhizome. **Leaves** 25–75 cm long, 0.5–1.5 cm wide, sword-shaped, usually dark green with a very pronounced mid-rib. **Stem** 50–90 cm long with 1–3 reduced leaves and occasionally a lateral branch; 1–3 flowers in the terminal head, 1–2 on a branch; spathes narrow, about 7.5 cm long. **Flowers** 8–15 cm diameter; red-purple through purple-blue bitone to white with pink and pure white; falls: haft 2.5 cm long, narrow, yellow with raised ridge, blade 7.5 cm long, oval to obovate, width variable, with yellow signal; standards 6–7 cm long, erect, narrow, oblanceolate; style rounded, narrow, with small erect crest nearly level with tops of standards. Filaments short, colour related to flower.

Anthers pale yellow, twice as long as filaments. Tube 1–2 cm. Ovary rounded, trigonal, with 6 grooves. **Capsule** short, tapering at ends, obtusely beaked at tip, rounded and grooved. Pedicel variable at all times. **Seeds** nearly circular thin discs. $2n = 24$ (Kazao 1928; Simonet 1928; Inariyama 1929).

Cultivation
These are plants of wet, grassy places up to 2400 m flowering between June and August. They can be grown in boggy ground and, if planted in clay puddled with old farmyard manure, in 5–15 cm of water, when the flower will be larger; also in a well-watered, enriched bed in a semi-shady position. Propagate true species by seed, or division in September, or after frost has ceased in spring.

Observations
This species has been known as *I. kaempferi* for many years and will possibly be called so for many more. Generally speaking, plants will not tolerate alkaline soils, but it is possible to raise lime-tolerant plants from seed as shown by Herr Steiger in the 1950s. Also it has differently shaped seeds from the other members of the series.

Subspecies and varieties

var. grandiflora Dykes
A larger form with yellow veining on the falls.

var. spontanea (Makino) Nakai
Reddish-purple flowers 10 cm diameter with yellow base to fall, from Hokkaido, Honshu and Kuosho, Japan.

Hybrids
These are known with *Ii. crocea* (syn. *aurea*), *laevigata, monnieri, pseudacorus, setosa, spuria, versicolor* and *virginica*.

I. laevigata Fisch.

Synonyms
I. aestiva Pall., *I. gmelini* Ledeb., *I. itsihatsi* Hassk., *I. mackii* Maxim., *I. mandschurica* Hort., *I. pseudacorus* Regel, *I. semperflorens* Hort., *Xyridion laevigatum* Klatt, *X. violaceum* Klatt.

Distribution
Central Russia eastward through China, Korea and Japan.

Description
Rootstock compact, with slender rhizomes covered with bases of old leaves. **Leaves** 45 cm long, 1.5–4 cm wide, pale green without prominent midrib. **Stems** 35–40 cm tall, stout, with 2 reduced leaves at base and a third part way up, giving an angled appearance; terminal head with 3–4 flowers, 2–3 on lateral branch; spathes keeled, green, outer up to 10 cm long, inner 6 cm. **Flower** blue-purple, sometimes white, 8–10 cm diameter; falls 6 cm long, 5 cm wide, blade obovate and twice as long as the narrow haft; standards up to 6 cm long, oblanceolate, erect, smaller than falls. Filaments short, white; anthers twice as long as filaments, narrow pollen sacs distinctly separated; pollen white. Tube up to 2 cm long. **Capsules** 5 cm long, oblong or much rounded trigonal with blunt ends. Pedicel up to 2 cm long. **Seeds** semi-circular or compressed; encased in thick, smooth, corky coat, brown. $2n = 32$ (Kazao 1928; Simonet 1934).

Cultivation
This is a species for the water garden and is easy to establish as a submerged plant up to 15 cm depth. Plant in the bottom of an enriched, puddled pool, or in a basket of compost enriched with well-rotted manure during September or after May frosts. Propagate by division for selected forms, or from seed for true species. Seed from selected forms will not come true, although a large percentage of the seedlings will be similar or close to the variety.

Observations
Dykes stated that the forms of this iris come true from seed. This may be so if only one form is grown, but the variation seen in nursery stocks suggests that this cannot be accepted today. The paler 'water marked' leaves make the plant quite distinct from *I. ensata* in the garden.

Subspecies and varieties

var. *alba* Wallace
A grey-white form with single flower.

var. *albo-purpurea* Makino
White mottled purple on falls.

var. *atro-purpureum* hort.
A deep blue form (possibly that grown as Christie-Miller's form).

var. *plena*
Double blue flowers.

var. *variegata* van Tubergen
White and green striped leaves, purple flower.

Hybrids
These are known with *Ii. brevicaulis, delavayi, ensata, orientalis* (syn. *ochroleuca*), *prismatica, pseudacorus, sanguinea, setosa, sibirica, versicolor, virginica* and Louisiana hybrids.

I. pseudacorus L.

Synonyms
I. caspica Rodion., *I. flava* Tornab., *I. longifolia* Bak., also Lam. & D.C., *I. lutea* Lam. & St. Lager, *I. pallidiflava* Sime, *I. pallidiflora* Sime, *I. pallidior* Hill, *I. paludora* Per., *I. palustris* Moench, *I. sativa* Mill., *I. strictum* Hort., *Limniris pseudo-acorus* (L) Fuss, *Limnirion pseudo-acorus* (L) Opiz, *Pseudo-iris palustris* Medic., *Xiphion acoroides* Alef., *Xiphion pseudo-acorus* Schrenk., *Xiphium pseudacorus* Parl., *Xyridion acoroideum* Klatt, *X. pseudacorus* Klatt.

Distribution
Throughout Europe, N Africa into SW Asia and China.

Description
Rootstock a stout rhizome with fibrous remains of old leaves; inside colour pink; robust, clump-forming. **Leaves** up to 100 cm long, 2–3 cm wide, deep green, slightly glaucous at base, with conspicuous midrib. **Stem** 60–160 cm tall, round, well branched, often with equal laterals on opposite sides of same point on stem, with a reduced leaf at each branch; carrying 4–12 yellow flowers; spathe oblong, green with papery edges, outer valve 6–9 cm long, sharply keeled, inner as long, no keel. **Flower** bright yellow, 7–10 cm diameter, usually with dark brown blotch and veining on falls; falls 5–7.5 cm long, 4 cm wide, usually veined brown on a yellow ground with deeper yellow patch on blade; standards 2–3 cm high, narrowly oblong and usually spoon shaped; style 3–4.5 cm long, broad, keeled; stigma with prominent tongue; filaments slightly longer than anthers; anthers cream to orange in colour edged with purple; pollen cream. Tube 1 cm long, conical. Ovary 1.25 cm long, trigonal with narrow groove at angles and concave sides. **Capsule** 3.5–6 cm long, oblong, narrowing to short beak at apex. **Seeds** flattened D-shape with light brown outer skin. $2n = 34$ (Simonet 1932a; Randolph 1934).

Cultivation
This is an ideal plant for most positions except dense shade. At its most robust in pond or bog garden, it will grow and flower in the herbaceous border although it will not reach its full height there. Like most aquatic irises it is a gross feeder and winter and spring mulching with well-rotted manure or compost is beneficial. Propagate by division from September to May, or from seed, which may not come true.

Observations
Iridin is obtained for use in the drug trade and has been used for tanning, dying and in snuff. It is a drastic purgative and irritant, so dead, or dying plant material should never be left where cattle may eat it accidentally.

Varieties and forms

var. *acoriformis* (Boreau) Dykes
Blades of falls are rounded.

var. *alba* hort.
Ivory white flowers.

var. *bastardii* (Spach) Dykes
Pale sulphur-yellow flowers.

var. *caspica* Rodion.
Paler green leaves and pure yellow flowers.

var. *flore plena*
Double-flowered.

var. *gigantea*
Over 2 m tall, collected in Italy by Mr Berry in 1922.

f. *longiacuminata* Prod.
Longer, narrower, acuminate leaves; collected in Bulgaria.

var. *mackii* (Maxim.) Y.T. Zhao
Small-flowered with 2 branches on an 80 cm stem, from NE China. Prof Zhao gives this specific status as smaller form of *I. pseudacorus*; the original herbarium material consisted only of seed pods and it was regarded as a synonym of *I. laevigata*. Grown in England, seedlings of this variety gave robust plants with flowers of either the standard yellow or the paler *bastardii* colour.

F. *nyaradyana* Prod.
Fruit wider than the type, long beak; short pedicel; collected in Romania.

Hybrids
These are known to exist with *Ii. ensata, fulva, laevigata, prismatica, sibirica, spuria, versicolor* and *virginica* and there are tetraploid forms induced with the aid of colchicine.

I. versicolor L.

Synonyms
I. boltoniana Hort. ex Regel, *I. caurina* Herb. ex Hook., *I. flaccida* Spach., *I. murrayana* Farn., *I. picta* Mill., *I. pulchella* Regel, *I. virginiana* sensu Jacq., *Xiphion flaccidum* Alef., *X. versicolor* Alef., *X. virginianum* Alef.

Distribution
Eastern N America from eastern Canada southwards to Texas, USA.

Description
Rootstock a stout, creeping rhizome forming large clumps. **Leaves** 20–80 cm high, 1–2 cm wide, erect or curved, mid-ribs prominent in mature leaves which are grey-green stained purple at base. **Stem** 25–80 cm high, arching, with 1–3 reduced leaves, usually with 1 or 2 branches on upper half; branches and main stem level with leaves, 2–4 flowers in terminal head, 2–3 flowers in lateral heads; spathes unequal, 3–8 cm long, outer shorter with papery edge. **Flowers** lavender to blue-purple, veined yellow or whitish, rarely white, 6–8 cm diameter; falls 4–7 cm long, 2–3.5 cm wide, haft yellowish green, blade ovate or suborbicular with yellowish, soft, hairy blotch by the haft surrounded by white heavily veined with lilac-purple; standards 2–4 cm long, narrow, erect, spathulate; style 3.5 cm long, crest 1.5 cm long, lilac purple; filaments purplish, longer than violet anthers. Tube 0.5–1 cm. Ovary small, oblong, triangular. **Capsules** up to 5 cm, ovoid to oblong–ellipsoid. Pedicel up to 5 cm. **Seeds** D-shaped, shiny, dark brown with thin, hard skin. $2n = 72, 84$ (Longley 1928); 108 (Randolph 1934; Anderson 1936); 72, 84, 105 (Simonet 1934).

Cultivation
Although a water plant, it is just as happy in a border, only requiring extra humus to stop the soil drying out too quickly. Propagate by seed, or division of especially good colour forms.

Observations
This species is very variable in colour; the varieties listed below seem only to be selected forms which may not come true from seed. This is another of the few irises used in the drug trade; irisin is still in the American Pharmacopoeia List.

Varieties

var. *arkansensis* Perry 1930
Light blue flowers spotted purple.

var. *columnae* Barr 1899
Velvety-purple flowers.

var. *kermesina* Barr 1901
Red-purple flowers with bold white markings.

var. *rosea* Cleveland 1927
Soft rose pink with white markings, flowers below leaf height (now often the term for any pinkish form).

Hybrids
These are known with *Ii. brevicaulis, ensata, fulva, laevigata, pseudacorus, sibirica* and *virginica*.

I. virginica L.

Synonyms
I. carolina Radiu (& of Watt), *I. carolina* S. Watson, *I. georgiana* Britton, *I. sasha* hort., *I. versicolor* Walt.

Distribution
Eastern seaboard of USA south from Virginia along the Atlantic coastal plain of Florida, Georgia, N and S Carolina and SE Louisiana.

Description
Rootstock a stout rhizome around 2.5 cm diameter. **Leaves** 40–90 cm long, 1–4 cm wide, with several prominent ribs when mature, curving over at tips; grey-green or bright green. **Stem** 50–90 cm tall, slender, with several long, linear, reduced leaves, which can overtop the terminal head; simple, rarely branched above the middle; may be stained purple or black; 1–4 flowers in terminal bud with 1–2 on branch level with those on main stem; spathes compact, inner valve 3–14 cm, outer usually longer with papery edges. **Flowers** 6–8 cm diameter, lavender through to violet, rarely white, often scented; falls 4–5.5 cm long, 3–4 cm wide with central yellow signal covered with fine, long hairs (which help to separate this plant from *I. versicolor*), yellowish haft streaked with orange, often slightly shorter than standards. Styles 3–5 cm long with lilac rib and paler edges; filaments same length as anthers; tube short. Ovary small, oblong. **Capsules** 4–7 cm long, ovoid; seeds in 2 rows to each carpel. **Seeds** round or irregularly D-shaped with pitted corky skin. $2n$ = 71 (Randolph 1934); 70–72 (Anderson 1936).

Cultivation
The type is not as hardy, nor as robust, as *I. versicolor*, and requires a wetter, more acid position. Leaf shoots, which may be well grown by winter, should have a loose mulch for protection. The variety *shrevei* is more adaptable in British conditions. Propagate by division in September or April; or by seed, which should be soaked for 10–14 days before sowing.

Observations
Two features distinguish this plant from *I. versicolor*: the longer signal hairs on the fall and the seeds.

Varieties

var. *shrevei* (Small) Anders.
A weaker, fragrant variant from the Mississippi Valley with long branches and capsules 7–11 cm in length and half that width.

var. *shrevei* var. *virginica*
The capsules are much thicker and 4–7 cm in length.

Hybrids
These are known with *Ii. ensata, fulva, laevigata, pseudacorus, sibirica* and *versicolor*.

SERIES HEXAGONAE (DIELS) LAWRENCE

Map 16, Figure 16.

A group of five irises popularly known as Louisiana irises from their concentration in that state of the USA although some of them spread further north or east. The large number of original descriptions has now been reduced to 5 accepted species with considerable internal variations, but in practice the measurements given below are really only a guide to the type specimens; and all are very variable in colour. They can be raised easily from seed, which should be as fresh as possible; dried seed may take some years to germinate. Generally speaking they flower at roughly fortnightly intervals with *Ii. fulva* and *giganticaerulea* followed by *nelsonii, brevicaulis and hexagona.*

New work is prompting serious reconsideration of this classification since it seems likely that it was based on an inadequate selection of plants from a limited area.

There has been much discussion about how far these species are inter-related. Bennet & Arnold (1989) have recently set out to try and establish this through molecular genetic analysis. This is a time-consuming operation which has produced some interesting results to date as well as the location of still more forms of the various irises.

Cultivation
This is much the same for all these species: they are plants of the wetlands and need ample water in the growing season and rich feeding. Many are regarded as too tender to be satisfactory in temperate climates, but this may be a question of finding suitable seedlings: *I. hexagona* is rated as frost-tender. E.G. Luscombe noted that *I. nelsonii* and *I. fulva* are suitable for a shady site, but the others prefer full

sunshine. Dykes suggested that where it is necessary to grow the plants in pots, a second pot should be provided for the current year's offset to root into so that it could be detached when the new plant was established. *I. brevicaulis* and *I. fulva* seem to grow well in Britain.

I. brevicaulis Rafinesque

Synonyms
I. coelestina Nuttall, *I. foliosa* Mack. & Bush, *I. lamancei* Gerard.

Distribution
South Louisiana; reported from Arkansas, Missouri, Indiana and Ohio.

Description
Rhizome 1–3 cm thick, annual growths 8–15 cm, creamy-green, branched. **Leaves** 2–5 cm wide, 30–40 cm long, floppy. **Stem** 25–30 cm, usually recumbent, sometimes slanted upwards; stem-leaves conceal flowers and overtop the terminal ones; flowers in axils; stem changes direction at nodes, resulting in zig-zag effect; spathes 5 cm long, lanceolate, green; pedicel 0.75 cm; tube 1–2 cm long; ovary 0.75 cm, triangular, double-ridged; tube 2 cm, yellowish, with green-yellow ribs. **Flower:** 9–11 cm diameter, rich blue, fragrant; falls 8–9 cm long, 2–3 cm wide, broadly ovate, reflexed, narrowing sharply to haft, which is whitish veined green, this colour spreading onto haft, the small, pubescent median ridge is yellow; standards about 6.6 cm long, 2 cm wide, held diagonally; styles 4.5 cm long, oblanceolate, paler with greenish haft; crests 1.25 cm, quadrate, very pale; stigma bilobed, conspicuous; filaments slightly shorter than anthers, greenish; anthers cream, long, narrow, clearly bilobed; pollen yellowish-white. **Capsule** 3 cm long, globular. **Seeds** D-shaped, thick, corky. **Flowering** June. $2n = 42, 44$.

Observations
This plant is much valued for its intense blue colouring, but paler forms exist as well as pure white.

Hybrids
These are known with *Ii. prismatica* and *versicolor*.

I. fulva Ker-Gawl.

Synonyms
I. cuprea Pursh, *Isis fulva* Tratt., *Neubeckia fulva* Alefeld.

Distribution
South Louisiana, Arkansas, Missouri and Ohio.

Description
Rhizome up to 2 cm diameter, with 10–13 cm annual growths, green colouring. **Leaves** 45–80 cm long, 1.5–2.5 cm wide, drooping at tips. **Stem:** 50–90 cm tall, erect, nearly straight, stem-leaves 45–60 cm; spathes 5–10 cm long, unequal, green, persistent; pedicel 4 cm long; ovary about 1.75 cm, with 6 ribs equally spaced, green; tube 2–2.5 cm, yellowish, hollow as far as ovary. **Flower** 10–11 cm diameter, usually shades of orange-red, veined darker; falls 5–6 cm long, 2–3 cm wide, oblanceolate, cuneate, with yellow veining into haft; standards 5 cm long, 2 cm wide, shorter than falls, narrowing gently to haft; styles around 2 cm long, convex, unkeeled; crests rounded, triangular, shallowly toothed, small; stigma 2-pointed; filament yellowish, short; anthers cream, as far as stigma; pollen cream. **Capsule** 2.75 cm wide, 5 cm long, 6-ribbed, 3 wide and 3 narrow sections all slightly ribbed, staying green even after seeds have ripened. **Seeds** 1 cm × 1.75 cm, flat-sided, with corky covering. **Flowering** June. $2n = 42$.

Observations
Although the terracotta-coloured form of this plant is the most widely known, lighter and darker shades are found as well as a rarer rich yellow. Shade-tolerant.

Hybrids
These are known with *Ii. aurea, delavayi, prismatica, pseudacorus, spuria, versicolor, virginica, wilsonii* and *spuria* cultivars.

I. giganticaerulea Small

Distribution
South Louisiana and, possibly, adjacent corner of south-east Texas.

Description
Rhizome: 2–4 cm thick, annual growths 13–30 cm. **Leaves** to 90 cm, but usually overtopped by flowers. **Stem:** 95–116 cm (even to 165 cm), nearly straight, 2-flowered at apex, with singles in axils; stem leaves 50–65 cm long, 2–3 cm wide; spathes 15–20 cm, green, membranous, keeled, herbaceous; pedicel around 2.5 cm; ovary trigonal, with grooved concave sides; tube to 5 cm, multi-grooved. **Flower:** 13–15 cm diameter, typically blue or blue-purple; falls 9.5 cm long, 4 cm wide, flaring, slightly reflexed at ends, the raised pubescent mid-rib varying from orange through yellow to pale green, haft greenish-white; standards 8.5 cm long, 2 cm wide, erect; styles narrow, raised clear of falls; crests triangular, coarsely toothed; stigma bilobed, with 2 triangular teeth; filaments green, short; anthers greenish, sacs narrow, much

longer than filament. **Capsule** 3.5 cm wide, 10 cm long. **Seeds** roughly D-shaped, very large, with corky covering. **Flowering** March–April. $2n = 44$.

Observations
Colours actually include bright blues, lighter shades, lilacs and pure whites. For a long time, this plant was taken as a larger version of *I. hexagona*, but for various reasons has finally been accorded independent status. The flowers have a curious smell, although nobody describes it.

I. hexagona Walter

Synonym
I. virginica Michaux.

Distribution
Florida, Georgia and the Atlantic coast areas of the Carolinas.

Description
Rhizome 2 cm or more across, annual growths to 30 cm or more, greenish. **Leaves** up to 90 cm long, 2.5 cm wide, slightly glaucous, upright, yellow-green. **Stem** 30–90 cm long, variably straight or zig-zag, 2-flowered at apex, with singles in axils; stem-leaves long, wrapping round nodes; spathes outer 15–20 cm long, inner 10 cm, membranous; pedicel 2.5 cm long; tube 3 cm long, funnel-shaped; ovary trigonal, concave-sided, grooved; tube to 5 cm, multi-grooved. **Flower**: 10–12 cm diameter, darkish blue, fragrant; falls 9–10 cm long, 4 cm wide, median ridge clear yellow dotted yellow or white changing to greenish haft, obovate–elliptic; standards 9–10 cm long, 2 cm wide, erect, oblanceolate, haft veined green; styles narrower than haft, ridged; crests triangular, serrated; stigma bilobed, 2-toothed, triangular; filaments green, short; anthers greenish, with narrow sacs, longer than filament. **Capsule** 4–6 cm long, ovoid, definitely hexagonal, ribs prominent. **Seeds** D-shaped, with corky coating. **Flowering** in June–July. $2n = 44$.

Observations
This plant is very similar to *I. giganticaerulea*, but not quite as large.

Hybrids
These are known with *I. pseudacorus*.

I. nelsonii Randolph

Distribution
Louisiana in a restricted area south of Abbeville.

Description
Rhizome 2–3 cm thick, annual growths 10–15 cm long. **Leaves** 80–90 cm long, about 2.5 cm wide, ends reflexed, pale green. **Stem** 80–105 cm long, with 4–6 branches, stem-leaves 50–75 cm long, 2–3 cm wide, not longer than bloom stalk; branches usually 2-budded; inner spathes membranous, outer partly so at flowering; pedicel very short; ovary about 2 cm long; tube about 2 cm long; 2-flowered at apex. **Flower:** 11–12 cm diameter, rich purple-red; falls 6–8 cm long, 3–4 cm wide, with inconspicuous signal sometimes missing, semi-flared, notched at apex, oblong, unguiculate; standards 5–5.5 cm long, 2 cm wide, reflexed, oblong, unguiculate; style 2–3 cm long, raised clear of falls, bluntly bilobed, ridged, relatively short; stigma rounded; anthers may show under style. **Capsule** 5–6.5 cm long, 2–2.5 cm wide, ellipsoid to oblong, tapering at both ends, apex often obtuse. **Seeds** 1.5 cm wide, D-shaped generally, light brown, with corky covering. **Flowering** a week or two later than *Ii. fulva* and *giganticaerulea*. $2n = 42$.

Observations
This species has been shown by Randolph (1966) to be a long-established hybrid between *Ii. fulva* and *giganticaerulea* and, to some extent, gives the impression of a larger *I. fulva*. The colour usually seen tends to have more blue in it, but there is a wide range of shades from purple through to clean reds and occasionally yellow-browns or the reverse.

For a long time this species was thought merely to be a recent hybrid or a giant form of *I. fulva* and was collectively known as the 'Abbeville Reds'.

Hybrids
Hybrids are known with *Ii. delavayi, pseudacorus, spuria, versicolor, virginica* and *wilsonii.*

SERIES PRISMATICAE (DIELS) LAWRENCE

Map 17, Figure 17.

The sole representative of this N American species was once classified with the Sibiricae which it does resemble to some extent, but all too often the plant on offer is not *I. prismatica*. In practice it serves to demonstrate that gardeners neglect to look at their plants sufficiently carefully.

I. prismatica Pursh ex Ker-Gawl

Synonyms
I. boltoniana Roem. & Schult.; *I. caroliniana* Radius; *I. gracilis* Bigelow; *I. trigonocarpa* Br. & Bouche; *I. virginica* A. Gray.

Distribution
Atlantic Coast of N America from Carolina north to Nova Scotia.

Description
Rhizome very slender with long stolons. **Leaves** 50–70 cm long, 0.2–0.7 cm wide, narrowly ensiform, acuminate, glaucous, finely ribbed. **Stem**: 30–80 cm tall, slender, wiry, tending to zig-zag at the nodes, with 1 reduced stem-leaf, usually 1 branch; spathe acuminate, scarious, narrow, about 2.5 cm long; 2–3 terminal flowers, 1 on branch; pedicels up to 4 cm; tube only 0.2–0.3 cm long, relatively wide; ovary trigonal, corners flattened and extended. **Flower:** falls pale violet with deeper veins, 4–5 cm long, to 1.5 cm wide, ovate, with base and long narrow haft marked violet on whitish ground; standards violet, 3–4 cm long, 0.5 cm wide, oblanceolate, with short caniculate haft, semi-erect; styles arched, narrow, violet; crests divergent, quadrate, upper edge serrated and recurved, stigma sharply triangular; filaments reddish-purple or light violet; anthers equal to filaments, deep bluish-purple; pollen cream. **Capsule** trigonal with sharply winged edges. **Seeds** approx 0.2–0.3 cm, pyriform, smooth, usually compressed or almost cubical, buff to dark brown. **Flowering**: June–July. $2n = 42$.

Cultivation
The plant seems to prefer light shade in a dampish, well-drained, rather acid soil. Its general resemblance to *I. sibirica* is misleading as it does not form compact clumps and the stoloniferous offsets should probably be liberally mulched with leaf mould. Any artificial fertiliser should be suited to acid-loving plants.

Observations
I. prismatica has a reputation for being difficult to keep, at least in Britain, but this may be partly due to its wandering habit. There are several rules of thumb for distinguishing it from *I. sibirica*, but the only sure ones seem to be the shape of the capsule, the shape of the seeds and the very loose clumps resulting from the stoloniferous habit. Flower colours vary from violet through lighter blue shades, and a full range of pinks to white.

var. *austrina*
An inland version of the type and rather less delicate in appearance.

SERIES SPURIAE (DIELS) LAWRENCE   MARIAN BOWLEY
Map 18, Figure 18.

The species of this series are easily distinguished by the trigonal capsule, which has six longitudinal ribs arranged in three pairs, one pair at each angle of the capsule, which ends in a sharp beak. The stigma is also characteristic: it always has two points and, finally, the seeds are enclosed in a loose papery envelope. The number of species and subspecies in the series is, to some extent, debatable; in the interests of gardeners as well as botanists I have included 20 species and subspecies for individual discussion.

The species vary considerably in size from giants with stems 90–125 cm to genuine dwarfs with stems under 30 cm and even less than 2.5 cm. The majority, however, are between these extremes. Whatever the size, practically all the spurias are elegant as well as hardy with plentiful leaves setting off the flowers, which have a distinctive purity of form. The falls are more or less panduriform and this means that the hafts and styles are conspicuous and contribute significantly to the appearance of the flowers.

There is always a terminal cluster of flowers at the top of the stem and also frequently lateral branches carrying one or two flowers. These lateral branches are always very close to the stem and usually very short so that the impression is given of a spike of flowers, which is enhanced by the reduced leaves more or less sheathing the stems.

The rhizomes of the spurias are hard, compact and woody, more or less wiry in the smaller species. The method of growth differs from that of many other irises. There is only one growth bud at the end of each rhizome instead of the usual trio of a central and two or more side buds; the side branches, also with only one growth bud, are thus the only means of increase of flower stems. Annual growth is in a straight line with the spaces gradually being filled by the branches. Thus the large species spread over a large space that is rather thinly covered. In the more moderate-sized species, however, the rhizomes are shorter and compact clumps form more quickly.

Increase and dispersion is facilitated by the seeds, for the papery envelope means they can float and so be dispersed by streams and ditches, a method particularly useful in some of the habitats described below. The seeds can also be wind-borne. These devices, coupled with the stalwart character of the species, facilitate the establishment of colonies over wide areas.

As the characteristics described in the preceding paragraphs apply to all the species, they will be assumed in the sections on individual species unless there is an unusual feature.

A group of irises so varied and so beautiful might be expected to be well known to gardeners, but in fact very few have been widely known as garden plants and, until the later nineteenth century, very few even to botanists. It might be supposed that this was due to rarity or inaccessible locations, but this was not so. The spuria species are widely distributed as a group and as individual species. They are found in a broad belt stretching from south western Europe, with outliers in the Baltic and, until recently, England, through central, southern and south-east Europe, with an outlier in N Africa, into SW Russia and Turkey through the south-eastern former USSR, e.g. Turkomania, into Siberia and W Mongolia; further south there are spurias in Iran, Afghanistan and Kashmir, even into SW China.

Although over this immense range individual species and subspecies have proved well adapted to the variations in climate and altitude from the mediterranean conditions of Europe to the much more extreme conditions of central Asia, the habitat requirements are fairly constant. For nearly all the spurias the essential requirement is ample moisture, particularly in the growing period, and absence of desiccating baking in the summer, but plenty of sunshine. The most common habitats are wet grassland or marshes, river valleys, and drainage and irrigation ditches, of which a very common character is salinity of soil and water. For this reason the irises are sometimes called 'the salt marsh irises' and one subspecies is actually named '*halophila*' to denote its affinity to salt. Although these are the general conditions there are a few species whose habitats are well-drained hillsides or even dryish scrub. They all require a generous food supply.

Unfortunately, urban expansion and modern technical developments and changes in agricultural methods may, sooner or later, lead to a serious destruction of some of these habitats. Some are indeed already diminished in western Europe. There is thus a conservation question involved for spurias. It is possible that they perform an ecological role where they grow in great colonies by helping to stabilise the soils and, on a small scale, may be useful, as the Romanian botanist Prodan suggested, by preventing erosion of the banks of drainage and irrigation channels.

Cultivation

The spuria species and subspecies are ideal plants for herbaceous borders as well as for gardens or parts of gardens too moist for bearded irises. They extend the flowering season into July.

Their needs are simple and they do not require a saline soil. They are usually considered to prefer a rather heavy clay loam, neutral or slightly acid, but they also seem indifferent to lime in the soil. However, the soil must be moisture-retentive and they require plenty of moisture particularly in the growing period. Although they nearly all like full sun they dislike being baked; a rather moist atmosphere seems to suit them best. Indeed I have grown, in a garden with these conditions, 11 species and subspecies without trouble and would grow more if I could obtain them. Spurias are adaptable plants and will grow in a great variety of soils including the chalk hills of Wiltshire, the acid soils of Cornwall and some sandy soils if sufficiently enriched and provided with enough moisture. There are, however, limits; and some species are more particular than others. For instance, large spurias could not be grown in a hot, walled garden with a thin soil in East Anglia, although *I. graminea* and its varieties could. Again, only *I. orientalis* 'Shelford Giant' would settle in a garden in an old sand pit, however much enriched.

There are reports of difficulties in a number of countries, e.g. Australia and the United States, where it seems that some of the spuria species will not grow or flourish in many places. On the other hand *I. orientalis* and *I. spuria* Linn. have naturalised in fields and ditches in one area in New Zealand. There are also disturbing reports of a mustard seed fungus disease occurring occasionally, but not as far as I know in the UK.

Unfortunately, all spurias resent disturbance; rhizomes deteriorate or die quickly if allowed to dry out while being moved even in moist weather. They will grow from seed, however, although this takes 2–4 years to flowering.

Where possible the seed should be sown as soon as the capsule starts to dehisce and in an emergency the seed is worth sowing while still green.

Classification

Several medium-sized spurias, often included under *I. spuria* Linn. (sensu lato), provide botanists with difficult classification problems. Recently Mathew has developed Dykes's idea of regarding them as a complex *without* a type species. Mathew treats each as a subspecies of an abstract entity called *I. spuria*. Some people may feel that this is 'Hamlet without the Prince of Denmark'; others that it will reduce the realisation of the beauty, variety and interest of these irises. The system is, however, very tidy. The species *carthaliniae, demetrii, musulmanica,* and *spuria* Linn. (sensu stricto) become subspecies of *I. spuria* and the varieties *halophila, maritima,* and *sogdiana* become subspecies. Mathew's system is used in this section.

I. crocea Jacquemont ex Baker

Synonym
I. aurea Lindley.

Distribution
Valley of Kashmir; 1600–2000m; particularly near cemeteries.

Description
Leaves 60–90 cm long, 2.5 cm wide, ensiform. **Stem:** up to 150 cm; spathes 2–3-flowered, narrow, stiff, green, lanceolate; pedicel 3.75–7.5 cm long; terminal cluster at apex and 3 lateral clusters on short, erect branches. **Flowers:** rich golden-yellow; falls 5 cm long, blade oblong, tapered, crimped at margin, narrowing to haft, which is shorter; standards 7.5 cm long, oblanceolate, waved at edges, upright; styles 3.8 cm; crest deltoid. **Flowering** June–July. $2n = 40$.

Cultivation
Not only does *I. crocea* resent disturbance, but like *I. orientalis*, once established it is difficult to move as the old rhizomes plunge deeper and deeper into the soil.

It thrives usually in the heavy loams common in England, flowering after *I. orientalis* and extending its flowering season into July. Sometimes it is temperamental; Dykes suggested that in gardens where it did not thrive the old hybrid between it and *I. orientalis*, called *I. ochroaurea*, should be grown instead as this is very robust and free-flowering.

I. crocea is an excellent garden plant if there is adequate space, for the tendency of spurias to concentrate new growth at the terminal shoot seems particularly marked with it as with the other larger species. As it makes branches rather slowly it does not make compact clumps like many of the smaller species, but rather spreads over a larger space.

Observations
I. crocea is perhaps the most striking of all spurias. It is a beautiful and stately plant and it is disturbing that it is only known to be growing wild in the Valley of Kashmir, for with so limited a location there must be some anxiety that it might disappear in the course of agricultural and other developments.

It was described in 1847 by J. Lindley and called *I. aurea* and introduced to Europe. In 1877 it was also described by Jacquemont and called *I. crocea*, the name by which it has recently become officially known.

Hybrids
Hybrids are known with *Ii. ensata, fulva* and *fulvala*.

I. graminea Linn.

Synonyms
I. colchica Kem. Nat., *I. sylvatica* Balbis.

Distribution
NE Spain, central and SE Europe, particularly Romania, Bulgaria and Turkey; former USSR: Caucasus and Crimea.

Description
Rhizomes slender, branching, forming dense mats. **Leaves** usually 40–90 cm, narrowly ensiform, acuminate, densely nerved, upright, bright green and glossy above, subglaucous and dull beneath. **Stem:** usually 7.5–10 cm, flattened with distinct flanges, with 1–2 sheathing leaves near base often rising above flowers; spathes sharply keeled, usually unequal, outer sometimes foliaceous and erect, extending well above the flowers; pedicel 3.75–5 cm long, 1–2 flowers at apex. **Flowers:** 7–8 cm diameter, strongly scented; falls 3.8–5 cm long, ground colour yellowish-white, veined red-purple on haft and blue-purple on blade, which is orbicular, separated from haft by very gradual constriction; standards broadly

lanceolate, shorter than falls, rich purple; styles pale red-purple, sharply keeled rich purple; crest broadly semi-ovate. **Flowering** May–June, occasionally again in September. $2n = 34$.

Cultivation
In the wild it grows in alpine and forest meadows and though evidently liking a humus-rich soil, it is tolerant of other soils and indifferent to lime. It will grow in partial shade and is an accommodating plant in gardens, but easier to see on a raised bed. The conditions of successful moving are the same as for other small spurias and it comes readily from seed.

Observations
I. graminea is a hardy dwarf spuria growing over a wide area. It has long been familiarly known as 'the narrow leafed plum scented iris' and was grown under that name in the Cambridge Botanic Garden as early as 1733. *I. graminea* forms dense upright clumps of narrow leaves, which conceal the small, violet, plum-scented flowers, although sometimes it is so floriferous that the flowers push the leaves out of the way round the edge of the clumps. It is variable, particularly in the width of the leaves and the length of the spathes.

var. *pseudocyperus* Schur.

Synonym
I. pseudocyperus Schur.

Distribution
Limited to Romania, Transylvania and Slovakia.

Description
Differences from type: larger and more robust in all its parts and forming lax clumps to 60 cm across. **Leaves** broader, thicker and lax rather than upright, grey-green in the wild, but deep bright green and strikingly glossy in cultivation. **Stem** short, but flowers show conspicuously well above and among the leaves. **Flowers:** larger than the type and brighter, richer colours, but not scented; falls have a medial yellow stripe; haft and blade more or less equal in size, and constriction between them very slight; standards stiffly erect. **Flowering** June.

Cultivation
A very good and attractive plant for the front of borders; the leaves remain decorative until the frosts. In the wild it grows in the same sort of conditions as the type; and cultivation requirements are also similar except that there is no advantage in growing it on a raised bed.

Observations
Var. *pseudocyperus* has a much more limited distribution than the species. Perhaps for this reason and because it was only described in 1866 it is much less well known to gardeners than the type.

Its status has been disputed by botanists. Though first described as a species, Dykes was doubtful whether it was sufficiently distinct even to be classed as a variety. However, it is now accepted as a variety and, certainly to gardeners, it is significantly different from the type.

I. kerneriana Ascherson & Sintensis ex Baker

Synonyms
I. haussknechtii Bornmuller, *I. graminifolia* Freyn. *I. gransaultii* Siehe,

Distribution
Endemic to N Turkey from Balikesir province to Erzincan in the east.

Description
Leaves about 45 cm, slender, ensiform, finely ribbed, slightly twisted, pale yellowish-green, produced in thick clusters. **Stem**: 15–30 cm, hidden by sheathing leaves, but internodes sometime bare; spathes 7.5 cm long, slightly keeled, inflated, pointed, scarious at tips and upper edge; 2-flowered at apex. **Flower**: 7–10 cm diameter; falls panduriform, blade lanceolate with slight ridge, edges wavy and crimped; colour rich yellow, the haft narrower and paler; standards linear, lanceolate, wavy edges tending to twist spirally, yellow; styles much arched longitudinally, pale yellow; crest very small, triangular, much recurved. **Flowering** May–June. $2n = 18$.

Cultivation
I. kerneriana is obviously a most attractive and desirable plant, but it is not always easy to establish in gardens, resenting disturbance and slow to come from seed. It prefers a slightly acid to neutral soil and plenty of moisture in the spring, but good drainage; even in summer it should not be allowed to dry out. It flourishes on sandy loam and seems to prefer it to a heavy clay-based one. It should always be planted out as a seedling or moved in moist conditions in autumn.

Observations
The species is endemic to Turkey, where it grows in the northern mountains in dryish oak scrub, open pine woods and sparse grassland. It was not known to botanists until late in the nineteenth century and is the only dwarf yellow-flowered spuria. *I. kerneriana* forms an upright clump with leaves growing in tufts; although some leaves are as tall as the scape, the two flowers show up well.

I. lilacina Borbas

Synonym
I. spuria var. *lilacina.*

Distribution
Not known, but suggestions are Kashmir, central Asia and Mongolia; or Romania.

Description
Leaves around 100 cm long, up to 5 cm wide, deep dull green, finely ribbed, ensiform, stiff and upright. **Stem:** 50–100 cm tall, rounded, fairly stout; 6–7 sheathing leaves of varying length (the lowest being 60 cm long) but stem visible between them; spathes both green, inflated, inconspicuously keeled in upper half; pedicels unequal; ovary the normal spuria shape with the usual prominent double ridges at the 3 angles, gradually tapering into a very narrow neck; terminal head of 2 flowers and 3 lateral buds. **Flower** lilac; falls have a broadly oval blade veined densely by rich lavender on the pale lavender or whitish ground, the median yellow stripe at the base continuing into the deeply caniculate haft, veined red-violet; standards 7.5 cm long, erect at first, becoming oblique, lilac-lavender; styles narrow, pale pinkish-violet, keel darker; crests deltoid, recurved. **Flowering** June–July. $2n = 44$.

Cultivation
There is no information about its cultural requirements in Britain, but it seems a plant grown as *I. lilacina* in New Zealand is difficult to establish there, like *I. spuria* subspp. *halophila* and *musulmanica.*

Observations
Mystery surrounds *I. lilacina,* for Borbas did not describe the plant for which he introduced this name nor where it grew. The chromosome number, 44, published in 1955, is consistent with its reported origin in Kashmir and the Turkestan–Mongolia area. It is included by Prodan (1964), however, so it could be Romanian!

Plants have grown under this name in recent years in Britain, New Zealand, Australia and the USA, but have not been formally compared. A would-be Australian grower complained in 1968 that *I. lilacina* was not standardised and that it varied greatly from garden to garden. She also quoted from a USA nurseryman's catalogue a description of a plant they were distributing, namely: 'This may be *spuria lilacina,* we don't know, but it is pretty . . .!' It seems that *I. lilacina* is not always easy to grow in the Antipodes. One commentator from near Wellington in New Zealand failed to establish it and commented that like '*I. halophila* and *I. musulmanica,* not being as robust as *I. orientalis* it is only found comparatively rarely in gardens'.

In Britain, E.G.B. Luscombe gave a detailed description (unpublished) of a plant he called *I. spuria* var. *lilacina* supplied by Perry's nursery in 1950. It has been found to fit plants grown from seed of that name supplied by Thompson and Morgan, the well-known seed firm. This was reported by Marchant (1968) together with E.G.B. Luscombe's description of the plant he had from Perry's. A summary of this description has been given above. The features of *I. spuria* var. *lilacina* do not suggest that it can be a synonym for *I. spuria* subsp. *sogdiana* as Mathew has suggested unless there is another iris more like *sogdiana* circulating under the name of *lilacina*.

Whatever the exact botanical status of the plant described by E.G.B. Luscombe as *I. spuria* var. *lilacina,* it was included in the Species Notes where it was concluded that it is a horticulturally, if not botanically, distinct plant. It would be a pity if knowledge of it were to disappear before its botanical status and its whereabouts were established.

I. longipedicellata Czeczott

Distribution
Turkey; Galatia, nr. Seraiky in the Eldiven–Cagh Mountains, near springs in marshy ground at about 1300 m.

Description
Leaves 35–40 cm long, 1–1.2 cm wide, ensiform, greyish-green, semi-evergreen. **Stem** about same length as leaves, terminal head with 2 flowers, occasionally 1 laterally. Spathes 11–13 cm, narrowly lanceolate, acuminate, greenish tending to greenish-purple at edge; straw-coloured when dry. **Flower**: falls entirely pale yellow with long slender haft, short ovate blade emarginate at tip and and shorter than haft; standards roundish, cuneate, emarginate, same yellow as falls; styles deep yellow, partly veined. $2n = ?$

Observations
Although *I. longipedicellata* was described as long ago as 1932, the only information available about it is that in the original very detailed description. This does not seem to me to support the statement by Mathew (1989, p.113) that this iris appears to be identical with *I. orientalis* Miller.

I. longipedicellata in its native marshy habitat in the mountains of C Turkey is a short, but not dwarf spuria. It is, for instance, rather taller than *I. kerneriana*, the only other short yellow spuria at present known to be native to Turkey. Its size and entirely pale yellow colour suggests that it would be a very welcome addition to our gardens if seed were available for its introduction. It may also have the merit of a long flowering period, for it was collected in flower in July, but was reported as also flowering in late March–early April.

The discovery of this iris in 1932 together with the rather more recent one of *I. xanthospuria* suggest that there may still be more spurias to find in Turkey.

I. ludwigii Maxim.

Distribution
Endemic to the Altai Mountains in E Kazakhstan.

Description
Rhizome creeping. **Leaves** up to 50 cm, linear, lanceolate, with 5–7 nerves. **Stem** not more than 3 cm, closely clasped by 2 reduced leaves like those of *I. ponticum*, spathes membranous, 2-flowered. **Flowers:** falls panduriform, blade ovate, half as long as violet haft, which is slightly winged, distinguished from *I. ponticum* by small longitudinal beards of dense hairs; standards shorter than falls, oblanceolate, violet blue; styles linear; tube almost as long as the segments. **Flowering** in May. $2n = ?$

Cultivation
I. ludwigii is of more interest to botanists than to gardeners, but it can be grown on rock gardens as can *I. pontica*.

Observations
I. ludwigii is a rare dwarf spuria. Practically stemless, it is very like the European *I. pontica* (syn. *I. humilis*) and for a long time was considered as possibly synonymous with the Eastern form. It is, however, differentiated from *I. pontica* by the presence of minute longitudinal beards on the falls, confirmed by G.I. Rodionenko from living material, and by the perianth tube, which is almost as long as, or equal to, the segments.

I. monnieri De Candolle

Distribution
Suggested, but not proved, Rhodes, Crete, Cilicia.

Description
Leaves 100 cm long, 2.5–4 cm wide. **Stem:** 100–125 cm long, with several reduced leaves; spathes lanceolate, firm, green; pedicel 2.5–4 cm long, becoming much longer; 2.3 flowers in terminal cluster and 1 or more lateral clusters. **Flower:** falls with orbicular blade, rather longer than 2.5 cm, haft bright lemon-yellow; standards 7 cm long, oblong, cuneate, bright yellow; styles 2.5 cm long, pale yellow; crests small, deltoid, sharply recurved. **Flowering** late June–July. $2n = 40$.

Cultivation
I. monnieri is an excellent garden iris needing the same conditions as its relatives *I. orientalis* and *I. crocea*.

Observations
I. monnieri was first described in 1808 by A.P. de Candolle. He found it growing in the garden of M Lemonnier at Versailles where it was called the Iris of Rhodes. De Candolle named it *I. monnieri* in compliment to M Lemonnier.

To the gardener this beautiful iris, together with *I. crocea* and *I. orientalis,* most satisfactorily completes a trio of tall white and yellow spurias which will grow under the same conditions.

Colour is not the only difference between *I. monnieri* and the other two. The standards and falls of *I. monnieri* lack the crimped edges of *I. crocea* and the blade of the fall is orbicular, not oblong as in *I. crocea*.

The style crests of all three irises differ: those of *I. orientalis* are triangular and more than 1.25 cm long, but those of *I. monnieri* are short and deltoid and very recurved, while those of *I. crocea*, though deltoid, are not recurved.

I. monnieri poses some awkward questions for the botanist. It has never been recorded in the wild and, worse still, when self-pollinated, the majority of seedlings resemble *I. orientalis*. This suggested to Dykes that *I. monnieri* might not be a true species. This question has been investigated more recently by Dr Lee Lenz. He received seed from near Ankara in Turkey in 1948 which produced deep golden-yellow-flowered spurias with short and extremely recurved style crests. Dr Lenz called this spuria 'Turkey Yellow'. This suggested to him that *I. monnieri* might be a natural hybrid between *I. orientalis* and a deep yellow spuria growing in Turkey. As the seed of 'Turkey Yellow' had been collected in the Mugla province of Turkey and as *I. orientalis* is reported in some nearby eastern Aegean islands, it seems possible the *I. monnieri* may have originated as a natural hybrid on the island of Rhodes, thus explaining the name used for it in M Lemonnier's garden in 1808. This would offer a neat solution to the problem. 'Turkey Yellow' is now officially called *I. xanthospuria*.

Whether or not *I. monnieri* is ultimately officially re-classified as a hybrid, it has itself been long used for hybridisation. For instance, Sir Michael Foster, who inspired so much of Dykes's work on irises, derived the well-known Monspur hybrids by crossing *I. monnieri* with *I. spuria* which resulted in irises of various beautiful blues; it is still involved in modern hybridisation, particularly, perhaps, in the USA.

Hybrid
A hybrid is known with *I. ensata*.

I. orientalis Miller

Synonyms
I. albida Davidoff, *I. gigantea* Carr., *I. longipedicellata* Czecz., *I. ochroleuca* Linn.

Distribution
Asiatic Turkey, probably not east of Kayseri, the Aegean Islands, Lesbos and Samos, NE Greece. Reports from Yugoslavia and Syria may relate to escapes from cultivation.

Description
Leaves 2.5 cm or more wide, 50–100 cm long, rather glaucous, stiff, harsh and fibrous with a slight spiral twist. **Stem** up to 125 cm, slightly flattened, with 2–3 reduced leaves; spathes green, lanceolate, acuminate; pedicel 2.5–7.5 cm.; 2–3 flowers at apex, 1–2 lateral clusters on short erect branches. **Flowers**: falls with narrow haft 3.8–5 cm long, the broad orbicular blade is shorter, white with deep yellow blotch on blade; standards erect, white; styles nearly 5 cm long, white; crests more than 1.25 cm long, triangular. **Flowering** June–July. $2n = 40$.

Cultivation
I. orientalis is a plant requiring conditions similar to those of its close relation *I. crocea*, but also flourishing in lighter soils. It can, however, be shy of flowering in some gardens, like *I. crocea*. Dykes suggested that in such cases the hybrid with *I. crocea* (*I.* × *ochroaurea*) should be used instead. Apart from its greater robustness, this differs only slightly from *I. orientalis* from a gardener's point of view in that a creamy yellow replaces the white of *I. orientalis*, but the characteristic deep blotch on the falls is retained.

I. orientalis has been used a great deal in hybridising and it is suggested that *I. monnieri* may be a natural hybrid between it and *I. xanthospuria* (see under *I. monnieri* above).

Observations
I. orientalis was generally known as *I. ochroleuca* before the recent change of name under the international rules of botanical nomenclature. It is perhaps the most familiar of the large spurias. Since its principal home is in western asiatic Turkey with outlying stations in the eastern Aegean and in Greece and possibly Yugoslavia, it was accessible to European botanists and gardeners relatively early. It was described by Miller as *I. orientalis* in 1768, but given its Linnaean name of *I. ochroleuca* in 1771. As far as is known the giant forms such as 'Shelford Giant', 'Ephesus', or simply 'Gigantea' are not found in the wild. Like so many of the spurias it enjoys saline conditions in the wild, growing in great sheets in marshes in Turkey and, also, often growing alongside drainage and irrigation ditches.

I. pontica Zapalowicz

Synonyms
I. humilis Bieb., *I. marschalliana* Bobrov.

Distribution
SE Romania, W Ukraine and the Caucasus, possibly extending into Russian Central Asia.

Description
Leaves 15–35 cm long, firm, ribbed, overtopping the flower. **Stem**: up to 4 cm, very short, with 2 long leaves immediately below the spathes; spathes 5–6 cm long; pedicel very short; usually 1-flowered. **Flowers**: relatively large; falls with winged haft separated by sharp constriction from almost orbicular blade, haft and base of blade veined and mottled red-brown on greenish-yellow ground, rest of blade deep blue-purple, central ridge slightly yellow; standards oblanceolate, unguiculate, blue-purple; styles slightly keeled; crests almost quadrate, sharply revolute. **Capsule** normal spuria type, but situated at ground level. **Flowering** end–May. $2n = 72$.

Cultivation
In the wild, *I. pontica* grows on rather dry grasslands among other grassland plants, but in gardens it can be grown on rockeries, or as a pot plant. Rodionenko considers that it could be much in demand as a rockery plant. A good clay loam with crushed limestone added and good drainage is recommended by Köhlein, who also regards it as a worthwhile garden plant, floriferous when established. Dykes, however, took a very poor view of it, finding it difficult to establish and persuade to flower. It seems to be a case in which the type of soil may be very important.

Observations
I. pontica is another of the hardy dwarf spurias growing in Romania and southwest Russia. Its very short stem and capsule at ground level distinguishes it from other dwarf spurias. The leaves are much longer than the flower stems and form tufts fairly effectively concealing the blue-purple flowers, although these are relatively large. They are sometimes slightly fragrant.

I. pseudonotha Galushko

Distribution
Caucasia.

Description

Leaves about 50 cm (–100 cm) long, up to 2 cm wide, blue-green, drooping. **Stem** about 60 cm (–85 cm) tall, 2-flowered at apex, with 1 lateral, slight zig-zag at nodes. **Flowers** 8–10 cm diameter, light lilac; standards nearly erect; falls slightly arching, slightly paler than standards; signal yellow, with faint veining from edge; style has slightly darker ridge. **Flowering** May–June. $2n = ?$

Observations

This plant was found in 1984 or 1985 by Dr Rodionenko. It appears to be hardy and sets seed freely according to Dr Maurice Boussard, France, and Mrs Revie Harvey, New Zealand. Mrs Ball, also from New Zealand, compares it in appearance to the illustration of *I. spuria* (Kashmir) in Dykes (1913a).

I. sintenesii Janka

Distribution

S Italy, Balkans, SW Russia and Turkey mainly in the NW and the Pontus mountains.

Description

Rhizomes hard, slender, wiry, clothed with remains of old leaves. **Leaves** 20–45 cm, narrow, linear, acuminate, evergreen. **Stem** 10–30 cm, rounded, almost entirely clothed in 2–3 reduced leaves; spathes 3.8–7.5 cm long, narrow, linear, acuminate, inner slightly longer, both sharply keeled; pedicel 2.5–7.5 cm long; 2-flowered terminal head. **Flowers:** falls slightly panduriform, haft separated by gradual constriction from the elliptical blade, which is densely veined blue-purple on a white ground, the haft being veined with red-purple; standards oblanceolate, deep purple; styles narrow with slight keels; crests small, triangular. $2n = 16, 32$.

Cultivation

I. sintenesii grows wild among dry scrub and grass. In gardens it will grow well on rockeries or in the front of the border if it has plenty of moisture at its roots, particularly in the growing season, and provided its shallow roots are not allowed to dry out, for it will not survive being baked dry in the summer. Like other spurias it resents disturbance and being split up into small sections and should only be moved in the moist autumn weather. Fortunately it comes easily from seed and should be planted out from pots.

A named form *'constantinopolitan'* Prodan is distinguished by 1 cauline leaf enfolding the stem and overtopping the flower; the basal leaves are equal to, or shorter than the stems. Other forms are distinguished, differing chiefly in length of leaf.

Observations
I. sintenesii is a charming, hardy, dwarf spuria; it forms low evergreen cushions 30 cm across. Although it has a fairly wide distribution and has been known to botanists since 1876, it does not seem to have been known well in Britain before the first World War as Dykes had a good deal of difficulty in obtaining living material.

var. *urumovii*

Synonyms
I. urumovii Vel.

Distribution
Bulgaria.

Description
Differences from type: an upright plant and more slender. **Leaves** very glaucous and rough to the touch, herbaceous, not evergreen; spathes also glaucous and rough to touch, not keeled. $2n = 20$.

Cultivation
I. var. *urumovii* is an elegant little plant that grows well in a rockery. Whatever may be the final classification by botanists, this iris could be of interest to gardeners, so it would be a pity if the description disappeared.

Observations
I. var. *urumovii* was not found until 1902. So far it has only been found in Bulgaria. Dykes thought it a very elegant little plant and that the glaucous leaves showed up the quiet colour of the flowers well. He noted that it is very floriferous and Köhlein (1981) has described it as more delicate in all its parts than subsp. *brandzae*. Dykes considered that *I. urumovii* was a distinct species closely related to *I. sintenesii*. He pointed out the significant differences listed above in 1914. Since then it has been re-classified as a variety of *I. sintenesii*. Mathew nevertheless, referring to the *Flora of Bulgaria*, adopts the view that it is actually synonymous with *I. sintenesii* and does not include any description of it; Köhlein, however, does not follow Mathew and includes a full description in his book.

subsp. *brandzae* Prodan

Synonym
I. brandzae Prodan

Distribution
Romania: Bessarabia and Moldavia.

Description
Differences from the type: **Leaves** narrower with fewer, but more prominent veins and rough to the touch; herbaceous, not evergreen. **Stems** taller in relation to leaves; spathes strongly inflated. **Flowering** in May. $2n = 20$.

Observations
I. subsp. *brandzae* was first described by Prodan in 1936. It is now regarded as a subspecies of *I. sintenesii*. It will be noticed that subsp. *brandzae* has a more limited geographical distribution than *I. sintenesii* and, unlike it, grows in saline marsh. There are a number of forms, differing mainly in the relative lengths of the leaves and stems. A named form, *topae* Prodan, has longer, more slender spathes and a longer, zig-zag stem with leaves attached at the angles.

I. spuria Linn.

subsp. *carthaliniae* Fomin

Synonym
I. carthaliniae Fomin.

Distribution
Caucasus, Transcaucasia, Georgia, West of Tiblis, Gruzia.

Description
Leaves 130–140 cm long. **Stem**: 90–100 cm, usually slightly shorter than leaves, with abbreviated stem-leaves; spathe green, leathery, acute, 4–5-flowered at apex. **Flowers**: usually sky-blue, but white is known; falls elliptical with deep indentation at tip, normal colour sky-blue with dark veins; a narrow line runs along the haft into which the blade narrows abruptly; standards upright, same colour as falls; styles slightly shorter than haft; crests erect or slightly curved. **Flowering** May to mid-June. $2n = 44$.

Cultivation
Although *I.* subsp. *carthaliniae* grows in wet marshy ground in the wild, it grows happily in ordinary deep, heavy loam in gardens in association with paeonies and other summer herbaceous plants, provided that it has plenty of moisture.

Observations
I. subsp. *carthaliniae*, now classified as *I. spuria* subsp. *carthaliniae*, is a beautiful, vigorous iris found in marshy areas in the Caucasus region of the former USSR. It grows up to 1 m and has very beautiful, large flowers, normally a clear sky-blue.

It varies considerably, though, and the Russian botanist Mrs L. Soboleva describes finding a form with 'great marble-white flowers' in Gruzia.

Although the flower stems may be slightly shorter than the leaves, which are broad and stiff, the flowers show up extremely well; it is a robust plant and forms handsome clumps. Dr G.I. Rodionenko and Mrs L. Soboleva both regard this as one of the finest spurias in the (former) USSR, closely related to the deep blue *I. musulmanica* (syn. *klattii*) with which it hybridises readily. Dykes thought that probably the fine blue iris that was sent to him from Kashmir (illustrated in *The Genus Iris*) might be *I. carthaliniae* although it was difficult to explain its presence there.

subsp. *demetrii*

Synonyms
I. demetrii Akhverdon & Mirsoeva, *I. prilipkoana* Kem.-Nat.

Distribution
Transcaucasia, foothills of Azerbaidjan etc., Soviet Armenia.

Description
Leaves 60–90 cm, stiff, dark green, not so wide as in *I. carthaliniae*. **Stem** to over 1 m, spathes slightly inflated; 2–5 flowers at apex well above leaves. **Flower**: falls blade width constant up to haft, which is longer than blade. $2n = 38$.

Cultivation
This plant requires a well-drained, but fertile clay. According to Köhlein it is susceptible to mustard-seed fungus.

Observations
I. subsp. *demetrii* is a relatively recent addition to the list of spuria irises found growing in Transcaucasia. Officially named in 1950, it was discovered to be identical with an iris named *I. prilipkoana* the previous year but not officially described.

It grows on the foothills, like its close relative, *I. spuria* subsp. *notha*, for, unlike most of the spuria subspecies, they both dislike wet soils. It is an attractive plant of modest height with flowers described by Rodionenko as 'big, finely patterned dark blue'. He considers that as the spathes are less inflated than in many spurias it should be useful for hybridisation. B.R. Hager considers that the deep violet colour is more valued in the USA.

subsp. *halophila* Pallas

Synonyms
I. desertorum Ker., *I. gueldenstadtiana* Lepechin, *I. halophila* Pallas, *I. stenogyna* Redouté.

Series Spuriae

Distribution
European Russia, Caucasus, Altai Mountains, Soviet central Asia, Siberia, W Mongolia; Iran and Afghanistan; China, Gansu and Xinjiang.

Description
A strong-growing plant. **Leaves** 75–100 cm, lanceolate, almost ensiform, longer than stem. **Stem**: up to 90 cm, sheathing leaves mostly lanceolate, acute, 3–4 flowers at apex, 1–2 in lateral clusters. **Flowers** white, dingy white-yellow or grey-purple, sometimes golden-yellow; falls elliptical, narrowing abruptly into haft, which is longer than blade; standards upright; crests small. $2n = 44$.

Cultivation
It grows easily and quickly in gardens with the usual moisture requirements, and sets seed plentifully. Except for the white and yellow forms it is not particularly attractive, but it is floriferous.

Observations
First identified by P.S. Pallas in 1773, the name refers to its habitat in saline, marshy ground. It was recognised as the eastern equivalent of *I. spuria* Linn. and regarded as a variety, being recently re-classified as a subspecies.

It has elegance, but is shorter than many other spurias; the clear yellow and rare golden-yellow forms are of particular interest as the only ones in these subspecies. Rodionenko describes how it makes colonies many kilometres long in river valleys where its roots can reach underground water. It is the hardiest and most northerly of spurias native to the former USSR.

subsp. *maritima* Lamarck

Synonyms
I. spuria var. *maritima*, *I. maritima* Lamarck, *I. spathulata* Lamarck.

Distribution
SE France, central Spain.

Description
Rhizome clothed with base of old leaves. **Leaves**: 30 cm at flowering, lengthening after, linear, ensiform, upright, stiff, ribbed, dark green, subglaucous. **Stem**: 25.5–30 cm long, sheathed with reduced leaves sometimes hiding internodes; spathes firm, green, somewhat inflated, lanceolate, outer slightly keeled; pedicel about 2.5 cm; 2-flowered at apex, with 1–2 lateral heads of 1 flower. **Flowers**: falls orbicular, blade blue-purple veined deep blue, haft twice as long, veined red-purple on white ground with central greenish-yellow ridge dotted purple;

standards shorter than falls, oblanceolate, deep violet-blue; styles narrow; crest small, triangular or subquadrate. $2n = 38$.

Cultivation
It grows well in gardens under the usual conditions.

Observations
This is the most southerly of the European complex of spurias related to, or part of, *I. spuria* Linn. It grows in wettish areas, often saline, usually by the sea. A rather small iris, the flowers are very effective growing wild in a mass as well as individually pretty.

There seems to be a good deal of variation in the length of the stem-leaves in relation to the internodes. The plants originally described from France had stem-leaves longer than the internodes; those found later in Spain had the leaves equal to or shorter than the internodes. The Spanish plants were distinguished as *I. maritima hispanica* and similar plants found in France were included with them. The distinction has now been given up as unreliable.

var. *reichenbachiana* Klatt

From Algeria and north Africa; this is a more robust plant, but the stem-leaves do not fully cover the internodes. Mathew suggests that it should be treated either as synonymous with subsp. *maritima* or as a variant of it.

subsp. *musulmanica* Fomin

Synonyms
I. daënënsis Kotschy, *I. klattii* Kem.-Nat., *I. musulmanica* Fomin, *I. violacea* Klatti.

Distribution
N Caucasus, Transcaucasia east to Azerbaijan; N and W Iran and E Turkey.

Description
Leaves wider than in subsp. *notha*, strong, substantial. **Stems** 40–90 cm, thick, strong; spathes broadly lanceolate, somewhat inflated; flowers at apex. **Flower**: falls fine, brilliant blue-violet; there is also a white form; the elliptical blade has a yellow-white area at base and is joined to a haft of about the same length by a tubular strip, which is a distinctive characteristic; standards same colour as falls; styles shorter or equal to hafts; crest reflexed. $2n = 44$.

Cultivation
This plant is very like subsp. *notha*, but stronger and more substantial. It is a beautiful companion to *I.* subsp. *carthaliniae,* contrasting well in colour and requiring similar conditions.

Observations
Found originally in E Transcaucasia by Klatt and later by Fomin in Iran and Turkey, it has only recently been recognised that they are the same plant. It is closely related to subsp. *halophila*, but has a more limited geographical distribution. It also grows in salt marshes and irrigation and drainage ditches in great masses and hybridises in the wild with subsp. *carthaliniae*. A yellow-flowered plant has recently been found in Turkey, which may be another form.

D. Niswonger after considerable experience using both '*musulmanica*' and '*klattii*' is convinced that these are separate species.

subsp. *notha* Bieb.

Synonym
I. notha Bieb.

Distribution
Caucasus, N Caucasia, Transcaucasia.

Description
Leaves 1.25 cm wide, 68–75 cm long, linear, smooth, dark green. **Stem:** 1–1.1 m long, slender, bent at nodes with stem-leaves on lower part; spathes linear, lanceolate and long, pointed at apex, not inflated; tube 2.5 cm, 3–5 flowers at apex. **Flowers** large, bright deep blue or violet-blue with broad median yellow stripe; rounded ovate blade narrows sharply into the haft, which is of equal or greater length; standards almost vertical, same colour as falls; styles same length as haft; crests subacute, reflexed. **Flowering** June–July. $2n = 38$.

Cultivation
This elegant plant is regarded as rather less hardy than its relations by Köhlein and by some gardeners in New Zealand. It certainly flowers and grows well in gardens in southern England without special care provided that its preference for dryish, well-drained soil and sunshine is respected.

Observations
It was described and named by Bieberstein in 1819. Unlike its close relations it grows on dry, well-drained hillsides requiring full sun and surviving drought. It is rather taller than subsp. *musulmanica;* Rodionenko considers it more elegant in all its parts.

subsp. *sogdiana*

Synonyms
I. sogdiana Bunge, *I. spuria* var. *sogdiana*.

Distribution
Central Asia and S Kazakhstan; NE Iran, Afghanistan, Pakistan, Kashmir, Kopet Dag and Tien Shan mountains; China, Gansu and Xinjiang.

Description
Leaves: dark green. **Stem:** 10–35 cm, slightly twisted, 3–4-flowered at apex. **Flowers:** falls short, pale blue varying to greyish-lilac, haft narrow and twice as long as blade; standards same colour. $2n = ?$

Cultivation
I. subsp. *sogdiana* requires the normal spuria conditions. It is a tough plant for it comes from areas of inhospitable climate where no other spurias will survive, but its green leaves will withstand very high temperatures.

Observations
Originally named by Bunge in 1847, it has been suggested that it is simply a mauve variety of subsp. *halophila*, but this is not accepted at present. It has had a poor press among botanists, but Mrs L. Soboleva, who has studied this iris in the wilds of South Turkmenia, states that the flowers vary greatly in size and colour. It is a small iris growing in clumps in marshy areas and irrigation ditches, frequently saline, sometimes forming large masses and favouring high ground.

subsp. *spuria*

Synonym
I. spuria Linn.

Distribution
Sweden, Denmark (Island of Saltholm), Germany, Austria, Hungary and Czechoslovakia, and formerly England.

Description
Leaves: less than 2.5 cm wide, nearly as long as stem, glaucous, green. **Stem:** up to 0.5 m, sheathing leaves shorter than upper internodes; spathes keeled, semi-transparent at apex; 2–4 flowers at apex, flowers at 3 nodes. **Flowers:** blades orbicular, elliptic, light violet-blue with darker veins, with pale yellow or white area at base, sometimes a yellow streak, hafts longer than blades, violet with yellow feathering; standards about as long as falls, medium violet; styles narrow. $2n = 22, 38$.

Cultivation
The requirements are those of a standard spuria. It is a decorative plant and well worth preserving in the wild and establishing in gardens.

Observations
The recent re-classification by Mathew has equated *I. spuria* subsp. *spuria* with what was known as *I. spuria* Linn. (sensu stricto). It is the most widespread of the European species of *I. spuria*. Unfortunately it is under severe ecological threat everywhere; in the past 20 years it has been lost as a wild plant in Britain, where its two colonies were destroyed. It now grows only in Cambridge University Botanic Garden and a few private gardens. It was reported in 1973 that the colony on the island of Saltholm was diminishing and that in Germany 'it is now very rare'.

It varies considerably in height, but is usually taller than subsp. *maritima*, shorter and less conspicuous than subsp. *notha*.

subsp. *spuria* var. *danica*
This the name given to the famous spuria iris growing on the island of Saltholm. It is a vigorous form of the central European stock to which it is morphologically very similar. It can be grown in gardens, but Köhlein states that it can be difficult to establish. It is to be hoped that the decline reported on the island and on the nearby mainland can be checked.

I. xanthospuria Mathew & Baytop

Synonyms
'Turkey Yellow' Lenz.

Description
Leaves 1–1.8 cm wide, equal to or shorter than stems, greyish-green, rigid, erect. **Stem** very variable, 50–100 cm; spathes green, margins membranous, with 1–2 erect branches each 2–5- flowered. **Flowers** entirely deep yellow, but smaller than *I. crocea*; falls with orbicular blade, broadly elliptical or ovate, equal in length to haft; standards oblanceolate; crests small, but very strongly recurved. $2n = 40$.

Cultivation
Standard spuria requirements.

Observations
I. xanthospuria was first grown from seed collected near Ankara in 1948 and was the first definite proof that there were tall yellow spurias in the wild in Turkey. Others have been found more recently at Mugla and Antakya. The resemblance between the strongly recurved crests of both *I. xanthospuria* and *I. monnieri* (q.v.) has led to the suggestion that the latter may be a hybrid between this, or another yellow Turkish spuria, and *I. orientalis*.

It is now being used in hybridisation.

I. farreri **Dykes** and *I. polysticta* **Diels**.
These were formerly included in this series, but recent work has indicated that they belong with Series Tenuifoliae.

Hybrids
Hybrids are known between members of this series and *Ii. ensata, fulvala, laevigata, pseudacorus* and *sibirica* as well as inter-specifics.

SERIES FOETIDISSIMAE   ANNE BLANCO WHITE

Map 19, Figure 19.

The sole representative of this series is unique among irises in having a fleshy covering to the seeds: red in the type specimen. Morphologically, it appears to be related to Series Spuriae.

I. foetidissima (Diels) Mathew

Distribution
North Africa and northwards over Europe to the heavy winter snow line; east from the British Isles at least to Malta.

Description
Rhizome to 1.5 cm across, compact, tough. Leaves 30 cm long or more, 3 cm wide or more, heavily veined, dark green, evergreen. **Stem** to 30 cm or more, several stem-leaves with flowering points in axils of upper ones, slightly flattened; spathes 7.5 cm, lanceolate, green, rigid, 2–3-flowered; pedicels variable, around 7.5 cm, lengthening as capsules grow; ovary trigonal, grooved at angles and on faces, tapering slightly at both ends; tube 1.25 cm long, rounded trigonal, constricted at ovary. **Flower** 5–7 cm diameter, colour variable; falls 5 cm long, 2 cm wide, blade orbiculate, haft narrow, whole frequently heavily veined; standards to 4 cm long, usually clear-coloured, oblanceolate or obovate, narrow, haft canaliculate; styles 2.5 cm long, widening sharply at ends, frequently same colour as standards; crests small, deltoid; stigma bilobed with pointed teeth; filaments short; anthers often projecting beyond stigma; pollen cream. **Capsule** to 8 cm, trigonal, with small beak at apex. **Seeds** usually appearing bright scarlet, attached to capsule until it rots. **Flowering** June–July. $2n = 40$.

Cultivation
An extremely tolerant species for neutral and alkaline soils, but not so happy on really acid ground. It will grow in heavy shade, but at the expense of poor plants with poor flowering and worse seeding. Best results will be obtained in the classical rich, loamy soil which is never allowed to dry out. Otherwise, a good garden plant in most sites since it is slow-growing.

Observations
This plant has been much maligned by having a really dull colour form as the type. There are forms with good blue falls, yellow or white flowers. The 'berries' too are bright red, yellow or white and of varying sizes. The red and yellow 'berry' forms breed true; the white rarely does, but Brian Mathew raised some white-berried plants in 1995. The yellow darkens over the winter to a dull, light orange, which can be confused with the normal ones, and the white discolour to buff. There is a form with variegated leaves which is slower-growing. It seems likely that the white berries and the variegated leaves are unstable chimaeras. The north African plants are usually larger than the European forms and have good yellow flowers, while a magnificent form with clear yellow flowers has been found by Richard Nutt in the Picos de Europa. There are curious reports of isolated plants well to the east of the Mediterranean basin far outside the normal range of this species.
Hybrids are doubtful.

SERIES TENUIFOLIAE (DIELS) LAWRENCE

Map 20, Figure 20.

A group of distinctive small irises from semi-desert and steppe lands which are not widely grown partly because of the difficulty of obtaining material and partly because of problems with humidity in temperate climates. They are notable for the tendency of the rhizomes to grow vertically and for the very tough remains of the old leaf bases, which give protection from predators; plants should be handled with care since the dead veins can be very sharp. They may be intolerant of transplanting.

I. anguifuga Y.-T. Zhao ex X.J. Xue

Distribution
Mid-China; Anhai, Hupei and Kuangxi.

Description
Rhizomes short, thick and, at the soil surface, having a bulbous appearance caused by the fibrous remains of old leaf bases; colour varies through brown-red, deep red and yellowish brown, but is creamy yellow at the growing tip. **Leaves** 20–30 cm long, 5–7 cm wide, ensheathing at base, tips pointed, with 3–5 main veins. **Stem** 30–50 cm, with 4–5 acuminate leaves 10–13.5 cm long, 0.8 cm wide; spathe single, acuminate, 10–13.5 cm long, 0.8 cm wide, pointed at tip, rather similar to stem-leaves; pedicel 2.5 cm; 1 flower. **Flower** 10 cm diameter, violet-blue or blue-purple; falls 5–5.5 cm long, 0.3 cm wide, haft about 2/3 length held almost horizontal, with brown veining and spots, widening sharply to rather convex blade; standards 4.5–5 cm long, 0.3 cm wide, oblanceolate, nearly erect,

with blue-brown veining; styles 4.5–5 cm long, 0.5 cm wide, flat, springing clear of falls, crests wide, shallow, triangular; anthers 2.5 cm, filaments flat, rather shorter than the anthers but about the same width, pollen bright yellow; tube about 3 cm long, narrow. **Capsule** 5.5–7 cm long, 1.5–2 cm thick, with long beak and sparse, soft, yellow-brown hairs on surface. **Seeds** roundish, 0.4–0.5 cm diameter. **Flowers**: end March–early April for about 1 week. $2n = ?$

Cultivation

The plant is found naturally on open, grassy hillsides and so would seem to need good drainage and plenty of light. Probably hardy given those conditions, but this perennial remains green over winter and dies down in summer so will need protection against summer rainfall. The stem-like rhizomes do not become woody and so do not form compact clumps. It has overwintered in the open at Kew.

Observations

This distinctive-sounding plant has been placed in Section Ophioiris by Prof Zhao, in which it would be the only known representative. However, its general characteristics suggest that it is a member of the Tenuifoliae. It appears to be extensively grown in China for use by herbalists and is traditionally used for the treatment of snake-bite, contusions, constipation and diarrhoea. A very wide range of local names indicate local uses and habits of growth.

I. *bungei* Maxim.

Distribution

Eastern Mongolia through northern China.

Description

Rhizomes woody, lumpy, held upright, tightly clumped; old leaf bases persist; roots thick, long. **Leaves** to 45 cm long, to 0.4 cm wide, with several marked veins. **Stem** 15–30 cm long, with 2–3 sheathing leaves at base; 3 spathes 8–10 cm long, inflated, membranous at margin, with close parallel ribs; pedicel about 1.5 cm long; 2 flowers. **Flower** 5–6 cm diameter, violet; falls pale blue, 5–6 cm long, to 1.5 cm wide; standards darker, to 5.5 cm long, to 1.0 cm wide, oblanceolate, erect; style about 5 cm long, narrow, crests triangular, sharply divided; stamens to 3 cm; tube to 7 cm long, filiform; ovary to 4.5 cm long, spindle-shaped. **Capsule** to 9 cm long, to 2 cm diameter, narrowly cylindrical, with 6 ribs, pointed at apex, dehiscing below apex. **Seeds** ? **Flowering** May-June. $2n = ?$

Cultivation

A plant of open, dry, sandy areas, possibly difficult to keep in temperate areas. It has survived in the open at Kew over recent winters so raising seedlings is important.

Observations
Under certain conditions the flower stem may be much shorter, to the point of not emerging from the soil.

I. cathayensis Migo

Distribution
China: Anhui, Jiangsu and Hubei.

Description
Rhizome apparently tough, almost erect with persistent, dark red leaf bases; knobbly, brownish. **Leaves** to 25 cm long, to 0.4 cm wide, rigidly erect and over-topping flowers, lengthening through season with tips arching, sheathing at base. **Stem** short, subterranean; 3–4 spathes to 12 cm long, 2 cm wide, margins membranous, apex acute; 2 flowers; pedicel to 2 cm, thin. **Flowers** violet, apparently heavily dotted on pale ground; falls to 5.5 cm long, 0.5 cm wide, ovate blade reduces sharply to very narrow haft; standards to 5 cm long, narrowly oblanceolate, with mid-rib and some single-celled hairs; style to 4 cm long, 0.3 cm wide; crests 1.0 cm long, narrow; tube to 9 cm long, slender, with slight widening at apex; ovary to 1.5 cm long, spindle-shaped. **Capsule** and seeds unknown. **Flowering:** April. $2n = ?$

Cultivation
A plant of open hillsides in grass. Nothing is known of the best conditions, but cultivation might be relatively easy for members of this group.

I. farreri Dykes

Distribution
SW China.

Observations
Originally classified in Section Spuria by Dykes, it now seems likely that this plant is just a variant of *I. polysticta* and so, like it, is better placed in this group pending more and better information about them both. Recent work in China leaves the problem unresolved.

I. kobayashi Kitagawa

Distribution
China: southern Liaoning.

Description
Rhizome short, lumpy, tough, with strong fleshy roots. **Leaves** 10–15 cm long, 0.2–0.3 cm wide, apex pointed, overtopping flowers, spiralling slightly. **Stem** very short, sometimes subterranean, 2-flowered, spathes to 8 cm long, 1 cm wide, lanceolate; pedicel to 1.5 cm; ovary around 1 cm long, cylindrical. **Flowers** about 3 cm diameter, densely spotted yellow on blue ground; falls to 3 cm long, 0.5 cm wide, blade horizontal narrowing sharply to haft; standards to 2 cm long, 0.3 cm wide, erect, narrowly oblanceolate; style shorter and narrower than fall, crests narrow, triangular, stigma obtusely bilobed; stamens to 1.8 cm long, anthers yellowish; tube to 5 cm long, slender. **Capsule** to 2 cm long, 0.8 cm diameter, with 6 conspicuous ribs and short point on apex. **Seeds** wrinkled. **Flowering:** May. $2n = ?$

Cultivation
A plant of dry hillsides; cultivation requirements unknown.

I. loczyi Kanitz

Synonyms
I. tenuifolia auct. non Pall., *I. tenuifolia* Pall. var. *thianschanica* Maxim., *I. tianshanica* (Maxim.) Vved.

Distribution
From Tien Shan and Pamir Alai through NE Iran and Afghanistan, to Baluchistan.

Description
Rhizome short, thickly covered with very persistent dead leaf bases, forming tussocks. **Leaves** 10–30 cm long, 0.2–0.4 cm wide, erect, pointed at apex, bases sheathing. **Stem** 2–2.5 cm, probably subterranean; basal leaves sheathing, membranous; 3 bracts to 15 cm long, 1.5 cm wide, with distinct mid-rib, pointed; pedicel?, 1–2 flowers. **Flower** 4–7 cm diameter; falls white or creamy with heavy purple-blue veining, 6 cm long, 1–2 cm wide, obovate or oblanceolate; standards pale blue or bluish-purple, to 5 cm long, 0.8 cm wide, oblanceolate; style same colour as standards, to 4 cm long, 0.8 cm wide, crests elliptic, stamens to 2.5 cm long; tube to 10 cm, filiform; ovary 1.2 cm long, spindle-shaped. **Capsule** 4–7 cm long, 2 cm diameter, with 6 veins, tube initially persistent. **Seeds** 0.5 cm, pyriform, wrinkled, dark brown. **Flowering** May–June $2n = ?$

Cultivation
This plant comes from open sunny slopes, either rocky or grassy, at high altitudes. Köhlein finds it more resistant to temperature and humidity differences than others of the group. Ordinary garden soil or sandy leaf mould and protection from winter damp.

Observations
This plant is ecologically and botanically distinct from *I. tenuifolia*, which it closely resembles.

I. polysticta Diels

Distribution
N Szechuan, China.

Description
Rhizome woody, with persistent brown leaf bases. **Leaves** up to 70 cm long, 0.7 cm wide, acuminate, rather rigid, strongly ribbed. **Stem:** 15–45 cm, with 1–2 leaves; 3 spathes to 20 cm long, 2.5 cm wide, herbaceous, inner longer and exceeding tube in length; pedicel very short; 1 flower. **Flower** to 9 cm diameter, single, yellow with purplish netted veining making the colour appear violet; falls 6.5 cm long, 1.2 cm wide, blade rounded, haft narrow, longer than blade, spotted along the ridge; standards to 4.5 cm long, 0.8 cm wide, subspathulate, obtuse, widespreading; styles to 4 cm long, crests elliptic; tube 0.3–0.5 cm long, 0.5–0.6 cm diameter; ovary narrow, beaked. **Capsule** 3.5–7 cm long, 1.5 cm diameter. **Seeds** ? **Flowering:** July. $2n = ?$

Cultivation
A plant of mountain meadows around 3200 m. It is probably attractive as well as botanically interesting, but there is no information about its needs.

Observations
This plant was difficult to classify as only incomplete herbarium material has been available. Recent work at Kew by Dr Cutler and Qi-gen Wu, *Botanical Journal of the Linnean Society* (1985), suggests that it belongs in this series, and this is supported by the Chinese description.

I. regelii Maxim.

Distribution
Originally described from Turkmenistan.

Observations
Virtually nothing is known about this plant, which appears similar to a miniature *I. tenuifolia* of 6–7 cm height with flowers about 3 cm in diameter.

Series Tenuifoliae

I. songarica Schrenk

Synonym
I. songarica Schrenk var. *gracilis* Maxim.

Distribution
Iran, Pakistan, Soviet central Asia, Mongolia and northern China.

Description
Rhizome woody, very dark in colour, permanently covered in dead foliage and forming large clumps. **Leaves** to 80 cm long, 0.3 cm (–1.0 cm) wide, with 3–5 veins, grey-green. **Stem:** to 50 cm long, with 3–4 leaves and 2 flowers, 3 spathes to 14 cm long, 2 cm wide, green, margins membranous; pedicel 4.5 cm long. **Flowers** to 9 cm diameter, violet through to white; falls to 5.5 cm long, 1.0 cm wide, blade ovate or rounded, hafts lanceolate; standards to 3.5 cm long, 0.5 cm wide, erect, oblanceolate; style to 3.5 cm long, 1.0 cm wide, crests narrowly triangular, stamens about 2 cm long, anthers brown; tube to 0.7 cm long; ovary to 2.5 cm long, spindle-shaped. **Capsule** to 6.5 cm, narrowing to long pointed apex, leathery, dehiscing from apex, with conspicuous ribbing; reticulate veining on the surface of the capsule distinguishes it from *I. ventricosa*. **Seeds** brown, wrinkled, flattened globular. **Flowering** June–July. $2n = ?$

Cultivation
This plant comes from high mountain grassland and stony areas where it is often the dominant vegetation. Köhlein recommends protection from winter damp and summer rain. Rodionenko comments that germination is poor and that seeds sown in pots germinated over 5 years.

Observations
Rodionenko says that this one is best seen from above. Dykes took his description of the plant from unpublished material by Foster. The two descriptions are quite incompatible; could Foster possibly have had *I. anguifuga*? In 1934, Edward K. Balls found the flower colour very variable and the plants attractive in the mass.

I. tenuifolia Pall.

Distribution
From the lower Volga valley up through Siberia to China, Tibet and Mongolia. Up to 4000 m in the Pamir mountains; more usually 1000–2000 m.

Description
Rhizome covered in dead reddish fibrous leaf bases and growing into dense tussocks; occasional stolons may be thrown out and the roots are hard and wiry.

Series Tenuifoliae

Leaves to 60 cm long, 0.2 cm wide, linear, twisted. **Stem** to 5 cm but usually very short, subterranean; 4 spathes to 10 cm long, 1 cm wide; pedicel very short; tube 5–8 cm long; 2-flowered. **Flowers:** 5–8 cm diameter, violet; falls to 5 cm long, 1.5 cm wide on blade, haft broadly wedge-shaped carrying a few hairs on mid-rib, blue-violet to lilac veining on cream ground with cream centre stripe; standards to 5 cm long, 0.5 cm wide, blue-violet to lilac, oblanceolate, erect; style 4 cm long, 0.5 cm wide, crests narrowly triangular, same colour as standards, stigma bilobed; stamens to 3 cm long, filaments and anthers equally long; tube to 6 cm long; ovary to 1.2 cm long, 0.2 cm diameter. **Capsule** about 4.5 cm including short beak, 1.8 cm diameter, plump, dehiscing from apex. **Seeds** cubical, dark brown, coarsely wrinkled. **Flowering** April–May. $2n = ?$

Cultivation
This plant comes from stony steppes, sand or sandy river banks. Mathew recommends a bank of sandy soil in full sun. Köhlein says seedlings grow very slowly. Both are agreed that the decorative value of the plant is not high.

I. ventricosa **Pall.**

Distribution
NE China, Inner Mongolia, Mongolia and Siberia.

Description
Rhizome wiry with thick fleshy roots and persistent old leaf bases. **Leaves** to 35 cm long, 0.3–0.7 cm wide, with exceptionally strong fibres. **Stem** 8–20 cm, 3 spathes to 8 cm long, 4 cm wide, inflated, clearly reticulated, margins membranous; pedicel to 1.5 cm; 1–2-flowered. **Flowers** 5–7 cm diameter, strongly purple in colour; falls to 5 cm long, 1 cm wide, blade edges turning up, pale base colour with heavy, dark veining; standards to 4 cm long, 0.8 cm wide, erect, oblanceolate; style about 3.8 cm long, 0.6 cm wide, flat, crest flatly triangular; stamens to 3.5 cm, anthers yellow-purple; tube to 4 cm long, very slender, enlarged at top; ovary 1.5 cm long, 0.3 cm diameter, narrowed top and bottom. **Capsule** to 4 cm long, 1.0 cm diameter, with 6 prominent ribs and long narrow beak (to 4.5 cm), dehiscing from apex. **Seeds** ? $2n = ?$

Cultivation
This plant comes from sandy plains and river banks. Köhlein finds it needs full sun and sandy soil. Worth growing for the remarkable bracts, which Rodionenko compares to 'the membranous wings . . . of a mantis', but the flowers are of no garden value. Nevertheless, this plant, too, has overwintered in the open at Kew so that seedlings are important.

Observations
The outstanding feature of this plant is the reticulate veining on the spathes, ovary and capsule, in which it differs from *I. songarica*.

SERIES ENSATAE (DIELS) LAWRENCE

Map 21, Figure 21.

The sole representative of this series must not be confused with *I. ensata* Thunb. Inevitably, with a plant as widely distributed as this, there are innumerable forms both wild and semi-cultivated. Very properly, when the finder was in doubt, a new specific name was attributed and it probably has more synonyms than any other iris. Recent work in China has suggested three distinct forms.

I. lactea Pallas

Synonyms
I. biglumis Val., *I. caespitosa* Pallas, *I caricifolia* Pallas, *I. doniana* Spach., *I. ensata* Dykes, *I. fragrans* Lindl., *I. graminea* Thunb, non Linn., *I. haematophylla* Link, *I. iliensis* Pol., *I. lactea* var. *chinensis* Fisch., *I. longifolia* Royle, *I. longispatha* Fisch., *I. moorcroftiana* Wall ex Don., *I. oxypetala* Bunge, *I. pallasii* Fisch., *I. triflora* Balbis. *Ioniris biglumis, I. doniana, I. fragrans, I. longispatha, I. pallassii, I. triflora,* Klatt, *Xiphion donianum* Alef., *X. pallassii* Alef.

Distribution
In a wide sweep over the mountain ranges from Altai to central China and Korea.

Description
Rhizome forming closely compacted tufts and retaining coarse remnants of old leaves; roots, adapted to extremes of aridity, are very long and complicated. **Leaves** up to 40 cm long, 0.3–0.5 cm wide, strongly ribbed, continuing to grow after flowering; one form spiralling slightly. **Stem:** 5–30 cm tall, oval cross-section; spathes 5–7 cm or more, green, with long points; pedicel lengthening from 2 to 4 cm as capsules ripen; tube 0.2–0.3 cm; ovary 2.5–5 cm long, roughly trigonal; 2–3-flowered. **Flowers**: 4–6 cm diameter, creamy white; falls oblanceolate, narrowing at haft, which has a low central ridge often marked with small purplish dots; standards oblanceolate, base of haft green, usually darker than falls; style narrow, rounded with triangular overlapping crests; stigma small, pendant, triangular; filament white; anther cream. **Capsule** 2.5–5 cm long, width very variable as is the length of the beak, with six ribs at equal intervals. **Seeds** roundish, smooth, dark brown, sometimes compressed. **Flowering** May–August. $2n = ?$

Cultivation

This plant is tolerant of most soils and may safely be planted in places liable to dry out in summer. It should never be left out of soil for any length of time because if the root system is dehydrated it is unlikely to recover. When leaves first start growing in spring, they may appear chlorotic, but this will pass. In temperate climates, leaves come into growth early and may well overtop the flowers; in more extreme areas where leaf growth is delayed the scented flowers will be displayed to better advantage. It has a reputation for poorly shaped flowers, but forms are very variable and inter-breed, as might be expected. The cultivated form known as 'chinensis' used to be considered the best as having large and well-shaped flowers.

Observations

Plants are found over an immense range of the Near and Far East and confusion over the various forms has been worse confounded by its use both as a fodder plant and for making string and rope from the leaf fibres, thus ensuring that a migrant community would take their favourite form with them. However, recent work in China has established variant forms in the wild.

var. *chinensis* (Fisch.) Koidz.

This, the most widely distributed form, has flowers in the blue-violet range and is extremely tolerant of saline, alkaline and degraded soils. It has considerable commercial use and can improve soil texture.

var. *chrysantha* Y.-T. Zhao

This variant from Tibet, only recently described, is similar to the type except for having yellow flowers.

SERIES SYRIACAE (DIELS) LAWRENCE   OFER COHEN

Map 22, Figure 22.

This is an isolated and distinctive series which is separated from all others by having short, massive and compact rootstocks, which grow somewhat vertically and are characterised by clusters of very spiny bristles at their upper side, around the leaf bases. The species are notable for their unbranched stem with only a single substantial flower in its apex.

Systematically, there has been much confusion over this series. Five species were described as belonging to the series, with their distribution area limited to the Middle East: Israel, Syria, E Turkey and N Iraq (Kurdistan). As a result of my studies, the series has been separated into two species only, which are closely related. This division is based on certain features, which were studied and found to coincide with both the distribution area and the flower's colour and shape. *Iris grant-duffii* (including *I. melanosticta* and probably *I. aschersonii*) is the yellow-flowered form with wide organs. It occurs in the western and southern parts of the section's distribution area: in the coastal area and slightly inland, both of which are areas of Mediterranean climate. *I. masia* (incl. *I. caeruleo-violacea*) is the purple-blue and narrow-flowered form, occurring inland in the northern and eastern parts of the distribution area in the drier areas.

Many morphological characteristics were given to identify the species, based mainly on leaf size and flower colour and tint. However, comparative studies on the Israeli coastal and inland populations show high variability in leaf width and leaf length, and also in the height of the stem in comparison to the leaves. These variable features probably derive from the microhabitat conditions and the difference in each site's rain and temperature situation each year. As opposed to this, there is only a little variability in the flower's colour within one population and a general similarity in colour among all Israeli populations. Hence, from the vegetative view, all the variants can actually be regarded as one variable species. Future field observations, especially of *I. masia* populations, together with cytological studies, will be able to revive the sensible possibility that both species, in fact, represent two colour forms of one variable species.

The following review is based on general information about all five presumed species. Some details were compiled by analogy using my experience of Israeli plants. These are plants of seasonally damp meadows or open turf fields, usually with no drainage, which are flooded from time to time in winter. They are distributed from sea level in the west, to 1100 m in the north-east. The habitat's climate is biseasonal: cold and wet in the short winter, hot and extremely dry during the other 7–8 months of the year. The habitat of *I. grant-duffii* has often been incorrectly cited as 'marshy'. This is inaccurate because the occurrence of water in the growth area is only a seasonal phenomenon. Also, the 'dry habitats' attributed to *I. masia* and to the eastern populations of *I. grant-duffii* appear to point to the aridity of the area and do not consider the occurrence of flooding during the short winter. Therefore, the considerable difference between the two species' habitats can be concluded to be only a natural gradient.

Soils are alkaline, deep and heavy, often basalt soils of volcanic origin, which are not calciferous at all but rich in potash. Because of their mineral compounds, montmorilonite in volcanic soils for instance, the soils become impervious to water when wet. The water is not free to drain into the ground and remains on the area's surface for anything from some days to a few weeks. The soils of the coastal area are usually heavy alluvial loam (Grumusol) formed from limestone, poor in calcium, and are populated by many calciphobic plants.

Leaf growth begins during the first rains (November–December). Flowers begin to bloom in the wild when the winter temperature begins to rise, from late January onwards, according to altitude. Israeli coastal populations bloom first from late January to mid-February; the inland Golan Heights plants in mid-March; the inland N Syrian plants in March–April; and those of the mountain areas in E Turkey and Kurdistan in May.

Areas of high ground dry out about a month after the last rains (May–June). In high summer, the leaves disappear, owing chiefly to grazing, but also to summer easterly wind storms and local human fires. The dry capsules, however, can persist for up to a year, during which time they eventually split and disperse the seeds. During this season, the sites are completely grassless and the ground is very dry and cracked.

One must be careful when handling the rootstocks: anyone who has ever tried to dig them out will never forget the sharp, fine and needle-like bristles, which inflict painful damage on fingers. These spiny bristles probably serve to protect the plants from drying during the hot season. They also protect them from marauding wild-life, especially from rodents whose many holes are distributed thoughout the population's area and are noticeable during the dry season. Unfortunately, the plants are not protected from human invasion. The intensive grazing in all of the distribution area, especially in the eastern parts, has greatly impoverished the populations. The damage is especially critical in developed areas. Throughout Israel, and also along the northern coastal area, almost all the heavy soil areas have been developed for agriculture. Most of the *I. grant-duffii* populations have become extinct in the past decades, while those of *I. masia* are poor in field information but seem to be endangered as well.

With regard to *I. grant-duffii*, Dykes commented that by the end of the first year, the seedlings from bulb-like rootstocks, covered with a reticulated coat, are very similar in appearance to plants of *I. reticulata*. Mathew also draws attention to the noticeable resemblance between *I. masia* and *I. pamphylica* (another member of the reticulata irises) in the extremely sharp fibres of the old leaf bases protecting the bulb. However, this resemblance does not necessarily attest to systematic relationship but is rather a protective adaptation feature which is known in bulbs of many arid zones, not only in irises.

Mathew also takes note of the similarity in the pollen structure between *I. pamphylica* and *I. masia* and of the pronounced difference from the other species of the reticulata irises. Moreover, he points out that 'within the subgenus *Hermodactyloides*, *I. pamphylica* is very distinct in having an aerial stem holding the flower above ground level rather than on a long perianth tube with a subterranean stem as in all other species'. In view of all these factors Mathew's suggestion, that there is a link between *I. pamphylica* (and possibly the entire reticulata group) and this series, seems to be a reasonable conclusion worthy of further study.

Cultivation

Plants require maximum sunlight for flowering. It is better to plant them in rich and heavy alkaline soils, poor in calcium and rich in potash. Turf areas, which can be flooded in winter but must remain dry in summer, are best. Despite their 'marshy' habitats in the wild, they suffer when waterlogged for long periods and are not suitable as candidates for 'the year-round water garden'. They are relatively resistant to dry conditions and can also be grown without excess water in winter. Most of the efforts to coax the species into bloom in Europe have been unsuccessful, probably because of wrong soil and water conditions.

Iris grant-duffii Baker

Synonym
I. melanosticta Bornm.

Distribution
Coastal regions of C and N Israel and W Syria (no records from the Lebanon coast). It penetrates inland from N Israel to S Syria and possibly adjoining areas in Jordan. First collected by the British General Grant-Duff from the banks of the Kishon River in the western Esdraelon Plain, it grows in the Acre and Esdraelon Plains, the Lower Galilee valleys and especially the basalt plateaus of the S Golan Heights, Hauran and Djabel Druz districts (plants that were defined as *I. melanosticta*). In the north it is known from the south of Latakiya in Syria and possibly in S Turkey too (*I. aschersonii*).

Description
Plant 30–70 cm in height. **Rootstock**, as described above, forms very large clumps. **Leaves** up to 70 cm long, 0.3–1 cm wide, stiff, flat, narrow, grey-green and strongly veined, frequently with a slight spiral. **Stem** 15–40 cm tall, round; spathes linear, greenish, papery at margin, tapering to apex. **Flower**: 6–8 cm diameter, solitary on each stem, fragrant; tube very short; falls 5–8 cm long, 0.8–1.2 cm wide in the haft, widening after constriction to obovate blade, 1.0–1.6 cm wide; standards 4–6 cm long, oblanceolate; styles (including lobes) 4–6 cm long, widening to apex; lobes 1–2 cm long, lanceolate acute with irregular edge; stigma bilobed; filaments about 1 cm long, anther 1–1.5 cm long, pollen yellowish. Flower basically bright greenish-yellow colour; the lower narrow parts of the falls, standards and styles are veined purplish on a pale yellow ground and spotted with linear purple-blackish dots; the blade of the falls has a single bright lemon patch in the centre, often with a few blackish dashes on it and sometimes with several merged dots in the throat. **Capsule** cylindrical–ellipsoid, up to 10 cm long, splitting when dry into 3 valves. **Seeds** rounded, tuberculate, red-brown, with defined neck. **Flowering**: late winter in wild (late January) February–March, according to altitude. $2n = 24$.

Habitat
Heavy soils in alluvial and volcanic areas; seasonal damp meadows and open turf fields, flooded in winter; sea level to 800 m.

Cultivation

As mentioned above, a heavy, degraded limestone soil or alkaline, uncalcareous soil (such as a basalt soil) seems best. Heavy feeding is suggested, especially with a low-nitrogen and high-potash diet. Maximum sunlight is needed in the flowering period. The plant must be protected from rain in summer (as with the high desert irises) although it should be winter-hardy in temperate climates. The species is easy to raise from seed but it will be ready to flower only after 6–7 years.

Observations

Plants from the coastal area which are grown in light soils have flexible reticulate rootstock fibres, whereas those of heavy soils, especially of mainland basalt origin, have rigid spiny spikes. It is clear that there is a gradient in this feature correlated with the soil structure and perhaps also with the amounts of rain (in *I. masia*).

Seeds are of normal fertility and the germination in the first year (in the nursery) is almost full.

Most of its habitats in Israel have been converted into agricultural land during the last decades, causing the extinction of most of the wild populations. The surviving large populations are restricted at present to only one area of the Golan Heights. The species is protected by law and is included in the *Red Data List of Israeli Rare and Endangered Plants*. The Golan sites have been designated as Nature Reserves and the conservation of the species in nature is assured.

I. aschersonii Foster (spec. dub.)

This species was described from the S Turkey coastal plain, near Adana in Cilicia. It is known only from the type collection and from the cultivated material originating from it.

Some characteristics of the flower colour and the rootstock's configuration were cited to recognise it as a separate species. According to Dykes, it was said to have 'bright yellow flowers veined and dotted with green' and falls edged with irregular, linear black-purple dots 'which merged into blotches on larger plants'. The growing point was cited as sheathed in flexible, reticulate (instead of rigid) spikes. However, all these variations are known in the Israeli populations of *I. grant-duffii*, at times even in one local stand.

I. aschersonii was the first of the series to flower in cultivation, many years before the other members. In the beginning of this century Dykes grew it in the Cambridge University Botanic Garden, and found that his specimens were easier to flower than *I. grant-duffii* specimens from the coastal plain of Israel. The plant seems to have been widely distributed and grown early in this century (in England), but now it appears to be extinct in cultivation.

Because of the great interest in *I. aschersonii*, and also because it has not been re-collected in the wild since its type collection, it continues to be cited in the iris literature. However, I believe it can be clearly identified with *I. grant-duffii* and is not merely a local variant. It is suggested that the plant should be looked for in the S Turkey coastal area near Adana, along the edges of fields of heavy soil which are flooded in winter.

Iris masia Stapf ex Foster

Synonym
I. caeruleo-violacea (Gombault) Mouterde.

Distribution
Boundary areas of the North Syrian Desert, from SE Turkey and NW inland Syria to NE Syria and adjacent Iraq (Kurdistan). Its first collection is unknown, but it was described in 1902 from 'Asia Minor', based on cultivated material which was sent to M. Foster by M. Leichtlin. Although the specimen has not been traced, its origin was probably the foot of the extinct volcano Karacadag (the Black Mountain) near Urfa, whose ancient name was Mons Masius. It is known from the Mesopotamia area, Gaziantep, Urfa and Diyarbakir regions in Turkey; in N Syria from many sites in the Halab and Homs-Hama regions and also from the Jira region near the Turkey–Iraq border; it was only found twice in the lower forest zone of Kurdistan.

Description
This resembles *I. grant-duffii* but it was said to have only narrow leaves (0.3–0.6 cm wide). As mentioned above, this is a very variable feature. Its flowers are violet to blue-purple, less substantial-looking than the yellow species. All the flower parts, chiefly the falls' blade, are narrower and slightly smaller than those of *I. grant-duffii*.
Flowering: late winter in the wild, March–May, according to site altitude. $2n = 24$.

Habitat
Fields and steppic plains, alluvial soil and often arid volcanic areas (flooded in winter?). Altitudes: Syria, 300–800 m; Turkey, 750–1050 m; Iraq (Kurdistan), 600–1100 m.

Cultivation
In my opinion, the best conditions are probably like those recommended for *I. grant-duffii*, with rather short and cold winter conditions. According to Mathew, who grew it in England in a number of different situations (pot, bulb frame, open sunny area), it only sprouted leaves and did not flower.

Observations
As mentioned above, the field information about this species is somewhat limited. Field observations are necessary in order to examine the possibility that it represents only a narrow, violet-blue colour form of the yellow species.

SERIES UNGUICULARES (DIELS) LAWRENCE   AARON DAVIS AND STEPHEN JURY

Map 23, Figure 23.

The Series Unguiculares was named because of the claw-like base of the perianth segment. It is a small and isolated group of just two species, *I. unguicularis* Poir. and *I. lazica* Albov. The unique features of this group are the stylar tube, formed by the fused stamen filaments, and the presence of peculiar yellow glands on the upper surface of the style branches. Dykes likened these glands to a 'sprinkling of gold dust'. In this series the nectarium is absent and instead nectar is secreted in drops at the base of the falls.

In *I. unguicularis* and *I. lazica* the stem is very short or virtually absent and the flowers are held above the leaves by a very long perianth-tube.

Many variants of this group are of great horticultural value because they produce large, attractive flowers at a time of the year when little else is in bloom. Most variants are slightly fragrant. *I. unguicularis* is a very variable species over its range particularly in the width of the leaves and the size, shape and colour of the perianth segments. However, it can be successfully classified into recognisable taxa using morphological and pollen characters. A full account of the taxonomy of this species is included in a paper in the *Botanical Journal of the Linnean Society* (1990).

I. unguicularis Poir.

Distribution
Algeria, Greece, Turkey, Syria, Israel and Lebanon.

Description
Rhizome tough with wiry roots, forming dense clumps. **Leaves** 12–85 cm long, 0.1–1 cm wide, linear to sword-shaped, strongly ribbed or fluted with distinct margins. Flowers solitary or rarely branched, fragrant. **Stem** very short or absent; spathe valves 5–13 cm, tapering, green and herbaceous. Perianth tube cylindrical, 5–26 cm long, 0.3–0.6 cm across. Ovary slender, oblong–ellipsoid, to 3.5 cm, more or less stalkless within the spathe. **Flower:** falls 4.3–8.7 cm long, 0.9–1.8 cm wide, erect, deflexed, oblanceolate to spathulate, claw-shaped at the base, apex acute to rounded, lavender-blue to deep lilac or white, veined darker on white region at base of blade, with a yellow median stripe; standards 4.3–8.6 cm long, 0.3–0.7 cm wide, erect, slightly reflexed, oblanceolate to spathulate, claw-shaped at the base, apex acute to rounded, colour as falls except for brownish-violet zone occurring at the base of the haft. Stylar tube 1.2–2.5 cm; style branches 2.3–3.5 cm long with acute lobes, yellow-glandular, with fine teeth pointing forwards. **Capsule** ellipsoid, 1.1–1.7 cm long, 2.4–4.0 cm diameter, carried on a short inflorescence stalk. **Seeds** brown to orange-brown, 0.45–0.65 cm, few, approximately spherical to wedge shaped, wrinkled when dry. **Flowering** February–June. $2n = 38, 40, 50$ (? 24, 36, 48).

Habitat
Rocky places, on banks and hillsides, and in dry scrub or open coniferous woodland.

subsp. *unguicularis*

Synonyms
I. stylosa Desf., *Neubeckia stylosa* Alef., *Ioniris stylosa* Klatt, *Siphonostylis unguicularis* W. Schulze.

Distribution
Algeria, Tunisia and possibly Morocco.

Description
Leaves 12–85 cm long, 0.3–1 cm wide, linear to sword-shaped. Spathe valves 6–13 cm long. Perianth tube cylindrical, 6–22 cm; ovary slender, oblong–ellipsoid to 0.35 cm. **Flower:** falls 5.9–9.7 cm long, 1.8–3.2 cm wide, oblanceolate–spathulate, lavender-blue to deep lilac or white, veined darker on white region at base of blade, with a yellow median stripe; standards 6.2–8.6 cm long, 1.7–3 cm wide, oblanceolate–spathulate, colour as falls except for brownish-violet zone occurring at the base of the haft. Stylar tube 2–2.5 cm long; style branches 2.7–3.5 cm long, with fine teeth pointing forwards. **Capsule** ellipsoid, 1.4–1.7 cm long, 0.3–0.4 cm wide, carried on a short inflorescence stalk. **Seeds** orange-brown, 0.65 cm. **Flowering** February–June. $2n = 38, 40, 48, 50$.

Habitat
This plant grows in rocky places, on banks in dry scrub or open coniferous woodland.

Observations
These N African plants are almost certainly the most amenable to cultivation in British, and similar, gardens.

In addition, there are some colour variants:

'Alba': white, greenish-orange-yellow median stripe.

'Marginata': lilac with white edging.

'Mary Barnard': violet-blue.

'Walter Butt': pale silvery lavender.

subsp. *cretensis* (Janka) Davis & Jury

Synonyms
I. cretensis Janka, *I. cretica* Janka [nomen nudum], *I. stylosa* Desf. var. *angustifolia* Boiss. & Heldr. p.p., *I. humilis* Sieber ex Baker [nom. in schaed.], *Siphonostylis cretensis* (Janka) W. Schulze subsp. *cretensis*.

Distribution
Greece: the Aegean islands of Crete and possibly Saria.

Description
Leaves 11–31 cm long, 0.1–0.33 cm wide. Spathe valves 6–12 cm long. Perianth tube 5–13 cm long. **Flower:** falls 4.3–5.7 cm long, 0.9–0.15 cm wide, oblanceolate, violet or deep lavender at apex only, rest of blade white with prominent veining and a thick orange median stripe not exceeding area of white coloration; standards 4.3–5.5 cm long, 0.7–1.25 cm wide, oblanceolate. Stylar tube 1.2–1.5 cm long; style branches 2.3–2.5 cm long. **Capsule** 2.4–3.2 cm long, 1.1–1.5 cm diameter. **Seeds** brown, 0.45 cm, few. **Flowering** February–May. $2n = 40$.

Habitat
Rocky places, on banks and in dry scrub.

Observations
The most distinctive flower patterning is found in this Cretan subspecies where white covers almost two-thirds of the blade and is strongly veined with violet. This character provides an easy means to separate subspecies *cretensis* from other narrow-leaved variants.

subsp. *carica* (W. Schulze) A.P. Davis & Jury

Synonyms
I. cretica Herb. ex Baker [nomen nudum], *I. stylosa* Desf. var. *angustifolia* Boiss. et Heldr. p.p.

Distribution
Greece, Peloponnisos, Turkey, Syria and Lebanon.

Description
Leaves 11–50 cm long, 0.18–0.5 cm wide. Perianth tube 6.5–26 cm long. **Flower:** falls 0.48–0.79 cm long, 0.104–0.22 cm wide, oblanceolate; standards 0.55–0.82 cm long, 0.12–0.21 cm wide, oblanceolate. Stylar tube 0.12–0.16 cm long; style branches 0.23–0.28 cm long. **Capsule** 0.24–0.34 cm long, 0.11–0.15 cm diameter. $2n = 40$. (?36)

subsp. *carica* var. *carica* (W. Schulze) A.P. Davis & Jury

Synonyms
Siphonostylis cretensis (Janka) W. Schulze subsp. *carica* W. Schulze.

Distribution
Greece: the east Aegean islands of Cos and Rhodes; SW Turkey.

Description
Leaves 0.1–0.35 cm wide. Falls 0.55–0.74 cm long, 1.05–1.7 cm wide, oblanceolate. **Flower**: standards oblanceolate. $2n = $?

subsp. *carica* var. *angustifolia* (Boiss. & Heldr.) A.P. Davis & Jury

Synonyms
I. stylosa Mazziari, *Siphonostylis* 'Hybridcomplex' W. Schulze.

Distribution
Greece; the Ionian islands (Kefallonia and Zakinthos), Peloponnisos, and the Aegean Islands (Paros).

Description
Leaves 0.18–0.5 cm wide. **Flower**: Falls 0.48–0.69 cm long, 1.05–2.2 cm wide, oblanceolate; standards 5.5–7.9 cm long, 1.6–2.1 cm wide, oblanceolate. $2n = 40$ (?36).

Observations
A very variable plant over its range: the flowers may approach the dimensions of the subspecies *unguicularis* and *cretensis*. Colour forms are known.

'Greek White': white, greenish-orange-yellow median stripe (collected by Dr. Elliott, ELLJ 8269, and distributed as 'Alba').

'Angustifolia': mostly white with bluish lilac margins.

'Speciosa': deep violet-blue.

subsp. *carica* var. *syriaca* (W. Schulze) A.P. Davis & Jury

Synonyms
I. cretica Herb. [nomen nudum]; *I. stylosa* Desf. var. *angustifolia* Boiss. & Heldr. p.p.; *Siphonostylis cretensis* (Janka) W. Schulze subsp. *syriaca* W. Schulze.

Distribution
E Turkey, Syria and Lebanon.

Description
Leaves 0.12–0.5 cm wide. **Flower**: Falls 5.5–7.9 cm long, 1.05–1.9 cm wide,

oblanceolate to lanceolate–rhombic; standards 6.5–8.2 cm long, 1.2–2.1 cm wide, oblanceolate to lanceolate–rhombic. $2n = ?$

I. lazica Albov

Synonyms
I. cretensis Janka forma *latifolia* Lipsky; *I. unguicularis* Poir. var *lazica* (Albov) Dykes; *Siphonostylis lazica* (Albov) W. Schulze.

Distribution
NE Turkey, Georgia (Adzhariya and Abkhasia) and Russia in the vicinity of the Black Sea Coast area.

Description
Rhizome tough with wiry roots, forming dense clumps. **Leaves** linear to sword-shaped, more or less arranged in two ranks, 12–85 cm long, 0.85–1.3 cm wide, weakly ribbed, with flat margins. Flowers solitary or rarely branched, fragrant; stem very short or absent; spathe valves 5–12 cm long, tapering, green and herbaceous. Perianth tube cylindrical, 5–12 cm long, 0.3–0.6 cm diameter. Ovary slender, oblong–ellipsoid, to 3.5 cm, more or less stalkless within the spathe. **Flower:** falls 7.1–8.7 cm long, 2–3 cm wide, erect, deflexed, oblanceolate or fiddle-shaped, claw-shaped at the base, apex acute to rounded; lavender-blue to deep lilac or white, veined or veined to spotted darker on white region at base of blade, with a yellow median stripe. Standards 6.9–8.7 cm long, 1.8–3.1 cm wide, erect, slightly reflexed, oblanceolate to spathulate, with a claw and acute, colour as falls except for brownish violet zone occurring at the base of the segment. Stylar tube 2.3–2.7 cm long; style branches 2–2.3 cm long with acute lobes, pale purple, yellow-glandular, with fine to coarse teeth; forward-pointing. **Capsule** 2.4–3 cm long, 1.2–1 cm diameter, ellipsoid, carried on a short inflorescence stalk. **Seeds** light brown, 0.35 cm, flattened-angled, finely wrinkled when dry. **Flowering** March–June. $2n = 32$.

Habitat
Wet, stony places; shady damp banks and open woodland (hazel groves). Sea level to approximately 2000 m.

Observations
This species is easily separated from *I. unguicularis* by its flat, wider, more or less unribbed leaves and larger fiddle-shaped falls. It also occupies a completely different ecological niche.

Cultivation
The wide-leaved variants of *I. unguicularis* present no real problems in cultivation, provided they are given a sunny, well-drained position. A south-facing border, preferably on an alkaline soil such as chalk, is ideal.

The narrow-leaved forms (subspp. *carica* and *cretensis*) need slightly more protection in the open garden and benefit from some overhead cover, particularly during a wet winter. Although these plants tend to be shorter lived in cultivation than the more robust, wide-leaved variants from N Africa (subsp. *unguicularis*), we have seen specimens that have remained in cultivation for over 20 years. They were grown in a raised border/rockery with their rhizomes level with the soil and given occasional dressings with a general purpose fertiliser.

Although *I. unguicularis* prefers a hot sunny position it must never be allowed to dry out completely; for this reason, we prefer to see plants grown out of pots with an unrestricted root run.

I. lazica grows in an area of much higher rainfall and lower winter temperatures than *I. unguicularis* and is therefore more tolerant of life in British gardens. This species can be grown in the open garden in any good, fertile soil.

Propagation
For both *I. unguicularis* and *I. lazica* this is by cuttings taken in autumn or spring. We have achieved almost 100% success by taking 4–8 cm rhizome cuttings in autumn. The wiry, non-fleshy roots need to be trimmed off, leaves cut down to half their length and any dry, non-living material removed. Cuttings placed in John Innes No.1. compost can be left in this medium until the following autumn, by which time they need potting on or planting out. Both species can also be propagated by seed, which should be sown as soon as possible after gathering.

Other cultivars, of uncertain affinities (mainly narrow-leaved variants), are as follows.

'Abington Purple': purple.

'Ellis's Variety': violet-blue.

'Oxford Dwarf': mainly white, violet lavender with purple veins.

'Starker's Pink': lilac/pink.

'Winter Treasure': white.

Further named cultivars do exist, but are either rare or no longer in cultivation.

Subgenus Nepalensis (Dykes)

CLIVE INNES

Map 24, Figure 24.

An awkward group where species have been in, and out, over many years. There is insufficient information about the individual plants, which need to be grown under conditions where they can be properly and thoroughly observed. Newly collected material is providing interesting results and there is still considerable debate about whether *I. collettii* and *I. decora* are in fact separate species.

I. barbatula Noltie & K.Y. Guan sp. nov.

Distribution
China: NW Yunnan.

Description
Rootstock a tiny growing point on a cluster of swollen roots giving rise to a clump of fans; old fibrous leaf bases persist. **Leaves** 3–5, to 25 cm long, 1 cm wide, slightly taller than flowering spike, matt surface, veins conspicuous, 1 on one side, 2 on the other, narrowing sharply to point at apex. **Stem** to 4 cm, sometimes none; spathes 2–6 cm long, pointed at apex, keeled, lanceolate, tightly clasping ovary and the base of the tube, with hyaline margins; tube 3–10 cm long, narrow, light green, with 2–4 flowers. **Flower** dark violet; falls 3–5 cm long, 1–1.5 cm wide at blade, crest densely covered with long dark yellow hairs, to mid-point, white ground with dark violet veining heavier towards apex; standards 2.7–4.9 cm long, 0.7–1.3 cm across blade, oblong, colour the same as the falls, sometimes with white marks at base; style 2–2.5 cm long, oblong, lobes 1.1–1.6 cm long, 0.6–0.7 cm wide, erect, triangular, irregularly toothed, stigma smoothly bilobed, narrow; filament 0.8–1.2 cm long, anther 0.8–1 cm long, pollen white. **Capsule** 2–2.5 cm long, tapering to both ends, sharply ribbed, trigonal, apex pointed. **Seeds** (unripe) about 0.2 cm, with white aril. $2n = ?$

Cultivation
So far, *I. barbatula* has grown well in the Royal Botanic Garden, Edinburgh, planted in loam-based John Innes No.2 compost with added grit. The pots are sunk in sand in frames, in full sun. Plenty of water is given in the growing season, but the frame is kept fairly dry after the plants die down for the winter.

Observations
This plant was named for its beard which distinguishes it from *I. collettii* and which somewhat resembles those of Section Pseudoregelia, thus emphasising the need for more detailed studies of both groups. Sadly, the short-lived flowers of this attractive plant will make it less popular with gardeners, but there is a good flowering sequence on an established plant.

I. collettii Hook. f.

Synonyms
I. duclouxii Léveillé, *I. nepalensis* for. *depauperata* Collett & Hemsley, *I. nepalensis* var. *letha* Foster.

Distribution
Mountainous regions of N Burma, Thailand and south-west China (Yunnan & Szechuan provinces) at altitudes from approximately 600 to 3400 m, growing in open areas of pine woods, scrub country and open grassland.

Description
Rootstock a small, relatively flat, disc-like rhizome densely covered with the bristly remains of leaf-sheaths and attached to which are slender, shortly elongating, fleshy, whitish roots, which are irregularly swollen. **Leaves**: short at flowering, later to 30 cm or more, 0.15–0.18 cm wide, smooth, sword-shaped, pointed at tips, dark green or slightly glaucous, prominently ribbed, 2 on upper surface, 3 beneath. **Stem**: variable from almost non-existent to about 5 cm or a little more, occasionally with 1–2 reduced leaves near base, frequently branched; spathes green, long, slender, keeled, with 1–2 or more flowers; pedicels 1–2.5 cm long; ovary 3-sided; flowers at apex and branches. **Flower**: 2.5–3 cm diameter, fragrant; tube 6–10 cm long, slender; falls 2.5 cm wide, 5 cm long, blade broadly lance-shaped, pale purplish-yellow, haft veined in deeper shade becoming yellowish-brown as it merges with blade; standards narrower than falls, tending to recurve at tips, bluish-violet with slightly deeper veining; style with pale orange edges, indented; crest orange-yellow, triangular, erect; stigma deeply lobed; filaments white, faintly tinged with violet; anthers cream, faintly tinged violet; pollen white. **Capsules** 2.5–4 cm long, 3-sided, pointed at tip. **Seeds** dark brown, about 0.3 cm long, with whitish appendage. $2n = 28$.

Cultivation
It would appear that this requires similar treatment to *I. decora*. A rich soil is certainly of importance and this must be consistently moist during the growing and flowering periods from April to late October. Once the foliage has died down in October a period of dormancy is necessary until mid-April–May. It would seem

appropriate to lift plants, taking precautions not to damage the rootstock, and store in dry sand or peat and keep cool throughout. In late April, or early May, they can be planted out again in rich, moist soil; a position of indirect light has been advised by Dykes.

Propagation
This is by seeds sown soon after harvesting in the autumn. Germination is not apparent until the following April–May, when the soil must be kept consistently moist.

Observations
Seeds of this plant have recently been collected again and plants have flowered. There is a possibility that, given really good drainage, such as that provided by tree roots, this plant may be reasonably hardy in temperate zones. Dykes stated that plants developed 2 or more stems from the same rootstock, but this has not yet been verified.

var. *acaulis* Noltie

Distribution
China: Szechuan, Yunnan. Shady areas in mixed oak and pine woodland.

Description
This variety has been distinguished from *I. collettii* because of the total absence of a stem. The flowers vary from pale to dark blue with a distinct white signal patch and yellow crests on the falls. **Capsule** under 2 cm long, rounded.

Cultivation
Plants do well under greenhouse and frame conditions in a loamy compost with added grit and full sunshine in Britain.

Observations
As a result of his work, Henry Noltie (1995) has separated this stemless form, found only in China, from the type.

I. decora Wallich

Synonyms
I. fasciculata Jacquemont, *I. nepalensis* D. Don, *I. sulcata* Wallich, *I. yunnanensis* Léveillé, *Evansia nepalensis* Klatt, *Junopsis decora* W. Schulze, *Neubeckia decora* Klatt, *Neubeckia sulcata* Klatt.

Distribution
Temperate Himalayan regions of Nepal, Kashmir, Sikkim, Bhutan; China (Yunnan, Szechuan), southern and western Tibet, India (Khasia Hills, Shillong) at altitudes from 1000 to 4300 m in open pastures and clearings in scrub and forest country.

Description
Rootstock a small, rather flat rhizome densely covered by remains of old leaves, attached to which are finger-like white, fleshy, slender roots, sometimes swollen near the tips. **Leaves**: 0.2–0.5 cm wide, varying from 15 to 30 and sometimes 45 cm long, strongly ribbed, sometimes streaked with purplish lines and dots. **Stem**: 10–30 cm high, rarely more, with reduced leaves towards base, often branched; spathes green, slender, acuminate, keeled, generally 2-flowered; pedicels to about 2.5 cm long; ovary 3-sided, these slightly concave; tube 2.5–5(–6) cm long, slender; flower cluster at apex. **Flower**: 4–5 cm diameter, fragrant; falls about 2.5 cm wide, 5 cm long, with broadly lanceolate blade, pale lavender-blue with violet and white veining, haft with deeper reddish-purple veining and pronounced central ridge, which can be wavy, often bearing a few thread-like processes, ridge brownish-yellow at base gradually becoming more yellowish then pale mauvish-white on blade; standards narrowly lanceolate, outward-curving in similar manner to falls, pale reddish-purple with deeper veins of almost violet suffusion; styles broadly lobed, pale violet, edges crisped and toothed; crests large, triangular, erect, exceeding falls and standards, yellowish-orange and whitish or purplish at tips; stigma deeply bilobed; filaments white, slightly suffused violet; anthers cream coloured, tinged violet at base; pollen white. **Capsule** 2.5–3.5 cm long, trigonal, grooved at side, with pointed tip. **Seeds** dark brown, small, round with an appendage as large as, or larger than, the seed. **Flowering** late June. $2n = ?$

Cultivation
Found on open hillsides, in scrub or clearings in rhododendron forest at altitudes between 1000 and 4300 m. This plant is generally considered most suited to alpine house culture, or at least where it need not be disturbed when it comes to the dormancy period; a location where they are able to have reasonably regulated seasons provided. Flowers are short-lived with the leaves beginning to die down in October when watering should be reduced to a minimum and then to almost complete dryness. Growth recommences in April from which time watering should be increased. Keep in a cool, but bright position. Propagation is easy from seeds, but capsule needs careful watching if the seeds are to be collected before the plant distributes them.

Observations
This species has for many years been better known as *I. nepalensis*, a title originally given by D. Don, but as Wallich had already used the name for another species

which has since been considered as a synonym of *I. germanica*, *I. decora* becomes the acceptable title.

Mathew comments that there appear to be several forms of *I. decora* in cultivation in Britain. Henry Noltie points out that the original description of *I. decora* states it to be a plant of 90 cm but that none of the old herbarium specimens or anything that has been in cultivation until recently approached this in height. However, some recent collections from the Ganesh Himal (Nepal) grown at Kew produce stems around 120 cm tall and might therefore represent the typical variety (this plant also differs in capsule shape). The question of how, and at what level, to recognise variability (e.g. the smaller forms) taxonomically remains to be resolved, as does the separation between *I. decora* and *I. collettii*.

I. leptophylla Lingelsheim

A plant originally considered to be a form, or perhaps a synonym, of *I. decora*. It is now classified among the Pseudoregelia.

I. staintonii Hara

Distribution
Ganesh Himal, Abuthum, Lekh, Nepal, in alpine meadows at altitudes of about 3800 m.

Description
Rootstock a small insignificant rhizome 0.2–0.3 cm long with a few fibrous roots and 1–2 ovoid or ellipsoid brown tuberous roots 0.6–0.8 cm thick and 0.1–0.15 cm long. **Leaves**: generally 4, short, 1–6 cm long, tips obtuse, bract-like. **Stem**: 3–8 cm long; single linear-ensiform stem-leaf 15–30 cm long; 2–3 thin bracts 1.5–3 cm long; spathe 2–3 cm long, tips more or less acute; tube 2–2.5 cm long, slender, widening gradually; single flowered at apex. **Flower**: about 3 cm diameter; falls 0.7 cm wide, about 2 cm long, spreading, obovate–spathulate, pale violet, no crest; anthers about 0.5 cm long. **Flowering** July. $2n = ?$

Cultivation
From alpine meadows at around 3800 m, so a cold greenhouse at present.

Observations
This very attractive miniature iris has just reached cultivation and little information is yet available. Flowers seen at Kew in 1994 were a rich blue and extremely close to the ground. The bracts had very narrow maroon-coloured edges and the leaves were single. The crest was only vestigial. When compared with *I. collettii* the stem is longer, but the flower tube shorter. It was named for the discoverer, J.D.A. Stainton.

Subgenus Xiphium (Miller) Spach

HENNING CHRISTIANSEN

Map 25, Figure 25.

The Xiphium subgenus, together with the 'juno' and 'reticulata' subgenera (Scorpiris and Hermodactyloides, respectively), contain the bulbous irises, but the three groups are not difficult to separate.
(a) The bulbs: the xiphiums have papery tunics, the reticulatas netted tunics and the junos swollen roots.
(b) the plants: the xiphiums are tall plants (25–90 cm) with channelled leaves, the reticulatas are low (10–15 cm) with leaves mostly square or cylindric in cross section, and the junos have channelled leaves arranged in a similar way to those of maize.

All the Xiphium species occur in southern Europe and N Africa (S France, S Italy, Sicily, Corsica, Spain, Portugal, Gibraltar, Tangier, Morocco, Algeria and Tunisia), and grow in soils which in most cases dry out during the summer. Only *I. boissieri* and *I. latifolia* come from areas with more humidity during the summer months and with colder, wet winters. The leaves of these two last species only appear in spring, whereas the other more southern ones already show up in the autumn.

Cultivation

Considering the natural habitats of the species, they can be roughly divided into two groups.
(a) The southern species should be kept fairly dry during the dormant period (summer) and planted in September, when they start growth, in an alkaline, well-drained soil.
(b) The northern species (*I. boissieri* and *I. latifolia*) should not be dried out so much during the summer months and can eventually be planted a little later as they only send leaves up in spring even in warmer climates.

Propagation

As most of the species produce offset bulbs around the mother bulb, these can be detached when they separate easily from the main bulb. After re-planting in September they will reach flowering size in 1–2 years. The species can also be raised from seed which should, for the best results, be sown in autumn or early spring.

I. boissieri Henriq.

Synonym
I. diversifolia Marino.

Distribution
N Portugal (Serra de Gerez) and adjacent Spain; a very limited distribution.

Description
Roots slightly fleshy. **Leaves** yellowish green, appearing in spring. **Stem** 30–40 cm tall. **Flower** usually solitary, deep purple, tube 3.0–4.5 cm long; yellow stripe on falls; sparsely bearded. **Flowering** May–June. $2n = 36$.

Cultivation
Not very much is known about this species, which is considered difficult outside the natural habitat. Coming from a region with cold winters, it should be hardy.

Observations
A very cold winter and cool summer may account for cultivation troubles. Perhaps bulbs should be refrigerated to prevent too early sprouting.

I. filifolia Boiss.

Distribution
SW Spain, Gibraltar, Morocco, Tangier.

Description
Bulb ovoid, invested in membranous leaf bases. **Leaves**: glaucous, 0.05–0.3 cm wide, appearing in autumn. **Stem**: 25–45 cm tall. **Flowers**: 1–2, dark reddish-purple; orange patch on falls; tube 0.12–0.3 cm long. **Flowering** June. $2n = 32$.

Cultivation
Probably not hardy and should be grown in a bulb frame or cold house.

var. *latifolia* Foster
A very wide-leaved variant; the leaves are even broader than those of *I. xiphium*.

var. *filifolia* Boiss.
Leaves thread-like.

I. juncea Poiret

Synonym
I. imberbis Asch. & Graebn.

Distribution
SW Spain, Sicily, N. Africa.

Description
Bulb ovoid, invested by coriaceous leaf bases becoming fibrous above. **Leaves** 0.1–0.3 cm wide, appearing in autumn. **Flower** bright yellow, scented; tube 3.5–5 cm long. **Flowering** June. $2n = 32$.

Cultivation
Probably not hardy and should be grown in a bulb frame or cold house.

var. *mermieri* hort.
Sulphur-coloured flowers.

var. *numidica* hort.
Lemon-coloured flowers from N Africa.

var. *pallida* Barr.
Large, soft canary-yellow flowers.

I. latifolia Miller

Synonym
I. xiphioides Ehrh.

Distribution
NW Spain (Picos de Europa), Pyrenees.

Description
Bulb ovoid, invested by thin, fibrous leaf bases. **Leaves**: 0.5–0.8 cm wide, silvery on upper surface; appearing in spring. **Stem**: 25–50 cm tall. **Flowers**: 2 to a spathe, violet-blue with yellow centre on falls, occasionally white; tube 0.05 cm long. **Flowering** June–July. $2n = 42$.

Cultivation
As mentioned above, this species is hardy and can be grown out of doors in a not too dry place.

Cultivars
These are the 'English Irises', which are selected forms and may be blue, violet, purple or white, but never yellow.

I. serotina Willk.

Distribution
SE Spain.

Description
Bulb ovoid, invested by membranous leaf bases. Stem 40–60 cm tall. **Leaves** 0.2–0.6 cm wide, appearing in winter, but dying away before flowering time. **Flowers** 2–3, bluish-violet with deeper violet veins and yellowish centre on the falls; standards very small; tube 0.5–0.7 cm long. **Flowering** August. $2n = ?$

Cultivation
Can be grown in bulb frame or cold house.

I. tingitana Boiss. & Reuter

Distribution
N Africa (Morocco, ?Algeria).

Description
Leaves silvery-green, thread-like, appearing in autumn. **Stem** 40–60 cm tall. **Flowers** 1–3, pale to deep blue; tube 1–2.5 cm long. **Flowering** February–May. $2n = 28, 42$.

Cultivation
Tender in colder climates, but can be lifted during the summer months; the natural habitat has dry, warm summers. In milder areas it can be grown outdoors all year on well-drained southern slopes.

var. *fontanesii* (Godron) Maire
A more slender plant which flowers later, with darker violet-blue flowers.

var. *mellori* C. Ingram
A very robust form (up to 100 cm) with purple flowers; the falls with very rounded blades.

I. xiphium L.

Synonym
I. battandieri Foster.

Distribution
Spain, SW France, Corsica, S Italy, Portugal, Morocco, Algeria, Tunisia.

Description
Bulb ovoid, invested by thin, membranous leaf bases. **Leaves** 0.5–0.8 cm wide, glaucous, appearing in autumn. **Stem** 25–60 cm tall. **Flowers** 2–3, in different shades of blue or violet, occasionally white, but yellow forms exist (see var. *lusitanica*); falls with a yellow centre; tube 0.1–0.3 cm long. **Flowering** April–May. $2n = 34$.

Cultivation
Easily grown out of doors in an open, sunny, well-drained place.

var. *xiphium*
The most common form with blue or violet flowers; occasionally white-flowering plants (albinos) appear mixed with the normal blue forms.

var. *battandieri*
White with orange centre on falls, from Morocco and Algeria.

var. *lusitanica*
Yellow flowers, often with yellow or yellowish-white standards. Pure yellow standards in W Portugal.

var. *praecox*
A large-flowering form appearing in early April; from Gibraltar.

var. *taitii*
A clear pale blue form, nearly lavender, from Portugal and Spain.

Cultivars
These are the 'Dutch Irises' and mostly hybrids between *I. xiphium* and *I. tingitana* (and, perhaps, *I. latifolia*) in a wide range of colours: white to yellow, bronze, pale blue to deep mauve. The yellow form 'Golden Harvest' seems to be a vigorous form of var. *lusitanica*. They are widely grown and sold as cut flowers, as they are easily forced into flower nearly all the year round.

Subgenus Scorpiris Spach

BRIAN MATHEW

Map 26, Figure 26.

This interesting group of irises from western and central Asia constitutes one of the larger groupings within the genus with over 50 species at present recognised. Unfortunately they are not among the easiest of irises to cultivate, although a few can be grown without protection in cooler northern climates.

The junos, to use their popular name, are bulbous plants, the bulb covered with papery tunics and consisting of a few fleshy scales attached to a basal plate, which gives rise to several thick fleshy roots; sometimes these are so swollen as to be radish-like. The channelled leaves are produced in one plane (i.e. they are distichous) and vary greatly in size, colour and texture from species to species. In many species they are glossy green on the upper surface and grey-green beneath, and they often have a white cartilaginous or transparent margin. The standards of junos are fairly characteristic: in nearly all the species they are much reduced, sometimes to mere bristles, and are held out horizontally, or downwards, so that the name 'standard' is not very apt in the case of this particular group. Only *I. cycloglossa* has standards that are held upright, or at least semi-erect. The outer three segments of the perianth, or falls, are beardless, but have a raised, often yellow, ridge or crest in the centre, which varies from being a very low ridge to a very prominent cockscomb-like crest. Sometimes the bracts (spathe-valves) enclosing the buds provide a useful means of distinguishing between the species; in some species they are green and rigid whereas in other species they may be transparent and soft. Going into finer detail, there are seed differences: some species have a whitish fleshy aril on the seeds and others do not.

Distribution

The species of subgenus Scorpiris are distributed mainly from central Turkey and the Caucasus south to southern Israel and Jordan and east to northern Pakistan and the Pamirs in Central Asia. One species, *I. planifolia*, lies outside this area in Mediterranean Europe and North Africa.

Cultivation

Most junos are plants of semi-arid steppe country and mountains in western and central Asia, receiving cold winters and hot dry summers. Correspondingly they are on the whole not very successful in northern countries in the open ground, for they either attempt to grow during muggy winter conditions and then rot off, or they do not receive the required warm dry period in summer. Most species, therefore, must be grown under glass: either in deep pots, since they have long and vigorous roots, in an unheated, well-ventilated greenhouse, or planted in a bulb frame. A few species grow perfectly well in the open garden, exactly which depending upon the prevailing conditions in the country concerned and on the local climate and soil of the particular garden; it really is best to experiment when enough plants are available. Since most species grow naturally on limestone-derived soils, it is advisable to raise the pH at least to neutral.

When growing them in pots or bulb frames, the chosen compost should be well drained, so I normally add about 25% by volume of extra grit to a loam-based compost. Planting or potting takes place in late August or September, taking care not to damage the roots if possible. A good thick layer of grit should be placed over the bulbs so that the neck is very well drained. After this the first watering can be given and then sparing amounts through the winter until they begin really active growth in spring. At no time should water fall onto the foliage: it is likely to rest in the leaf bases and cause rotting of the neck. Throughout their growth period they should have as much air flowing through the house or frame as is possible and it is advantageous to have a small fan to blow air over them if there is insufficient natural ventilation. After flowering, watering is continued for as long as the leaves stay green, but as soon as they begin to turn yellow and the seed pods are ripe, water can be withheld until the following September.

Propagation

Irises of the juno group are best grown from seed if at all possible. This gives rise to strong bulbs and reduces the risk of virus diseases spreading. However, some vigorous species do increase vegetatively and these can be divided in early autumn when re-potting or re-planting. It is best to hand-pollinate the flowers to encourage seeds to form, preferably cross-pollinating with another individual of the same species. Seed should be sown in pots in autumn and left outside in the cold until germination takes place in winter or spring; the pots can then be moved to a frame or greenhouse and given the same treatment as the mature bulbs. There is a tendency for seedlings to remain in growth longer than older bulbs and this should be encouraged by continuing to water them for as long as they show any green on the foliage; this gives them as long a growing season as possible and a better chance to develop a reasonable size of bulb. After a short summer rest they can be re-potted at the same time as the mature bulbs. If germination does not occur in the first season it is advisable to keep them for at least one more year since they do not necessarily come up straight away.

Bulbs of the juno group should never be collected from the wild except by qualified specialists in order to introduce a small 'nucleus' into cultivation to study and propagate either a species which is not already represented in gardens, or a particularly interesting form. Many of the Central Asiatic species are very poorly known botanically and do require studying, but it is unnecessary to collect more than about five bulbs in order to establish the species in cultivation. Seeds, of course, represent a much better way of introducing a species and their collection does little harm to a population if carried out rationally.

I. aitchisonii (Baker) Boiss.

Synonyms
I. aitchisonii var. *chrysantha* Baker, *Juno aitchisonii* (Baker) Klatt, *Xiphion aitchisonii* Baker.

Distribution
Pakistan, E Afghanistan.

Description
Bulb tunics papery, dark brown and developed into a neck; storage roots long, not markedly swollen. **Leaves** 3–6, the lower ones up to 40 cm long and 0.4–0.8 cm wide at the base, well developed at flowering time, suberect and rather flaccid; margin smooth, not conspicuously white. **Stem** up to 35 cm in height, with clearly visible internodes, unbranched or with 1 branch, 1–3-flowered. Bracts 5–6 cm long, green with a membranous margin, acute. **Flowers** about 5–6 cm in diameter, either yellow or violet or sometimes bicoloured yellow and brown; tube 2.5–3.5 cm long; falls 3.5–5.5 cm long, the claw 2–3.5 cm long with a conspicuous wing about 1.3 cm wide, the blade oblong–elliptic, acute or acuminate, about 0.8–1.5 cm wide, yellow, violet or brownish with an entire or crenate deep yellow ridge; standards reflexed, 1.3–2.6 cm long, linear to oblanceolate, subentire to distinctly 3-lobed with the middle lobe acuminate to caudate and the lateral lobes obtuse and much shorter. Style branches 3.5–4.5 cm long, the lobes about 1–1.5 cm long, 0.3–0.6 cm wide, acute. Anthers 1.2–2 cm long, filaments 0.5–1.2 cm long. **Flowering** March–April. **Capsule** not seen. $2n = 34$ (M. Gustafsson, unpublished data).

Habitat
Seasonally moist areas on grassy slopes, about 900 m.

Cultivation and propagation
I. aitchisonii has proved to be fairly amenable to cultivation in deep pots in an alpine house or planted out in a bulb frame. It does, however, have a tendency to grow too tall and lanky so must be kept as cool as possible during its growing season. Seeds are produced in cultivation and there seems to be no particular

problem in germinating and growing these on to flowering size. All the plants in cultivation appear to be the same yellow form and it is probable that the violet and bicoloured ones have not yet been introduced.

Observations
This is the most easterly occurring species in the juno group, easily recognised by its tall, slender habit and branched inflorescence. The existence of yellow and violet colour forms, a feature so common in, for example, the bearded irises, is unusual in the junos.

I. albomarginata R.C. Foster

Synonyms
I. coerulea B. Fedtsch, *Juno coerulea* (B. Fedtsch.) Poljak.

Distribution
C Asia: Tien Shan and Fergana Mountains.

Description
Bulb tunics papery; storage roots long and thick. **Leaves:** lower ones 1–2(–3) cm wide at the base, clustered together at the base or separated on the stem, falcate, tapering rather abruptly at the apex, dark green; margin distinct, smooth or scabrid. **Stem** up to about 20 cm long, either hidden by the leaves or with visible internodes in the flowering stage, unbranched with (1–)2(–5) flowers. Bract and bracteole details unknown. **Flowers** variable shades of bright blue with a white crest; tube about 4 cm long; falls about 4 cm long, the claw not conspicuously winged; blade blue, sometimes with yellow marks at the side of an entire white crest; standards spreading or deflexed, 2–2.5 cm long, oblanceolate. Style branches with lobes obliquely triangular, subacute. **Flowering** March–April. **Capsule** and seeds not known. $2n = ?$

Habitat
Stony and clayey slopes in foothills, about 2000 m.

Cultivation and propagation
I. albomarginata is a fairly easily cultivated species in a bulb frame or alpine house, but is nevertheless still rather a rare plant in gardens.

Observations
This is not a well-known species in the wild and there are marked discrepancies between the description in the *Flora of the USSR* (Komarov 1935) and the material which is in cultivation. It is clear that further studies are required in the field. The

new name *I. albomarginata* was provided by Sir Michael Foster because the epithet *coerulea* had already been used at an earlier date by E. Spach for an entirely different iris, not a juno.

I. atropatana Grossheim

This is almost certainly a synonym of *I. caucasica* Hoffm. The difference was said to lie in the shape of the falls, which in *I. atropatana* have a narrowed portion between the blade and the haft. However, *I. caucasica* is quite variable in the shape of its falls and this slight narrowing seems to constitute an acceptable amount of variation for the species. *I. atropatana* was described in 1936 from specimens gathered over quite a wide area in southern Transcaucasia at Nakhichevan, and at Oltu and Kagyzman in northeastern Turkey.

I. aucheri (Baker) Sealy

Synonyms
I. assyriaca Hort. ex Irving, *I. fumosa* Boiss. & Hausskn., *I. sindjarensis* Boiss & Hausskn., *Juno aucheri* (Baker) Klatt, *Xiphion aucheri* Baker.

Distribution
SE Turkey, N Syria, N Iraq, W Iran.

Description
Bulb tunics brown; storage roots long and fleshy, but not markedly swollen. **Leaves** 5–12, the lower ones up to 25 cm long and 2.5–4.3 cm wide at bases, clustered together and concealing the stem at flowering time, well-developed, becoming spaced out up the stem later on; upper surface glossy green, lower surface paler, matt grey-green; margin smooth, not conspicuously white or thickened; upper leaves progressively smaller and bract-like. **Stem** up to 35 cm in height at flowering time, hidden by leaves at first, then with visible internodes, unbranched with 3–6 flowers. Bracts 7–9 cm long, pale green and usually somewhat swollen, with conspicuous veins. **Flowers** 6–7 cm in diameter, usually pale blue, but varying from deep violet to almost white with a yellow crest; tube about 6 cm long; falls 4.5–5 cm long, the claw 3–3.5 cm long with a conspicuous wing about 2 cm wide, the blade nearly orbicular or elliptic, undulate at the margins, about 1.5 cm long, 1.3–1.7 cm wide, violet blue to pale blue or nearly white with a yellow crinkly crest; standards spreading or slightly reflexed, about 2.5 cm long and 1 cm wide, obovate with a narrow claw. Style branches 4–5 cm long, the lobes about 1 cm long, 0.6 cm wide, obtuse. Anthers about 1.5–1.7 cm long, filaments 1.7–2 cm long. **Capsule** about 6–8 cm long, cylindrical; **seeds** brown, without an aril. **Flowering** March–April. $2n = ?$

Cultivation and propagation
I. aucheri has been cultivated for a long time and is relatively easy in the alpine house or frame, or, in a well-drained sunny position, in the open ground. It tends to increase into clumps, which can be divided; seeds are also produced in cultivation.

Observations
Although having a relatively 'western' distribution, this plant in fact more closely resembles the large Central Asiatic species such as *I. graeberiana*.

I. baldschuanica B. Fedtsch.

Synonyms
I. rosenbachiana var. *baldschuanica* (B. Fedtsch.) Dykes, *Juno baldschuanica* (B. Fedtsch.) Vved.

Distribution
C Asia: Pamir–Alai Mountains.

Description
Bulb tunics brownish-grey, thinly papery; storage roots much swollen. **Leaves:** the lower ones 1–2 cm wide at first, but expanding after flowering, clustered together, sheathing the stem, much overtopped by the flowers, greyish-green, slightly falcate, margin smooth. **Stem** very short, carrying 1–2 flowers. Bract and bracteole greenish, membranous, sheathing the tube. **Flowers** described as being yellow; tube 8–11 cm long; falls 4–5 cm long, the claw not winged, but with downturned margins, 0.7 cm wide, the blade obovate, obtuse and sometimes notched at the apex, with a yellow undissected ridge; standards oblong–ovate, obtuse or notched at the apex. Style branch lobes obliquely triangular, obtuse, toothed, 0.6–1.5 cm long. **Capsule** not known. **Flowering** March-April? $2n = ?$

Cultivation and propagation
For cultivation methods, see *I. nicolai*.

Observations
This is very closely related to *I. nicolai* and *I. rosenbachiana* and differs from these mainly in the flower colour. It is a poorly known species and was described from cultivated material of which the exact provenance was not recorded. It is possible that the juno collected by Paul Furse in Afghanistan under the number 8206 belongs to this species. This was introduced as living material and showed variation in flower colour: some creamy, some with yellowish falls and some with pale pinkish suffusion.

I. bucharica Foster

Synonyms
I. orchioides Dykes, not of Carrière, *Juno bucharica* (Foster) Vved.

Distribution
NE Afghanistan, Tadjikistan.

Description
Bulb tunics brown; storage roots long and fleshy but not markedly swollen. **Leaves** (4–)6(–8), the lower ones up to 20 cm long and 3.5 cm wide at the base, clustered together and well developed at flowering time, becoming well spaced out on the stem in the fruiting stage, slightly falcate, glossy green above, paler beneath; margin whitish. **Stem** about 10 cm long at flowering time and hidden by leaves, elongating to up to 23(–30) cm in fruit with clearly visible internodes, unbranched with (1–)2–4(–6) flowers in the leaf axils. Bract and bracteole 7.5–9 cm long, green, rather narrow and pointed. **Flowers** 5–6.5 cm in diameter, deep yellow or bicoloured, the blade of the falls yellow and the styles and standards white; tube 4.5–5 cm long; falls about 4 cm long, the claw 2–2.5 cm long, not winged, 0.7–0.8 cm wide with the margins parallel, the blade broadly elliptic to obovate, 1.5–1.7 cm long, 1.2–1.3 cm wide, yellow, often with a brownish mark or blotch on each side of the deeper yellow crenate crest; standards speading to deflexed, yellow or white, 1.5–2 cm long, lanceolate, acuminate, sometimes 3-lobed. Style branches about 3.6 cm long, the lobes 1 cm long, 0.5 cm wide, obtuse, usually toothed. Anthers about 1.2–1.5 cm long, filaments about 1 cm long. **Capsule** 5–6 cm long, ellipsoid. **Seeds** light brown with an aril. **Flowering** April–May. $2n = ?$

Habitat
Stony hills and edges of fields, 800–2500 m.

Cultivation and propagation
This is one of the easiest of junos to cultivate, requiring only good drainage and a sunny position. It can be grown outside without difficulty and does not require lifting or protecting during the summer months.

Observations
The wholly golden yellow form of this species was, for many years, cultivated under the name of *I. orchioides*. However, the true *I. orchioides* of Carrière has distinct wings on the haft of the falls and the flowers are pale translucent yellow suffused with lavender in the late stages of flowering.

I. cabulica Gilli

Synonym
Juno cabulica (Gilli) Kamelin.

Distribution
E. Afghanistan.

Description
Bulb tunics brownish, thinly papery; storage roots much swollen. **Leaves** 3–4(–6), partly developed at flowering time and about 2 cm wide, then expanding to up to 25 cm long and 4.5 cm wide, clustered together, sheathing the stem, greyish-green, falcate; margin smooth, white. **Stem** very short, concealed by the leaves, carrying 2–4 flowers. Bract and bracteole membranous, 7.5–8.5 cm long, narrow and long-pointed at the apex. **Flowers** white or very pale lilac with a yellow crest; tube 5–8 cm long, falls 3.5–4.5 cm long, the claw not winged but with down-turned margins, 1.3–1.5 cm wide, the blade elliptic and about the same width as the claw, with a yellow crenate ridge; standards lanceolate to elliptic, deflexed, 1.3–2 cm long, 0.5–0.7 cm wide. Style branches 3.8–4.2 cm long, the lobes subacute or obtuse, toothed, about 1.5 cm long, 0.6–0.8 cm wide. Anthers 1.1–1.7 cm long, filaments slightly shorter. **Capsule** about 5 cm long. **Seeds** reddish-brown with a white aril. **Flowering** March–April. $2n = ?$

Habitat
Dry slopes; 1800–2000 m.

Cultivation and propagation
Although there is no information on the cultural requirements of *I. cabulica* it seems likely that it will need the same treatment as *I. nicolai*.

Observations
This is closely allied to *I. nicolai* and *I. rosenbachiana* and differs from these mainly in the shape of the falls, which have scarcely any difference in width between the haft and the blade. It has possibly never been in cultivation, although dried specimens have been collected on several occasions in the Kabul area of Afghanistan.

I. capnoides (Vved.)–

Synonym
Juno capnoides Vved.

Distribution
C Asia: W Tien Shan mountains.

Description
Bulb tunics greyish, membranous; storage roots thickened in the middle. **Leaves** lower ones 0.8–1.6 cm wide, clustered together at flowering time and conspicuously sheathing at the base, falcate, margin distinctly thickened. **Stem** carrying 1–3 flowers. Bract details not known. **Flowers** described as 'smoky'; tube 4.5–5.5 cm long; falls 3–4 cm long, the claw strongly winged, 1.7–2.2 cm long, with a pale yellow median vein, the blade smoky coloured with a crinkly yellow ridge; standards 1.3 -1.7 cm long, pale violet, trilobed. Other details unknown. **Flowering** May. $2n = ?$

Habitat
Mountain slopes.

Cultivation and propagation
This, as far as known, is not in cultivation at the present time.

Observations
A little-known species, said by A.I. Vvedensky to differ from *I. parvula* by having conspicuously sheathed leaves, a winged claw to the falls and a different flower colour. Features such as this must, however, make it fairly unrelated to *I. parvula*.

I. carterorum **Mathew & Wendelbo**

Distribution
E Afghanistan.

Description
Bulb unknown. **Leaves** 4, lower ones 14 cm long and 2.4 cm wide at the base, clustered together and well-developed at flowering time, narrowing rather abruptly and then long-acuminate in the upper part, falcate, greyish-green, densely papillose; margin distinctly white. **Stem** hidden by leaves, unbranched with 2 flowers. Bract and bracteole 5–5.5 cm long, green with a transparent margin. **Flowers** 4–5 cm diameter, yellowish-green, spotted purple-black in the lower parts, with a yellow crest; tube 3.2 cm long; falls 3.5–3.7 cm long, the claw without a wing, 0.7 cm wide, the blade elliptic–lingulate, emarginate at apex, 1.2 cm long, 0.7 cm wide, with a serrated deep yellow crest; standards spreading or deflexed, 0.4 cm long, bristle-like, shortly lobed at the base. Style branches 2.8–3 cm long, the lobes 0.8–1 cm long, 0.2 cm wide, narrowly triangular, subacute. Anthers 0.9–1 cm long, filaments 1.2–1.4 cm long. **Capsule** (details taken from unripe fruit) ellipsoid, about 2.5 cm long. **Seeds** with a distinct aril. **Flowering** April–May. $2n = ?$

Cultivation and propagation
In the absence of any living material it is impossible to say how this is likely to behave in cultivation.

Observations
This is a little-known species, collected only twice as herbarium material and never introduced into cultivation as far as is known. In its structure it resembles *I. wendelboi*, which also has rather small flower parts, but the colour is quite different. It is named in honour of the British Ambassador to Afghanistan (1967–72) and his wife, Mr & Mrs P. Carter, who collected material of this species in 1971.

I. caucasica Hoffm.

There are two subspecies.

subsp. *caucasica*

Synonyms
Juno caucasica (Hoffm.) Klatt, *Xiphion caucasicum* (Hoffm.) Baker, ?*Iris atropatana* Grossh., but see comments on above on this species.

Distribution
NE Turkey, Caucasus.

Description
Plant about 8–18 cm at flowering time. Bulb tunics brown; storage roots fleshy but not markedly swollen. **Leaves** (4–)5–7, the lower 5–15 cm long and (0.5–)0.8–2 cm wide at the base, clustered together and concealing the stem at flowering time, well-developed, sometimes becoming spaced out up the stem later on, lanceolate, falcate; upper surface green, lower surface grey-green; margin smooth, white. **Stem** hidden by leaves or with internodes just visible, unbranched, with 1–4 flowers. Bract and bracteole almost equal, 6–7.5 cm long, pale green and usually slightly swollen and tightly sheathing the perianth tube at the apex. **Flowers** 5–6 cm in diameter, greenish-yellow with a darker yellow crest; tube about 3–4(–4.7) cm long; falls (3–)3.5–4cm long, the claw (1.1–) 1.5–2.5 cm long with a conspicuous wing about 1–2 cm wide, the blade elliptic, slightly narrower than the claw, (0.9–)1.1–2 cm long, 1–1.6 cm wide, yellowish-green with a dark yellow or orange crinkly crest; standards spreading or reflexed, 1.3–2.5 cm long and 0.2–0.8 cm wide, linear, oblanceolate or 3-lobed. Style branches 2.3–3.5 cm long, the lobes about 1 cm long, 0.5–0.8 cm wide, obtuse. Anthers about 1.2 cm long, filaments about 0.8 cm long. **Capsule** about 3 cm long, ellipsoid; **seeds** brown, without an aril. **Flowering** May–June. $2n = ?$

Habitat
Rocky slopes and mountain steppe, 2200–2300 m.

Cultivation and propagation
I. caucasica is one of the easiest of the smaller junos to cultivate and makes an excellent subject for the alpine house or frame in a gritty compost. It sets seeds fairly readily and increases slowly by bulb division.

Observations
The flower colour is normally a greenish yellow, but occasionally, especially south of Lake Van in Turkey, the plants may be found with dirty bluish-green flowers, which have a darker blotch on the blade of the falls; these are probably hybrids with *I. persica*.

subsp. *turcica* B. Mathew

Distribution
C & NE Turkey.

Description
As for subsp. *caucasica*, but with the leaf margins and veins smooth. **Flowering**: April–June. $2n = ?$

Habitat
Bare stony slopes, scree and steppe, 1200–3500 m.

Cultivation and propagation
As for subsp. *caucasica*.

Observations
Although differing only slightly from subsp. *caucasica*, this also has a more western distribution in Turkey, the two meeting in the border areas between the Caucasus and Turkey.

I. cycloglossa Wendelbo

Distribution
W Afghanistan, known only in the neighbourhood of Herat.

Description
Bulb tunics blackish-brown; storage roots long and slender. **Leaves** about 6, lower ones up to 30 cm long and 1.5 cm wide at the base, well spaced out and well devel-

oped at flowering time, tapered gradually to the pointed apex; upper surface green, lower surface slightly grey-green; margin smooth, not conspicuously white. **Stem** 20–40 cm in height, with clearly visible internodes, unbranched or with 1 to few branches, each branch with 1 or 2 flowers. Bracts adjacent to branches 7–9 cm long, long-tapering, green; floral bracts 4.5–6.5(–7) cm long, unequal, rather leathery in texture, green. **Flowers** large, up to 10 cm or more in diameter, fragrant, primarily lavender-blue with a large white patch on the falls; tube 3.3–4 cm long; falls 6–7 cm long, the claw 2.5–3 cm long with a conspicuous wing about 2 cm wide, the blade nearly orbicular, about 3.5–4 cm diameter, blue with a large white zone in the centre and smaller yellow blotch around a raised yellow ridge; standards spreading obliquely upwards, about 4 cm long, 1.3 cm wide, widest towards the apex and with a wedge-shaped base. Style branches 4–4.5 cm long, 0.6–0.8 cm wide, the lobes about 0.8 cm long, 0.3–0.5 cm wide. Anthers yellowish, about 1.5 cm long, filaments 1.5 cm long. **Capsule** 5–6 cm long, cylindrical; seeds brown without an aril. **Flowering** May. $2n = 28$ (Gustafsson & Wendelbo 1975).

Habitat
Wet grassy areas, drying out somewhat in summer, 1450–1700 m.

Cultivation and propagation
I. cycloglossa has proved to be a beautiful and easily cultivated species in the open ground, although it is also very successful in a bulb frame or deep pot. It increases naturally by vegetative means so that division of established clumps is possible. Seeds are also produced in cultivation. All the plants in cultivation are almost certainly derived from the collection Wendelbo & Eckberg 7727.

Observations
A botanically rather isolated species, instantly recognised by the large blue flowers in which the standards are inclined upwards, quite unlike all other species in which they are small and deflexed or horizontal. The pollen is quite different from that of other species in the subgenus Scorpiris. The moist habitat is also rather unusual.

I. doabensis Mathew

Distribution
NE Afghanistan.

Description
Bulb tunics brownish, thinly papery; storage roots much swollen. **Leaves** 5, expanding to up to 20 cm long and 4 cm wide after flowering, clustered together, sheathing the stem and partly developed at flowering time, glossy bright green, glabrous, falcate; margin not conspicuous. **Stem** very short and completely concealed by the leaves at flowering, extending in the fruiting stage to up to 12 cm

with clearly visible internodes. Bract and bracteole 5–6 cm long, greenish, membranous, sheathing the tube. **Flowers** yellow with an orange-yellow crest; tube 7–8 cm long; falls about 3.5 cm long, the claw not winged but with down-turned margins, about 1.6 cm wide, the blade broadly elliptic, about 1.3 cm long, 1.2 cm wide, with a prominent deep yellow or orange-yellow crest; standards about 0.8 cm long, rhomboid–ovate, acute. Style branches 3.3 cm long, the lobes obtuse, about 1 cm long, 0.4–0.5 cm wide. Anthers 1.2 cm long, filaments 1 cm long. **Capsule** narrowly ellipsoid, acuminate. **Seeds** brown, with an aril. **Flowering** March ? $2n = $?

Cultivation and propagation
I. doabensis was cultivated by Dr J.G. Elliott in the 1960s. It was given alpine house conditions and appears to be similar in its requirements to *I. nicolai*.

Observations
This is very closely related to *I. nicolai* and *I. baldschuanica* but has a markedly different flower colour. It has been collected only once, by Paul Furse, and is not known to be in cultivation at present.

I. drepanophylla Aitch. & Bak.

There are two subspecies.

subsp. *drepanophylla*

Synonym
Juno drepanophylla (Aitch. & Bak.) Rodion.

Distribution
E Iran, N & W Afghanistan, S Turkmenistan.

Description
Bulb tunics brown; storage roots rather short and swollen. **Leaves** (4–)6–9, up to 20 cm long and 2.5 cm wide, well developed at flowering time, clustered together at first, but becoming spaced out on the stem in the fruiting stage, strongly falcate and long-tapering at the apex, glossy green above, green or slightly greyish-green beneath; margin often undulate, minutely hairy, white. **Stem** more or less concealed by the leaves at flowering time, elongating to up to 20 cm with clearly visible internodes, unbranched, with 2–8 flowers in the leaf axils. Bract and bracteole green and membranous, 5–6.5 cm long, minutely papillose. **Flowers** 4–5 cm in diameter, yellow with a deeper yellow crest; tube 3.5–4 cm long; falls 3.5–4.5 cm long, the claw 2.2–3.5 cm, not conspicuously winged but widening gradually from the 0.7 cm wide base to about 1.1 cm wide in the upper part, the margins folded downwards, the blade lingulate, 1.2–1.5 cm long, 0.7–1.1 cm wide,

with a very prominent toothed crest; standards spreading or deflexed, very narrowly lanceolate or subulate, about 0.3 cm long. Style branches 3–4 cm long, the lobes obtuse 0.8–1.2 cm long, 0.2–0.5 cm wide. Anthers about 1.6 cm long, filaments about 0.8 cm long. **Capsule** 2.5–3 cm long, ellipsoid. **Seeds** reddish-brown with a white aril. **Flowering** April. $2n = ?$

Habitat
Dryish slopes, 600–1700 m.

Cultivation
I. drepanophylla has always been very rare in cultivation so there is not much information available, but it seems that it presents no great difficulties given standard juno treatment in an alpine house.

Observations
Although this species is very similar in appearance to *I. kopetdaghensis,* the two can be distinguished easily in the living state by observing the margins of the haft of the falls. In *I. drepanophylla* the margins of the haft are turned downwards whereas in *I. kopetdaghensis* they are curved upwards to give the claw a gutter shape.

subsp. *chlorotica* Mathew & Wendelbo

Distribution
NE Afghanistan.

Description
As for subsp. *drepanophylla* but with green flowers with a very pale yellow crest on the falls. **Flowering** April. $2n = ?$

Habitat
Loess hills in sparse grass, 700–1400 m.

Cultivation
As for subsp. *drepanophylla.*

Observations
This subspecies occupies the eastern part of the overall distribution of the species and has been recorded only three times. It was introduced into cultivation by Paul Furse in the 1960s under the name of 'Lime Green', but is probably no longer in existence in gardens.

I. edomensis Sealy

Distribution
Jordan, Edom.

Description
Bulb tunics membranous, dark or greyish-brown, extended into a neck; storage roots not markedly swollen. **Leaves** 4–5, lower ones 0.5–1.4(–1.6) cm wide at the base, clustered together at base at flowering time, falcate and very undulate in the wild but less so in cultivation, long-tapering to the very acute apex, prominently close-nerved, green above, greyish beneath; margin conspicuous, white and thickened. **Stem** short and hidden by the leaves, carrying 1–2 flowers. Bract and bracteole 5–6 cm long, membranous, green. **Flowers** conspicuously blotched violet on a white ground with a yellow, purple-spotted, stripe in the centre of the falls; tube 5–6.5 cm long; falls white, heavily blotched violet with a yellow signal stripe; claw wedge-shaped, not prominently winged, the blade ovate; standards spreading, 2.2–2.5 cm long, 0.4–0.5 cm wide, ovate with a narrow claw. Style branches about 3–3.4 cm long, with lobes about 1 cm long, 0.2–0.3 cm wide, acuminate. **Capsule** cylindrical, about 1.6 cm long. **Seeds** brown, without an aril. **Flowering** April. $2n = ?$

Habitat
Rocky situations, probably sandstone formations, about 1220–1525 m.

Cultivation
I. edomensis is a fairly difficult plant to cultivate and is only suitable for the alpine house or bulb frame. It has been cultivated in Britain: in fact it was first described from plants grown by Mrs Gwendolyn Anley from bulbs collected by Dr P.H. Davis. Recently, it has been re-introduced.

Observations
This is most probably related to *I. postii*.

I. fosteriana Aitch & Bak.

Synonyms
I. caucasica var. *bicolor* Regel, *Juno fosteriana* (Aitch. & Bak.) Rodion.

Distribution
NW Afghanistan, NE Iran, Turkmenistan.

Description
Bulb slender and elongated, tunics dark brown, forming a long neck; storage roots long and slender, not markedly swollen. **Leaves** 4–5, the lower ones up to 17 cm long and 0.4–0.8 cm wide at the base, clustered together and well developed at flowering time, becoming spaced out on the stem in the fruiting stage, falcate and glossy green above, greyish-green and densely papillose beneath; margin silvery-white. **Stem** 4–9 cm long at flowering time, hidden by leaves, elongating to up to 23 cm in fruit with clearly visible internodes, unbranched, with 1(–2) flowers. Bract and bracteole 4–6 cm long, green and papillose. **Flowers** 4–5 cm in diameter, bicoloured, the falls yellow and standards violet; tube 3.5–4 cm long; falls 3.5–4.5 cm long, the claw 2–2.5 cm long, not winged, 0.5–0.7 cm wide with the margins parallel, the blade broadly elliptic to suborbicular, 1.5–2 cm long, 1.2–2 cm wide, yellow with a few brownish veins, pale creamy margins, a conspicuous darker yellow blotch in the centre and a prominent yellow crest; standards sharply deflexed, violet, 2–3 cm long and 0.6–1.2 cm wide, elliptic–obovate with a narrow claw. Style branches about 3.5 cm long, the lobes 1–1.3 cm long, 0.5–0.6 cm wide, obtuse, usually toothed. Anthers about 1–1.4 cm long, filaments slightly longer. **Capsule** 3.5–4.5 cm long, narrowly ellipsoid. **Seeds** light brown without an aril. **Flowering**: March–April. $2n = 18$ (Gustafsson & Wendelbo 1975).

Habitat
Dry steppe, 750–2000 m.

Cultivation
This is a striking species, but is rare in cultivation and not at all easy to grow. The best chance of success seems to be in an alpine house or bulb frame with plenty of ventilation, planted in long pots to accommodate the long slender bulbs and deep roots. When growing well, it is a clump-forming species.

Observations
This is a very distinctive species in its flower colour and is unlikely to be confused with any other, except perhaps *I. narbutii*, which also has bicoloured flowers.

I. galatica Siehe

Synonyms
I. eleonorae Holmboe, *I. purpurea* (hort.) Siehe.

Distribution
C & CE Turkey.

Subgenus Scorpiris

Description
Bulb tunics brown, papery; storage roots long and slender, not markedly swollen. **Leaves** 3–4(–5), the lower ones expanding later to about 10–12 cm long and 0.6–1.2 cm wide at the base, clustered together and poorly developed at flowering time, remaining basal, even in the fruiting stage, falcate, glossy green above, greyish-green beneath; margin thickened and white, ciliate or scabrid. **Stem** more or less absent at flowering time, hidden by leaves and not elongating in fruit, unbranched, with 1–2 flowers. Bract and bracteole about equal, usually rather stiff and erect, often slightly swollen, green. **Flowers** 5–6 cm in diameter, reddish-purple, greenish-yellow or pale silvery purple, usually with a darker purple or violet blade to the falls and with a yellow or orange crest; tube (4–)6–8 cm long; falls (2.6–)3.2–5 cm long, the claw 2–3.5 cm long with a conspicuous wing 1.2–2.7 cm wide, the blade obovate, (0.9–)1.2–2 cm long, (0.5–)0.7–1.2 cm wide, with a very prominent yellow or orange crest; standards spreading or slightly deflexed, (0.9–)1.1–2.8 cm long and 0.4–0.7 cm wide, spathulate or oblanceolate, obtuse, sometimes 3-lobed. Style branches (2.5–)3–5 cm long, the lobes large and conspicuous, obtuse and usually toothed. **Capsule** 2.8–4 cm long, cylindrical. **Seeds** brown without an aril. **Flowering** March–April. $2n = ?$

Habitat
Rocky places, in steppe vegetation or sparse oak and juniper scrub, (400–)900–1700 m.

Cultivation
This requires similar treatment to *I. persica* although it does not appear to be quite so tricky to grow and is more persistent in cultivation.

Observations
This is very variable in its flower colour and may be wholly purple, a silvery-purple with darker falls, or greenish-yellow with a purple blade to the falls. Some populations in central Turkey approach *I. persica* in their coloration and bract characters and it seems likely that hybridisation is taking place in this region.

I. graeberiana Sealy

Distribution
C Asia, precise area not known.

Description
Bulb tunics dark grey, papery; storage roots long and thick, not markedly swollen. **Leaves** about 8, lower ones 1.3–4 cm wide at the base, clustered together, becoming spaced out on the stem later on after flowering, falcate, shiny

green above, greyish-green beneath; margin distinct, white. **Stem** 10–20 cm long, extending to 40 cm later on, usually completely hidden at flowering time, with internodes becoming visible after flowering, unbranched with 4–6 flowers. Bract and bracteole green, with membranous margins. **Flowers** lavender-blue with a darker violet-blue apex to the falls and with a large white, blue-veined zone on either side of the white crest; tube about 5.3–6 cm long; falls about 5–6 cm long, the claw with a wing 2–2.3 cm wide; blade oblong, about 2–2.6 cm long, 1.2–1.6 cm wide, dark violet-blue at the apex with a white, blue-veined area around an undulate white papillose crest; standards spreading, 2–2.6 cm long, ovate with a narrow claw, obtuse. Style branches with lobes about 1.5–1.7 cm long, 0.7–0.8 cm wide, ovate, obtuse. **Capsule** not seen. **Flowering** April (not known in the wild). $2n = ?$

Habitat
Unknown.

Cultivation
This is an attractive and easily cultivated species for the bulb frame or alpine house and will also do well planted out in a sunny well-drained position.

Observations
A species described from material collected by Paul Graeber in central Asia, distributed by the firm of van Tubergen. It is very closely related to *I. albomarginata* and *I. willmottiana*, but little can be said about the relationships of these until there has been a thorough field survey to determine the range of variation.

I. hippolyti (Vved.)–

Synonym
Juno hippolyti Vved.

Distribution
C Asia: Kyzyl Kum Desert.

Description
Bulb tunics papery; storage roots swollen. **Leaves:** lower ones 1–1.5 cm wide at the base, clustered together, falcate, margin distinct, smooth. **Stem** about 10 cm long, hidden by the leaves, unbranched, with a solitary flower. Bract and bracteole details unknown. **Flowers** pale violet with a yellow signal patch in the centre of the blade of the falls around a white crest; tube about 4 cm long; falls about 4–4.5 cm long, the claw with a wing 2 cm wide; blade oblong, about 1.5 cm long, 1 cm wide, pale violet with a yellow area around an entire crest, which is white with a yellow base; standards spreading or deflexed, about 1.5 cm long, rhombic

or weakly 3-lobed, acute. Style branches with lobes about 1.3 cm long, 0.4 cm wide, obliquely triangular, subacute. **Capsule** and seeds unknown. **Flowering** April. $2n = ?$

Habitat
Arid rocky places with *Amygdalus*, *Zygophyllum* and *Atraphaxis*.

Cultivation
This is unknown in cultivation.

Observations
A very little-known species, probably collected only once. It was said by A.I. Vvedensky to be related to *I. willmottiana*, but differing in having narrower leaves and by having a yellow zone in the centre of the falls.

I. hymenospatha Mathew & Wendelbo

There are two subspecies.

subsp. *hymenospatha*

Distribution
C & S Iran.

Description
Bulb tunics brown, forming a long neck reaching to soil level; storage roots long and slender, not markedly swollen. **Leaves** 3–4(–5), lower ones 4–10 cm long and 0.4–0.6 cm wide at flowering time but becoming broader and longer in fruit, clustered together and only partly developed at flowering time, remaining basal, even in the fruiting stage, slightly falcate, glossy green above, greyish-green beneath; margin silvery-white and papillose-scabrid. **Stem** more or less absent at flowering time, hidden by leaves and not elongating in fruit, unbranched with 1–3 flowers. Bract and bracteole similar, about 6 cm long, white and membranous. **Flowers** 5–6 cm in diameter, silvery-white with a violet-blue zone on the blade of the falls around a yellow crest; falls 3–4.5 cm long, the claw 2.5–3 cm long with a conspicuous wing 2–2.4 cm wide, the blade ligulate, 0.5–1.5 cm long, 1–1.2 cm wide, with a low yellow crest; standards spreading, 1.5–2 cm long and about 0.5 cm wide, oblanceolate or obovate. Style branches 2.5 cm long, the lobes 1–1.3 cm long, 0.5 cm wide, obtuse. Anthers about 1.2 cm long, filaments 1.2–1.4 cm long. **Capsule** about 3.5 cm long, ellipsoid; **seeds** light brown without an aril. **Flowering** February–May. $2n = ?$

Habitat
Dry stony hills in open ground and in sparse scrub, 1500–2000 m.

Cultivation
I. hymenospatha is probably one of the trickiest of the junos to cultivate since it originates from a very hot dry region and when brought into a damper climate comes into growth very early in the year when the atmosphere is moist and the light intensity rather poor. The best chance of success is to grow it in a well-aired alpine house with as much light as possible, perhaps supplemented by artificial lighting.

Observations
This is almost certainly the same plant as Foster's *I. persica* var. *isaacsonii.*

subsp. *leptoneura* Mathew & Wendelbo

Distribution
C & W Iran, NE Iraq.

Description
As for subsp. *hymenospatha,* but with slightly broader leaves, which are usually more strongly falcate and lack the very prominent ribs on the underside which are a feature of subsp. *hymenospatha.* **Flowering** February–May. $2n = ?$

Habitat
Dry slopes, 1200–2250 m.

Cultivation
Subsp. *leptoneura* seems to be slightly easier to grow than subsp. *hymenospatha,* but neither has been tried to any extent so information is rather sparse.

Observations
This subspecies has a slightly more north-western distribution than subsp. *hymenospatha* and differs most obviously in its leaf characters.

I. inconspicua (Vved.)–

Synonym
Juno inconspicua Vved.

Distribution
C Asia: W Tien Shan Mountains.

Description
Bulb tunics not known; storage roots 'funiculiform'. **Leaves** clustered together at flowering time, falcate; margin thickened. **Stem** about 5 cm, hidden by the leaf

bases, carrying 1–3 flowers. Bract details not known. **Flowers** pale lilac; tube 4.5 cm long; falls 4–4.5 cm long, the claw not winged, with almost parallel margins, about 0.6–0.7 cm wide, white spotted dull green, the blade lilac with a dissected white crest; standards about 1 cm long, lilac, trilobed, acute. Other details unknown. **Flowering** March. $2n = ?$

Habitat
Gypsum hills.

Cultivation
This is, as far as is known, not in cultivation at the present time.

Observations
A.I. Vvedensky notes that this is similar to *I. kuschakewiczii* but that the spots on the falls are dull green, not deep violet.

I. kopetdagensis (Vved.) Mathew & Wendelbo

Synonym
Juno kopetdagensis Vved.

Distribution
NE Iran, NW Afghanistan, Turkmenistan (Kopet Dag).

Description
Bulb tunics brown; storage roots rather short and swollen. **Leaves** 4–10, 10–25 cm long and 0.9–2.5 cm wide, well developed at flowering time, clustered together at first but becoming spaced out on the stem in the fruiting stage, strongly falcate and long-tapering at the apex, glossy green above, green or slightly greyish-green and papillose-puberulent beneath; margin smooth or minutely hairy, white. **Stem** more or less concealed by the leaves at flowering time, elongating up to 28 cm with clearly visible internodes, unbranched with (1–)3–9 flowers in the leaf axils. Bract and bracteole green and membranous, about 5 cm long, minutely papillose. **Flowers** 4–5 cm in diameter, greenish-yellow with a deeper yellow crest; tube 4–5(–6) cm long; falls 4–5 cm long, the claw 2.8–3.5 cm, not winged, 0.6–0.8 cm wide with parallel margins, the margins curved upwards, the blade elliptic-ovate, 1.2–1.4 cm long, 1 cm wide, with a very prominent toothed deep yellow crest; standards deflexed, very narrowly lanceolate or subulate, about 0.5–0.7(–1.2) cm long. Style branches 3.5–4.5 cm long, the lobes acute, 1–1.1 cm long, 0.5 cm wide. Anthers 1.2–1.6 cm long, filaments 1–1.6 cm long. **Capsule** 3 cm long, ellipsoid. **Seeds** brown, with a small aril. **Flowering** March–May. $2n = ?$

Habitat
Dryish hillsides and edges of fields, 1000–3000 m.

Cultivation
This appears to be not too difficult to cultivate in an alpine house or bulb frame, but is very rare in cultivation at present.

Observations
I. kopetdagensis has a more slender, greener flower than the similar *I. drepanophylla*. The two may be easily distinguished in the living state because in the former the haft of the falls has upturned margins while in *I. drepanophylla* they are folded downwards; this feature is, however, difficult to observe in dried specimens.

I. kuschakewiczii B. Fedtsch.

Synonym
Juno kuschakewiczii (O. Fedtsch.) Poljak.

Distribution
C Asia: Tien Shan Mountains.

Description
Bulb tunics papery; storage roots long and thick. **Leaves** 4–5, the lower ones 1–1.5 cm wide at the base, clustered together at the base, straight or slightly falcate, tapering gradually to the apex, dark green; margin distinct, scabrid. **Stem** up to about 5 cm long, hidden by the leaves, unbranched with 1–3(–4) flowers. Bract and bracteole green with membranous tips and margins. **Flowers** blue or lilac-blue, the falls with a white signal patch and dark violet blotches and streaks around the white crest; tube about 3.5–4.5 cm long; falls about 3.5–4 cm long, the claw slightly winged, up to 1 cm wide, narrowing into the blade; blade oblong–obovate, about 1.2 cm long, 0.7 cm wide, pale violet with darker blotches and with an entire or toothed white crest; standards spreading or deflexed, 1–1.5 cm long, often 3-lobed, acute. Style branches with lobes 1 cm long, 0.4 cm wide, obliquely triangular, subacute. **Capsule** and seeds not known. **Flowering** April–May. $2n = ?$

Habitat
Rocky slopes in foothills.

Cultivation
I. kuschakewiczii is proving to be reasonably easy to cultivate, given the standard juno treatment in an alpine house or cold frame.

Observations
The plants that have been introduced into cultivation under this name are very similar to *I. willmottiana* and it seems that the two species might not be distinct; the main difference is based on the width of the claw of the falls.

I. leptorrhiza (Vved.)–

Synonym
Juno leptorrhiza.

Distribution
C Asia: Tabakcha Mts nr. R. Vakhsh.

Description
Bulb tunics brown, conspicuously veined; storage roots slender. **Leaves** 3–4, lower ones 0.7–1 cm wide, clustered together at flowering time, falcate; margin distinctly thickened, scabrid. **Stem** carrying 1 flower. Bract details not known. **Flowers** described as 'violet-green' from the dried material; tube about 4 cm long; falls about 3.5 cm long, the claw not winged, with parallel nargins, 0.5 cm wide, the blade obovate, notched at the apex, with a smooth ridge; standards about 1 cm long, linear–lanceolate, acute. Style branch lobes obliquely triangular, acute, entire, 0.3 cm wide, 1 cm long. Other details unknown. **Flowering** March. $2n = ?$

Habitat
Arid mountain slopes.

Cultivation
This is almost certainly not in cultivation at the present time so it is impossible to comment on its needs.

Observations
A little-known species: as far as is known only collected once. The slender roots make it rather uncharacteristic of the juno group, in which they are normally swollen to varying degrees.

I. linifolia (Regel) O. Fedtsch.

Synonyms
I. caucasica var. *linifolia* Regel, *Juno linifolia* (Regel) Vved.

Distribution
C Asia: Tien Shan and Tadjikistan, Pamir–Alai Mountains.

Description
Bulb tunics papery; storage roots short and swollen. **Leaves:** lower ones 0.4–0.7 cm wide at the base, separated up the short stem and showing short internodes, slightly falcate, tapering gradually to the apex; margin distinct, smooth or scabrid. **Stem** 5–10 cm long, with short internodes visible, unbranched, with 1(–2) flowers. Bract and bracteole details unknown. **Flowers** small, pale yellow with a white ? crest; tube about 4 cm long; falls 3.5–4) cm long, the claw without a wing, with almost parallel margins, 0.6 cm wide, the blade broadly oblong, 1.2–1.5 cm long, 0.9–1.2 cm wide, pale yellow with a whitish crenate to rarely dissected crest; standards spreading, about 1 cm long, bluntly 3-lobed. Style branches with lobes about 1.1 cm long, obliquely triangular, obtuse. **Capsule** and seeds not known. **Flowering** May–June. $2n = ?$

Habitat
Stony and gravelly slopes in the upper mountain zone, about 2500 m.

Cultivation
No information is available, but it seems likely that since it is a fairly high mountain species it would not be too difficult with the standard juno treatment in an alpine house or cold frame.

Observations
This is a little-known species, very rare in cultivation.

I. magnifica **Vved.**

Synonyms
I. amankutanica O. Fedtsch., *Juno magnifica* (Vved.) Vved.

Distribution
Tadjikistan, Pamir–Alai Mountains.

Description
Bulb tunics papery; storage roots long and thick, not markedly swollen. **Leaves:** lower ones 3–5 cm wide at the base, spaced out on the stem, falcate, light green; margin indistinct, scabrid. **Stem** 20–40 cm long, with clearly visible internodes, unbranched, with 2–7 flowers in the upper leaf axils. Bract and bracteole green, membranous. **Flowers** pale lilac (or white in 'Alba') with a yellow zone around a white crest; tube about 4.5–5 cm long; falls about 4–5.5 cm long, the claw widely winged, the wing 2–2.5 cm wide; blade oblong, 2–2.5 cm long, 1.3–2.2 cm wide, pale lilac with a yellow area around an entire, white or partly yellow crest; standards spreading or deflexed, 2.2–2.7 cm long, obovate, obtuse. Style branches with

lobes 1–1.1 cm long, 0.4–0.5 cm wide, obliquely triangular, obtuse. **Capsule** cylindrical. **Seeds** without an aril. **Flowering** April–May. $2n = ?$

Habitat
Rock crevices and stony slopes.

Cultivation
This, perhaps the most garden-worthy of all the junos, is an attractive species and easy to cultivate in any open, sunny, well-drained situation. It increases vegetatively into clumps, which can be divided in early autumn, and also sets seeds quite freely. The albino version 'Alba' is a delightful plant, as easy to cultivate as the typical form.

I. maracandica **Vved.**

Synonym
Juno maracandica Vved.

Distribution
Tadjikistan, Pamir–Alai Mountains.

Description
Bulb tunics membranous; storage roots thickened in the middle. **Leaves:** lower ones 1.5–2 cm wide at the base, clustered together at flowering time, falcate; margin conspicuous, smooth, thickened, whitish. **Stem** 10–15 cm, carrying 1–4 flowers. Bract and bracteole membranous, green. **Flowers** pale yellow; tube 3–4.5 cm long; falls 3.5 cm long, the claw with a conspicuous wing about 2 cm wide, the blade yellow with a darker yellow toothed crest; standards 1–1.5 cm long, narrowly rhombic, acute. Style branches with lobes about 1 cm long, 0.5 cm wide, obliquely triangular, subacute. **Capsule** and seeds unknown. **Flowering** March–April. $2n = ?$

Habitat
Stony slopes in foothills.

Cultivation
Plants in cultivation as *I. maracandica* indicate that this is a fairly easy juno for the alpine house or bulb frame.

Observations
This is clearly very similar to *I. svetlanae*, but has paler yellow flowers. It seems likely that the two are not distinct enough to be maintained as separate species, but only field studies can shed further light on this question.

I. microglossa Wendelbo

Synonym
Juno microglossa (Wendelbo) Kamelin.

Distribution
NE & Central Afghanistan.

Description
Bulb tunics blackish-brown; storage roots long and slender. **Leaves** 4–6, lower ones up to 20 cm long and 1.5 cm wide at the base, spaced out and well developed at flowering time, tapered gradually to the pointed apex; upper surface bluish-green or greyish-green, lower surface paler grey-green; margin ciliate. **Stem** 5–16 cm in height at flowering time, with visible internodes, unbranched with 1–4 flowers. Bracts and bracteoles 5.5–6.5 cm long, greyish-green with ciliate margins. Flowers 4.5–5.5 cm in diameter, pale lavender-blue or nearly white with a very pale yellow crest; tube 3–4 cm long; falls 4–4.5 cm long, the claw 2.5–3.5 cm long with a conspicuous wing about 1.8 cm wide, the blade lingulate, 1–1.7 cm long, lavender-blue or whitish with a white or very pale yellow, coarsely toothed crest; standards spreading, 1.5–1.8 cm long, 0.4–0.6 cm wide, oblanceolate or lanceolate with a narrow claw, toothed at apex. Style branches about 3.5 cm long, 0.5–0.9 cm wide, the lobes about 1.3 cm long, 0.6 cm wide, irregularly toothed. Anthers pale violet-blue, about 1.3 cm long, filaments 1.3 cm long. **Capsule** about 4 cm long, ellipsoid. **Seeds** brown, without an aril. **Flowering** April–June. $2n = ?$

Habitat
Dry slopes, sometimes among junipers, 1700–2100 m.

Cultivation
I. microglossa was brought back by Paul Furse under the name of 'Salang Blue' in the 1960s and has also been collected by P. Wendelbo, T. Hewer and C. Grey-Wilson. However, it remains a very rare plant in gardens. It appears to be reasonably easy in alpine house conditions in deep pots.

Observations
The silvery-lavender flowers and well-developed greyish leaves with ciliate margins make this a rather distinctive species.

I. narbutii O. Fedtsch.

Synonyms
I. dengerensis B. Fedtsch., *I. hissarica* O. Fedtsch., *Juno narbutii* (O. Fedtsch.) Vved.

Distribution
C Asia: Syr-Darya and Tadjikistan, Pamir–Alai Mountains.

Description
Bulb tunics papery; storage roots swollen. **Leaves:** lower ones 0.5–2.5 cm wide at the base, clustered and usually remaining basal but sometimes becoming slightly spaced out in the fruiting stage, falcate, tapering gradually to the apex, dark green; margin distinct, smooth or very rarely scabrid. **Stem** about 5–10 cm long, hidden by the leaves, unbranched, with 1–2(–6) flowers. Bract and bracteole details unknown. **Flowers** bicoloured, the falls with a dark violet blade margined white and a yellow signal patch surrounding a white crest, the standards bright violet; tube about 4–5 cm long; falls about 3.5–4(–5) cm long, the claw without a wide wing, with almost parallel margins, about 0.4–0.7 cm wide, the blade obovate, 0.9–1.6 cm long, 0.7–1.2 cm wide, dark violet with white margins and with an entire or toothed white crest; standards deflexed, 2.5–3.5 cm long, bright violet, oblanceolate or obovate, acute, obtuse or emarginate. Style branches pale violet or yellowish, with lobes 1.1–1.2 cm long, obliquely triangular, subacute. **Capsule** and seeds not known. **Flowering** March–April. $2n = ?$

Habitat
Rocky slopes in foothills.

Cultivation
I. narbutii has proved to be reasonably easy to cultivate given the standard juno treatment in an alpine house or cold frame.

Observations
A rather distinct species with bicoloured flowers, noted by A.I. Vvedensky as being variable in the wild, the plants from western Tien Shan differing in having paler colouring than those farther to the east and possibly deserving separate status.

I. naryensis O. Fedtsch.

Synonym
Juno naryensis (O. Fedtsch.) Vved.

Distribution
C Asia: Tien Shan Mountains.

Description
Bulb tunics papery; storage roots short and much swollen. **Leaves:** lower 0.5 cm wide at the base, clustered, the internodes scarcely visible but the leaves sometimes

becoming slightly spaced out in the fruiting stage, falcate, tapering gradually to the apex; margin distinct, ciliate. **Stem** about 5 cm long, more or less hidden by the leaves, unbranched with 1(–2) flowers. Bract and bracteole details unknown. **Flowers** pale violet with a white crest; tube about 4.5–5 cm long; falls about 4 cm long, the claw without a wide wing, with almost parallel margins, about 0.8 cm wide, the blade about 1 cm long, 0.9 cm wide, violet with a white dissected crest; standards spreading, about 2 cm long, linear–oblanceolate, acute to obtuse. Style branches with lobes about 1 cm long, obliquely triangular, acute. **Capsule** and seeds not known. **Flowering** March–April. $2n = ?$

Habitat
Gravelly slopes in lower mountain zone.

Cultivation
No information is available but it is likely to require the standard juno treatment in an alpine house or cold frame.

Observations
A poorly known species, probably never introduced into cultivation.

I. nicolai Vved.

Synonyms
I. rosenbachiana var. *albo-violacea* Regel, *Juno nicolai* Vved.

Distribution
C Asia: Pamir-Alai Mountains; Afghanistan, Kataghan province.

Description
Bulb tunics brownish-grey, thinly papery; storage roots much swollen. **Leaves:** lower 1–2 cm wide at first but expanding to 5–6 cm after flowering, clustered together, sheathing the stem and poorly developed at flowering time, greyish-green, slightly falcate; margin thickened, minutely scabrid. **Stem** very short, carrying 1–3 flowers. Bract and bracteole greenish, membranous, sheathing the tube. **Flowers** with a white or pale lilac ground colour and a deep dull purple blade to the falls; tube 8–11 cm long; falls 4–5 cm long, the claw white or lilac streaked with purple, not winged but with down-turned margins, 0.9–1 cm wide, the blade dark purple, oblong, obtuse and sometimes notched at the apex, 1–1.2 cm wide, with a prominent yellow or orange ridge blotched with purple; standards about 2–3.5 cm long, oblanceolate or obovate, obtuse or notched at the apex. Style branch lobes obliquely triangular, obtuse, toothed 0.6–0.8 cm wide, 18–20 cm long. Other details unknown. **Capsule** broadly ellipsoid, subacute. **Seeds** large, with an aril. **Flowering** February–April. $2n = ?$

Habitat
On clayey slopes, usually 1000–2000 m.

Cultivation
I. nicolai is a marvellous plant for growing in a bulb frame or alpine house and is not too difficult if given plenty of light and air during the growing season. It is very important not to water from overhead so that water cannot lodge in the rosette of leaves; if this happens the plants can die off overnight.

Observations
This is very closely related to *I. rosenbachiana* and *I. baldschuanica* and appears to differ mainly in flower colour. Until extensive field studies are made it is best to uphold the separate names although it seems likely that there are intermediates connecting these three.

I. nusairiensis Mouterde

Distribution
Syria.

Description
Bulb tunics brown; storage roots long and fleshy but not markedly swollen. **Leaves** about 6, lower ones about 1.5 cm wide at the base, clustered together and concealing the stem at flowering time, well-developed, sometimes becoming slightly spaced out up the stem later on, falcate, upper surface glossy pale green; margin smooth, not conspicuously white or thickened. **Stem** up to 10 cm in height at flowering time, hidden by leaves and either remaining short or developing visible internodes in the fruiting stage, unbranched, with 1–3 flowers. Bracts and bracteoles pale green and usually somewhat swollen in the lower part. **Flowers** pale blue, sometimes almost white, with a yellow crest; tube about 5–6 cm long; falls with a conspicuously winged claw, the blade nearly orbicular or elliptic, undulate at the margins, pale blue to nearly white with a yellow crest; standards spreading or slightly reflexed, obovate with a narrow claw. Style branches with obtuse lobes. **Capsule** and seed details not observed. **Flowering** April. $2n = ?$

Habitat
Rocky places, between 1400 and 2000 m.

Cultivation
I. nusairiensis is a very easily cultivated species in the alpine house or bulb frame, arguably one of the easiest of all junos in these conditions. I have not tried it outdoors in the open garden, but it seems likely that it would perform reasonably well in a raised sunny position.

Observations
This is very closely related to *I. aucheri* and requires further study in the wild to ascertain whether or not the two are specifically distinct.

I. odontostyla Mathew & Wendelbo

Distribution
NW Afghanistan.

Description
Bulb tunics brown; storage roots long and tough, not markedly swollen. **Leaves** 4–5, lower ones up to 18 cm long and about 1.5 cm wide at the base, clustered together and well developed at flowering time, probably becoming rather more spaced out up the stem later on, nearly straight or slightly falcate, glossy green above, paler below; margin white. **Stem** more or less absent at flowering and having visible internodes, unbranched, with 1 (possibly more flowers). Bract and bracteole 5.5 cm long, green with a narrow membranous margin. **Flowers** 5–5.5 cm in diameter, greyish or silvery-violet with an orange-yellow crest; tube 4 cm long; falls 4.5 cm long, the claw about 2.7 cm long with a conspicuous wing 1.8 cm wide, the blade suborbicular, 1.8 cm in diameter, pale silvery-violet with an orange-yellow crest; standards spreading, 1.5 cm long and 0.6 cm wide, spathulate, bluntly toothed. Style branches 4 cm long, the lobes 1.3 cm long, 0.4 cm wide, obtuse and toothed. Anther 1.3 cm long, filaments 1.3 cm long. **Capsule** and seeds unknown. **Flowering** April ? $2n = ?$

Habitat
Stony slopes and rock ledges, about 1500 m.

Cultivation
I. odontostyla was cultivated in the 1960s but has probably now been lost to cultivation. It was apparently given the standard treatment for junos, grown in pots in an alpine house or frame in a well-drained gritty compost.

Observations
This is not a well-known species, having been collected only twice by Paul and Polly Furse in 1966 and cultivated for a short time afterwards in Britain. It is probably most closely related to *I. stocksii* but can be distinguished by the orbicular blade to the falls and the toothed style lobes. Unfortunately, seeds have not been available to check if they have an aril or not, so it is not possible to be more precise about its relationships.

I. orchioides Carrière

Synonym
Juno orchioides (Carr.) Vved.

Distribution
C Asia: Pamir–Alai; Syr-Darya and Kara-Tau Mountains.

Description
Bulb tunics papery; storage roots long, not markedly swollen. **Leaves:** lower ones 1.5–3(–4.5) cm wide at the base, spaced out on the stem, especially towards the end of flowering time, falcate or straight, light green; margin distinct, slightly scabrid. **Stem** (10–)20–30 cm long, with visible internodes, unbranched, with (1–)3–4(–8) flowers in the upper leaf axils. Bract and bracteole green, membranous. **Flowers** pale yellow or pale greenish-yellow, tinged with pale purple especially as they age, with a yellow area in the centre of the falls and with a yellow crest; tube about 3–6 cm long; falls 3–4.5 cm long, the claw winged, the wing 1.5–2.5 cm wide; blade elliptic, 1.2–1.8 cm long, 0.8–1.2 cm wide, pale yellow or greenish-yellow with a darker yellow zone in the centre around a dark yellow toothed crest, the whole flower becoming tinged with pale purple in the later stages; standards spreading or deflexed, 0.7–1.5 cm long, usually 3-lobed, acute. Style branches with lobes 0.9–1 cm long, 0.4–0.6 cm wide, obliquely triangular, subacute. **Capsule** cylindrical. **Seeds** without an aril. **Flowering** March–May. $2n = ?$

Habitat
Stony slopes in foothills.

Cultivation
This is not a colourful species, but is nevertheless quite attractive and is not difficult to cultivate in the alpine house or bulb frame.

Observations
For many years this name was used in cultivation for a deep yellow-flowered juno, which has been shown to be a form of *I. bucharica*. The true *I. orchioides* has a wide wing to the haft of the falls whereas *I. bucharica* does not.

I. palaestina (Baker) Boiss.

Synonym
Xiphion palaestinum Baker.

Distribution
Israel, Lebanon, Jordan and (?) Syria.

Description
Bulb tunics membranous, brown, extended into a short neck; storage roots long and thickened, but not markedly swollen. **Leaves** 5–7, lower ones 1–2 cm wide at the base, clustered together at base, falcate and often undulate, glossy light green; margin conspicuous, white and ciliate. **Stem** very short, hidden by the leaves and not elongating in the fruiting stage, carrying 1–3 flowers. Bract and bracteole membranous, greenish-yellow. **Flowers** fragrant, green, greenish-yellow or sometimes pale blue with yellow central stripe on the falls; tube (4–)5–7 cm long; falls 4.5–5.5 cm long, the claw with a conspicuous wing about 2.5–3 cm wide, the blade ovate, about 1–1.5 cm long, green or greenish-yellowish or occasionally pale blue, with a low yellow ridge in the centre and veined with greyish-blue; standards spreading or deflexed, 1.5–2 cm long, linear–oblanceolate. Style branches with oblong, acuminate lobes. **Capsule** oblong. **Seeds** without an aril. **Flowering** January–February. $2n = 24$ (Feinbrun 1986).

Habitat
Open stony areas at fairly low altitudes.

Cultivation
Like *I. planifolia* this is not too difficult, although it is only suitable for the alpine house or bulb frame where it will get some protection in winter, for it begins to grow very early on and usually has well-developed foliage by November.

Observations
Closely related to *I. planifolia* and sharing with it the character of spiny pollen grains; the colour seems to be the main distinguishing feature of these two species.

I. parvula Vved.

Synonym
Juno parvula Vved.

Distribution
C Asia: Pamir–Alai Mountains (Gissar and Zeravshan ranges).

Description
Bulb tunics brown, papery; storage roots short and swollen. **Leaves** about 3, lower ones 5–7 cm long, 0.5–0.8(–1.5) cm wide at the base, slightly separated up the short stem and showing short internodes, straight or slightly falcate, with nearly

parallel margins and tapering only at the apex, bright green above, grey-green beneath, smooth; margin indistinctly white, scabrid. **Stem** up to 10 cm long, with short internodes visible, unbranched, with 1–2(–5) flowers. Bract and bracteole 2–2.5 cm long, green with a transparent margin, slightly swollen in the lower part, narrowed and clasping the perianth tube at the apex. **Flowers** 3.5–4 cm in diameter, very pale green slightly tinged pale violet on the style branches and base of the segments, with a pale greenish-white crest; tube 3–4 cm long; falls 2.8–3.5 cm long, the claw without a wing, with almost parallel margins, 0.5–0.8 cm wide, the blade oblong to oblong–ovate, 1–1.2 cm long, 0.6–0.8 cm wide, pale green with a slightly darker yellowish-green stain in the centre and a few dark green spots on either side of a very prominent cockscomb-like whitish-green crest; standards spreading, 0.5–0.7 cm long and 0.1 cm wide, becoming wider to the 3-lobed apex, the central lobe bristle-like. Style branches pale green flushed violet, about 2.5 cm long, the lobes about 0.5–0.7 cm long, narrowly triangular, acute. Anthers about 0.6 cm long, filaments 0.7 cm long. **Capsule** and seeds not known. **Flowering** May–June. $2n = ?$

Habitat
Stony slopes, 2500–3000 m.

Cultivation
From the small amount of material that has so far been available, it appears that this is reasonably easy when given the standard juno treatment in an alpine house or cold frame.

Observations
This is an interesting little species, only recently introduced into cultivation. The very small, pale green flowers and few, narrow, parallel-sided leaves make it quite distinctive.

I. persica Linn.

Synonyms
I. bolleana Siehe, *I. hausknechtii* Siehe, *I. issica* Siehe, *I. sieheana* Lynch, *Juno persica* (L.) Tratt.

Distribution
S and SE Turkey, NE Iraq, N Syria.

Description
Bulb tunics dark brown, forming a long neck reaching to soil level; storage roots long and slender, not markedly swollen. **Leaves** 3–4 (–6), lower ones up to 10 cm long and

(0.3–)0.4–1.2(–1.8) cm wide at the base, remaining basal, even in the fruiting stage, falcate, glossy green above, greyish-green beneath; margin thickened and translucent or white. **Stem** more or less absent at flowering time, hidden by leaves and not elongating in fruit, unbranched, with 1–4 flowers. Bract and bracteole dissimilar, the outer bract stiff and erect, green and leaf-like, the inner bracteole membranous and semi-transparent. **Flowers** 5–6 cm in diameter, silvery-grey, dull straw-yellow, brownish or pale greyish-green, usually with a darker blade to the falls and with a yellow crest; tube (4–)6–8 cm long; falls 3.5–4.5 cm long, the claw 2–3 cm long with a conspicuous wing 2–2.5 cm wide, the blade broadly elliptic, 1–1.8 cm long, 0.9– 1.4 cm wide, with a very prominent yelllow crest; standards spreading or deflexed, 1.5–2.5 cm long and 0.3–0.6 cm wide, oblanceolate or 3-lobed, sometimes toothed. Style branches 3.5–4.5 cm long, the lobes 1.3–1.7 cm long, obtuse, usually toothed. Anthers about 1.2 cm long, filaments 1–1.2 cm long. **Capsule** about 4–5 cm long, ellipsoid. **Seeds** light brown without an aril. **Flowering** February–April. $2n = ?$

Habitat
Dry stony hills, oak and pine scrubland, 600–1350 m.

Cultivation
I. persica is a well-known species and has been cultivated for centuries. In the eighteenth century bulbs were placed in glass jars on windowsills in much the same way as Hyacinths! However, it is not an easy plant to grow and the best chance of succeeding for any length of time is to grow it in long pots in a very well aired alpine house or cold frame. It will not tolerate overhead water since the water collects in the leaf bases and causes it to rot off. The summer rest period should be warm and dry.

Observations
This is a variable species in its flower colour and several of the variants have been given specific names. *I. bolleana* had yellowish flowers with a violet blotch on the falls; *I. haussknechtii* (syn. *I. sieheana, I. persica* var. *magna)* and *I. persica* var. *mardinensis* were silvery-grey with purplish falls; *I. issica* had flowers of an overall straw colour. Although the name implies that it is a native of Iran (Persia), it has not actually been found there as far as can be ascertained; its place is taken in Iran by the related *I. hymenospatha*.

I. planifolia (Miller) Fiori

Synonyms
I. alata Poir., *Xiphion alatum* Baker, *X. planifolium* Baker.

Distribution
Crete, Greece, Sicily, Sardinia, Spain, Portugal.

Description
Bulb tunics membranous, brown; storage roots long and thickened but not markedly swollen. **Leaves** 5–7, lower ones 1–3 cm wide at the base, clustered together at base, falcate and often undulate, long-tapering, glossy light green; margin conspicuous, white. **Stem** very short, hidden by the leaves and not elongating in the fruiting stage, carrying 1–3 flowers. Bract and bracteole about 10 cm long, membranous, green. **Flowers** pale to deep blue or violet, rarely white, with yellow central crest on the falls; tube (7–)10–20 cm long; falls 4.5–8 cm long, the claw with a conspicuous wing about 1.7–3 cm wide, the blade ovate or oblong, about 2–3 cm long, varying shades of blue or violet with a low minutely pubescent yellow ridge; standards spreading or deflexed, 1.5–2.5 cm long, (0.3–)0.7–1 cm wide, linear–oblanceolate to obovate. Style branches with oblong, acuminate lobes. **Capsule** oblong. **Seeds** without an aril. **Flowering** November–February. $2n = ?$

Cultivation
This species has been cultivated for hundreds of years, but has never been very common since it is not a long-lived plant. It begins to grow early in the season, often as early as the autumn, and is likely to be in flower by mid-winter. It is thus not suitable for outdoor cultivation and must be given the protection of a cold greenhouse or frame; if grown under these conditions in pots, they must be plunged in sand to prevent the bulbs from freezing in very cold weather. This aside, it is relatively easy to grow and will increase into clumps when doing well. It sets seed fairly readily with hand-pollination.

Observations
The only member of the juno group represented in Europe. It and *I. palaestina* are very closely related and differ from all the others in having pollen grains that are covered with minute spines. For this reason Rodionenko placed them in a separate section, *Acanthospora*.

I. platyptera Mathew & Wendelbo

Distribution
E Afghanistan and adjacent Pakistan.

Description
Bulb tunics chestnut brown, forming a long neck; storage roots short and swollen. **Leaves** (3–)4–6, lower ones up to 18 cm long and 1–3 cm wide at the base, clustered together and well developed at flowering time, remaining basal, even in the fruiting stage, usually strongly falcate, greyish-green, glabrous; margin smooth, distinctly white. **Stem** more or less absent at flowering time, hidden by leaves and not elongating much in fruit, unbranched, with 2–3 flowers. Bract and bracteole 5–5.6 cm long, membranous. **Flowers** 4–5 cm in diameter, a semi-transparent pale or dirty purplish or brownish-violet with a yellow crest; tube about 4 cm long;

falls about 4 cm long, the claw with a conspicuous wing 1.6–2.2 cm wide, the blade lingulate, about 1.2 cm long, 0.7–1cm wide, with a very prominent yellow crest; standards spreading or reflexed, 0.8–1.2 cm long and 0.3–0.8 cm wide, lanceolate or 3-lobed, the middle lobe caudate. Style branches about 4 cm long, the lobes about 1.2 cm long, obtuse, sometimes toothed. Anthers about 1.4 cm long, filaments 1.2 cm long. **Capsule** about 4 cm long, ellipsoid. **Seeds** red-brown with a distinct aril. **Flowering** March–April. $2n = ?$

Habitat
Dry stony slopes, 1800–700 m.

Cultivation
Although this was introduced to Britain in the 1960s by Paul Furse, under the name of 'Old Smoky', it is almost certainly not now represented in cultivation. Under the circumstances it is not possible to comment upon its requirements, but it will probably need the same careful treatment as *I. nicolai*.

Observations
This is a very dull species in its flower colour, in some of the paler forms reminiscent of *I. persica*. The short swollen storage roots and the presence of an aril on the seeds suggest that it is, however, more closely related to species such as *I. rosenbachiana* and *I. nicolai*.

I. popovii (Vved.)–

Synonym
Juno popovii Vved.

Distribution
C Asia: Pamir–Alai Mountains.

Description
Bulb tunics brownish-grey, thinly papery; storage roots swollen, thickest in the middle. **Leaves:** lower 2–2.5 cm wide, clustered together and almost fully developed at flowering time, falcate; margin thickened. **Stem** carrying 2–4 flowers. Bract details not known. **Flowers** thought to be blue or pale lilac on the basis of dried type material; tube 6–9 cm long; falls 3.5–4.5 cm long, the claw not winged, with almost parallel margins, 0.6–1 cm wide, the blade oblong, notched at the apex, with a toothed ridge; standards about 2 cm long, obovate, acute. Style branch lobes obliquely triangular, acute, crenate, 1.3 cm long, 0.5 cm wide. Other details unknown. **Flowering** July. $2n = ?$

Habitat
In clayey soil by melting snow, 3600 m.

Cultivation
This is, as far as known, not in cultivation at the present time but is likely to require similar growing conditions to *I. nicolai*.

Observations
A little-known species, related to *I. nicolai* and *I. rosenbachiana*. It is said to differ in having leaves that are almost fully developed at flowering, a different flower colour, and standards that are narrower and acute.

I. porphyrochrysa Wendelbo

Distribution
C Afghanistan.

Description
Bulb tunics brown; storage roots long, fairly thick. **Leaves** (3–)4(–6), 10–17 cm long and 0.6–1.1 cm wide, well developed at flowering time, clustered together, falcate, papillose-puberulent, greyish-green; margin distinctly white. **Stem** hidden by the leaves, unbranched, with 1–3 flowers. Bract and bracteole greyish-green, 5–6.5 cm long, minutely papillose. **Flowers** 4–5 cm in diameter, the blade of the falls deep yellow with an orange crest, the claw, style arms and standards bronze (purplish-brown); tube 3.5–4 cm long, bronze in upper part; falls 4–4.5 cm long, the claw 2.5–3 cm long, not winged, about 0.5 cm wide at the base, widening towards the apex, the blade broadly elliptic–ovate to suborbicular, 1–1.5 cm long, 0.9–1.1 cm wide, with a very prominent toothed orange crest; standards spreading to deflexed, linear–subulate, sometimes weakly 3-lobed, 0.5–1.1 cm long. Style branches 3.5–4 cm long, the lobes subacute, 1.1–1.3 cm long, 0.25–0.3 cm wide. Anthers 1.5–1.9 cm long, filaments about the same. **Capsule** 2–2.5 cm, oblong–ellipsoid. **Seeds** red-brown, with a linear white aril. **Flowering** May–June. $2n = ?$

Habitat
Dry slopes with spiny cushion plants, 2700–3000 m.

Cultivation
This is probably not in cultivation at the present time and very little information is available about its cultivation, but it is likely that it would not be too difficult to cultivate in an alpine house or bulb frame.

Observations
I. porphyrochrysa is a rather distinctive species with its yellow and bronze coloration. In fact Paul Furse, who introduced material into cultivation in the 1960s, referred to it as 'Shibar Bronze' in his writings.

I. postii Mouterde

Synonym
I. palaestina var. *caerulea* Post.

Distribution
Iraq, E Jordan and E Syria.

Description
Bulb tunics membranous, brown, extended into a long neck; storage roots long and rather wiry, not markedly swollen. **Leaves** (3–)4–6, lower ones 0.8–1.2 (–1.6) cm wide at the base, clustered together at base at flowering time but becoming slightly spaced out on the stem by fruiting time, falcate and long-tapering to the apex, prominently close-nerved, green above, greyish and papillose beneath; margin conspicuous, white and scabrid-ciliate or rarely smooth. **Stem** short and hidden by the leaves, elongating somewhat in the fruiting stage, carrying 1–3 flowers. Bract and bracteole subequal, 5–8 cm long, membranous, green, papillose. **Flowers** blotched and veined violet or brownish-violet on a pale lavender ground with a yellow crest on the falls; tube 2.5–4.5 cm long; falls 3–4 cm long, the claw with a conspicuous wing about 1.5–2 cm wide, the blade lingulate, 1–1.4 cm long, 1–1.2 cm wide, lavender, conspicuously blotched dark violet or brownish-violet, with a low yellow papillose crest in the centre; standards spreading or deflexed, 1.2–1.8 cm long, 0.5–0.6 cm wide, obovate, with a narrow claw. Style branches with lobes 0.7–1 cm long, 0.3–0.5 cm wide, acute. **Capsule** narrowly cylindrical, acuminate. **Seeds** dark brown, without an aril. **Flowering** February–April. $2n$ =?

Habitat
Sandy and gravelly soils in semi-desert, 175–650 m.

Cultivation
I. postii is a fairly difficult plant to cultivate, requiring plenty of light and ventilation during the winter and early spring. It occurs wild in desert regions at low altitudes and, although probably reasonably hardy, it is inclined to come into growth rather early and become etiolated in the poor light conditions of northern countries. It is suitable for the alpine house or bulb frame where it will get some protection in winter, for it begins to grow very early on and usually has well-developed foliage by November.

Observations
This is probably related to the semi-desert junos from farther to the east, such as *I. stocksii* and its relatives. The prominently blotched flowers, narrowly tapering leaves and long neck to the bulb make it easy to recognise.

I. pseudocaucasica Grossh.

Synonym
Juno pseudocaucasica (Grossh.) Rodionenko.

Distribution
SE Turkey, N Iraq, N and NW Iran, Armenia.

Description
Bulb tunics brown; storage roots fleshy but not markedly swollen. **Leaves** 4–6, lower ones up to 18 cm long and 2.8 cm wide at the base, clustered together and well developed at flowering time, becoming spaced out up the stem later on, strongly falcate; upper surface glossy green, lower surface paler matt grey-green; margin scabrid-ciliate, conspicuously white. **Stem** more or less absent at flowering time, hidden by leaves, then elongating with visible internodes, unbranched, with 1–4 flowers. Bract and bracteole almost equal, 4–5 cm long, green, slightly swollen and clasping the perianth tube at the apex. **Flowers** 5–6 cm in diameter, pale yellow, greenish or pale blue with a yellow crest; tube 3–4 cm long; falls 3.5–4 cm long, the claw 2–2.5 cm long with a conspicuous wing 1.6–2 cm wide, the blade lingulate, 1–1.2 cm wide, yellow, greenish or pale blue, with a low yellow undulate crest; standards spreading or slightly reflexed, 1.5–2.2 cm long and 0.2–0.4 cm wide, lanceolate, acute to trilobed. Style branches 3.5 cm long, the lobes about 1.1 cm long, 0.6 cm wide, obtuse to subacute. Anthers 1.3–1.4 cm long, filaments about 1 cm long. **Capsule** about 3.5 cm long, ellipsoid. **Seeds** brown, without an aril but with a small wing at one end. **Flowering** April–June. $2n = ?$

Habitat
Screes and rocky slopes, 600–3450 m.

Cultivation
I. pseudocaucasica is of relatively easy culture compared with, for example, *I. persica*. Nevertheless, it is best grown in the alpine house or frame in a well-drained gritty compost. When growing well it will increase vegetatively; seeds are also produced in cultivation.

Observations
Although rather similar to *I. caucasica*, the two can be distinguished by the shape of the haft of the falls. *I. pseudocaucasica* has a very wide conspicuous wing, much

wider than the blade whereas in *I. caucasica* the blade and wing of the claw are almost the same width.

I. regis-uzziae Feinbrun

Distribution
Israel, Jordan.

Description
Bulb tunics membranous, greyish-brown, extended into a long neck; storage roots fleshy but not markedly swollen. **Leaves** 5–6(–7), lower ones 1.7–4 cm wide at the base, clustered together at base at flowering time but becoming slightly spaced out on the stem by fruiting time, falcate and tapering gradually to the acute apex, prominently close-nerved, pale green above, greyish-green beneath; margin inconspicuous, very narrow and translucent-white. **Stem** short and hidden by the leaves, elongating somewhat in the fruiting stage, carrying 1–3 flowers. Bract and bracteole subequal and somewhat leaf-like, pale green, about 8 cm long, membranous, with a finely scabrid keel. **Flowers** pale blue to almost white, pale lilac or pale yellow, more or less translucent, with a yellow crest on the falls; tube length unrecorded; falls 3.5–5 cm long, the claw with a wing about 2.5–2.7 cm wide, the blade lingulate, 1–1.2 cm wide, very pale blue, lilac or yellow, with a raised yellow undulate crest in the centre; standards spreading or deflexed, 2–2.5 cm long, 0.6–0.8 cm wide, spathulate, truncate and apiculate or erose at the apex. Style branches with broadly ovate, subacute or obtuse lobes about 0.9 cm long, 0.9 cm wide. **Capsule** 5–7 cm long, ellipsoid. **Seeds** brown, without an aril. **Flowering** January–February. $2n = 22$

Habitat
Rocky places, 500–1000 m.

Cultivation
I. regis-uzziae has been cultivated successfully at Kew but is probably slightly tender and should be given some protection during severe winter weather; otherwise it requires the standard juno treatment.

Observations
This is a comparatively recently discovered species, which requires further investigation in the wild; it was described in 1978 from the Negev area of Israel. A similar but not identical plant occurs in the Ras-en Naqb area; of the few plants from there that I have seen, it appears to be very variable and to differ slightly from the Israeli material. It seems that those from Israel may vary from pale blue to near-white whereas those from Jordan are blue or yellow; this in itself is prob-

ably not of great significance, but it appears that there are also some differences in the dimensions of the wing of the haft of the falls.

I. rosenbachiana Regel

Synonym
Juno rosenbachiana (Regel) Vved.

Distribution
C Asia: Pamir–Alai Mountains.

Description
Bulb tunics brownish-grey, thinly papery; storage roots much swollen. **Leaves:** lower 2–3 cm wide at first but expanding to 5–6 cm after flowering, clustered together, sheathing the stem, developing at flowering time, but not overtopping the flower, slightly falcate; margin scabrous. **Stem** very short, carrying 1–3 flowers. Bract and bracteole greenish, membranous, sheathing the tube. **Flowers** with a purple ground colour and a deeper purple blade to the falls; tube 8–11 cm long; falls 4–5 cm long, the claw pale purple streaked with darker purple, not winged but with a down-turned margins, 0.9–1 cm wide, the blade dark purple, oblong, obtuse and sometimes notched at the apex, 1.2–1.7 cm wide, with a prominent yellow or orange ridge blotched with purple; standards about 2.5–3 cm long, obovate, obtuse or notched at the apex. Style branch lobes obliquely triangular, obtuse or subacute, toothed, 0.5–0.1 cm wide, 1.5–2.2 cm, long. **Capsule** broadly ellipsoid, subacute. **Seeds** large, with an aril. **Flowering** March–April. $2n$ = ?

Habitat
On stony slopes, usually 1000–2000 m.

Cultivation
For cultivation methods, see *I. nicolai*.

Observations
This is very closely related to *I. nicolai* and *I. baldschuanic*. Further comments are made under the former species.

I. schischkinii Grossheim

I have seen only a photograph of the type specimen of this, but it appears to be the same as *I. caucasica* although a very robust individual about 20–30 cm in height with correspondingly large flowers. It was described from the Transcauscasian region of Nakhicheven near the town of Bitschenach.

I. stenophylla Hausskn. & Siehe ex Baker

There are 2 subspecies.

subsp. *stenophylla*

Synonyms
I. heldreichii Siehe, *I. tauri* Siehe ex Mallet.

Distribution
S Turkey.

Description
Bulb tunics brown; storage roots fleshy but not markedly swollen. **Leaves** 4–5, short and narrow at first but becoming 10–20 cm long and 0.5–1 cm wide, clustered together and only partly developed at flowering time, remaining basal, even in the fruiting stage, straight or slightly falcate becoming more strongly falcate with age, glossy green above, green or slightly greyish-green beneath; margin minutely hairy, transparent; veins on underside smooth or very slightly scabrid. **Stem** more or less absent at flowering time, hidden by leaves and not elongating in fruit, unbranched, usually with 1 flower. Bract and bracteole similar, green and membranous, somewhat ventricose. **Flowers** 5.5–6.5 cm in diameter, violet-blue or lilac-blue, usually with a darker blade to the falls and a whitish, dark-blotched, zone surrounding a yellow or orange crest; tube 6–9 cm long; falls spreading nearly horizontally, (3–)3.5–5.2 cm long, the claw 2.3–3.5(–4) cm long with a conspicuous wing 1.7–3 cm wide, the blade oblong, (0.8–)1–1.7 cm long, (0.6–)1.4–1.7 cm wide, with a prominent, very finely pubescent, yellow or orange crest; standards spreading or deflexed, spathulate, obovate or oblanceolate, sometimes shallowly 3-lobed or toothed, 1–2.5 cm long and 0.6–0.7 cm wide. Style branches (2.5–)3–4.5(–5) cm long, the lobes obtuse. Anthers about 1–1.5 cm long, filaments 1–1.5 cm long. **Capsule** 3–3.5 cm long, ellipsoid. **Seeds** brown, without an aril. **Flowering** (March–)April(–May). $2n = ?$

Habitat
Open rocky slopes and in macchie, 400–2000 m.

Cultivation
I. stenophylla has been in cultivation for a long time (sometimes as *I. tauri* or *I. heldreichii*), although it has always been a rare plant in garden collections. It appears to be one of the easier of the dwarf junos but nevertheless requires alpine-house conditions if it is to thrive for long. Because it occurs wild in a part of southern Turkey that has a relatively high rainfall, it does not rot off quite so readily as *I. persica*, which inhabits a much drier climate.

Observations
Although closely related to *I. persica* this can be distinguished by the blue flower colour and by the way in which the falls are carried almost horizontally; in *I. persica* they are suberect forming a funnel shape.

subsp. *allisonii* Mathew

Distribution
S Turkey.

Description
As for subsp. *stenophylla* but with (4–)6–10 leaves which are 1.5–1.8 cm broad when mature; veins on underside of leaves strongly scabrid-papillose; claw of falls with a central band of multicellular hairs; perianth tube about 5.5 cm long. **Flowering** March–April. $2n = ?$

Habitat
Stony slopes at edge of pine woods, 850–1500 m.

Cultivation
As for subsp. *stenophylla*.

Observations
This subspecies has a much more restricted distribution than subsp. *stenophylla*; apart from the differences given above it usually has paler blue flowers with very conspicuous darker blotches on the falls.

I. *stocksii* (Baker) Boiss.

Synonyms
Juno stocksii (Baker) Klatt., *Xiphion stocksii* Baker.

Distribution
Afghanistan, Pakistan.

Description
Bulb tunics dark brown, forming a long neck reaching to soil level; storage roots long and slender, not markedly swollen. **Leaves:** lower up to 16 cm long and 0.6–1.6 cm wide at the base, clustered together and well developed at flowering time, becoming spaced out up the stem later on, falcate, greyish-green; margin conspicuously white. **Stem** more or less absent at flowering time, hidden by leaves, then elongating with visible internodes, unbranched, with 1–4 flowers. Bract and

bracteole 4.5–6 cm long, green with a membranous margin. **Flowers** 5–6 cm in diameter, lavender or bluish-violet with a yellow crest; tube 3–3.5(–5) cm long; falls 3.5–4.5 cm long, the claw about 2.5 cm long with a conspicuous wing 2–2.2 cm wide, the blade lingulate, 1–1.5 cm long, 0.8–1 cm wide, lavender or pale bluish-violet, with a yellow crest; standards spreading, 1–1.8 cm long and 0.4–0.6 cm wide, narrowly obovate. Style branches 3 cm long, the lobes about 1 cm long, obtuse. Anthers 1.1–1.3 cm long, filaments about 1 cm long. **Capsule** about 3.5–5 cm long, ellipsoid. **Seeds** dark brown, without an aril. **Flowering:** March–April. $2n = ?$

Habitat
Dry stony hillls, 1150–2700 m.

Cultivation
I. stocksii is one of the species that occur in very dry habitats and is not an easy plant to maintain in cultivation. It should be grown in the alpine house and given as much light and air as possible during the winter and spring months; on no account should overhead water be given, and the foliage must be kept dry at all times. The dormant period in summer needs to be warm and dry. In view of the fact that the bulbs are rather elongated and have a long neck it is necessary to grow them in 'long tom' pots.

Observations
I. stocksii is related to *I. odontostyla,* also from Afghanistan, but differs in flower colour, style branches that are not noticeably toothed, and the shape of the blade of the falls, which in *I. odontostyla* is rounded, in *I. stocksii* lingulate.

I. subdecolorata Vved.

Synonyms
Juno almaatensis Pavl., *J. subdecolorata* (Vved.) Vved.

Distribution
C. Asia: Syr-Darya Mountains.

Description
Bulb tunics papery; storage roots long and thick. **Leaves** 4–6, lower ones 0.8–2 cm wide at the base, clustered together at the base, falcate, tapering gradually to the apex, dark green; margin distinct, scabrid. **Stem** up to about 5 cm long, hidden by the leaves, unbranched, with 1–3 flowers. Bract and bracteole details unkown. **Flowers** translucent pale dirty green or tinged with lilac with dirty green blotches around a white crest; tube about 4.5 cm long; falls about 4–4.5 cm long, the claw

slightly winged, 0.7–0.9 cm wide; blade oblong or oblong–obovate, 1.4–1.6 cm long, 0.7–1 cm wide, pale green with darker dirty green blotches at the side of a dissected white crest; standards spreading or deflexed, 1.5–2 cm long, rhombic or often 3-lobed, acute. Style branches with lobes 1.1 cm long, 0.4 cm wide, obliquely triangular, subacute. **Capsule** and seeds not known. **Flowering** March–April. $2n = ?$

Habitat
Loess hills in foothills.

Cultivation
No details are available.

Observations
As far as known this has not been introduced into cultivation and the description is based on that in the *Flora of the USSR* (Komarov 1935). It is said to be closely related to *I. kuschakewiczii*.

I. svetlanae (Vved.)–

Synonym
Juno svetlanae Vved.

Distribution
C Asia: Pamir–Alai Mountains.

Description
Bulb tunics greyish, membranous; storage roots thickened in the middle. **Leaves:** lower 3–4 cm wide at the base, clustered together at flowering time, falcate; margin smooth, thickened, whitish. **Stem** 10–15 cm, carrying 1–2 flowers. Bract details not known. **Flowers** yellow; tube 4.5–6 cm long; falls 4–5 cm long, the claw with a conspicuous wing about 2–2.2 cm wide, yellow with 2–4 dull green veins, the blade yellow with an intense yellow ridge, which is slightly crinkled; standards 1.2–1.6 cm long, yellow, narrowly rhomboidal, acute at the apex. Other details unknown. **Flowering** March(–April). $2n = ?$

Habitat
Gypsum hills.

Cultivation
Plants cultivated at Kew as *I. svetlanae* indicate that this is not a particularly difficult juno for pot cultivation in an alpine house or frame.

Observations
In the original description, A.I. Vvedensky commented that this is similar to *I. maracandica*, but has deeper yellow flowers. Plants cultivated in Britain under both these names suggest that they are not distinct enough to be maintained as separate species, but only field-work can elucidate this question.

I. tadshikorum Vved.

Synonym
Juno tadshikorum Vved.

Distribution
C Asia: Pamir–Alai Mountains.

Description
Bulb tunics brownish-grey, thinly papery; storage roots much-swollen, thickest in the middle. **Leaves:** lower 0.6–1 cm wide, clustered together, falcate; margin thickened, minutely scabrid or smooth. **Stem** about 5 cm, carrying 2–4 flowers. Bract details not known. **Flowers** thought to be pale violet on the basis of dried type material; tube 3.5–4 cm long; falls 4–4.5 cm long, the claw not winged, with almost parallel margins, 0.7–0.8 cm wide, the blade ovate, acute or sometimes notched at the apex, 1.2 cm wide, with a crested white ridge; standards about 1.5 cm long, trilobed. Style branch lobes obliquely triangular, obtuse, crenate, 0.5 cm wide, 1.2 cm long. Other details unknown. **Flowering** June. $2n = ?$

Habitat
Stony slopes at about 2500–3000 m.

Cultivation
Plants under this name have been cultivated from time to time and it seems that these have been given pot cultivation in a frame or alpine house.

Observations
This is said to be related to *I. linifolia* but differing in flower colour. Unfortunately, both of these are so little known at present that it is impossible to discuss their relationships.

I. tubergeniana Foster

Synonym
Juno tubergeniana (Foster) Vved.

Distribution
Uzbekistan, Syr-Darya Mountains.

Description
Bulb tunics membranous; storage roots long and only slightly thickened. **Leaves:** lower 1.5–2.5 cm wide at the base, clustered together at flowering time, falcate, light green; margin conspicuous, slightly scabrid. **Stem** 10–15 cm, carrying 1–3 flowers. Bract and bracteole membranous, green. **Flowers** yellow with a deeper yellow crest; tube 4.5–5 cm long; falls 4–4.5 cm long, the claw with a conspicuous wing about 1 cm wide, the blade obovate, about 1.5 cm long, 1 cm wide, yellow with a darker yellow conspicuously dissected crest; standards about 1.5 cm long, 3-lobed, acute. Style branches with lobes about 0.9 cm long, 0.4 cm wide, obliquely triangular, subacute. **Capsule** and **seeds** unknown. **Flowering** March–April. $2n = ?$

Habitat
Clayey slopes in foothills.

Cultivation
Although there is no modern account of its cultivation, the indications are that *I. tubergeniana* is not too difficult although probably best suited to the alpine house or bulb frame.

Observations
This is very rare in cultivation at the present time and is not well known. The very prominent beard-like crest makes it rather distinctive, but it is presumably related to *I. svetlanae* and *I. maracandica*.

I. vicaria Vvedensky

Synonym
Juno vicaria Vved.

Distribution
C Asia: Pamir–Alai Mountains.

Description
Bulb tunics papery; storage roots swollen. **Leaves:** lower 1.5–4 cm wide at the base, spaced out on the stem, especially towards the end of flowering time, falcate, tapering gradually to the apex, light green; margin distinct, scabrid. **Stem** 20–40 (–50) cm long, with clearly visible internodes, unbranched, with (1–)2–4(–8) flowers. Bract and bracteole details unknown. Flowers pale lilac or almost white

with a yellow blotch around a yellow or white crest; tube about 4–4.5 cm long; falls about 4–4.5 cm long, the claw not winged, 0.5–1 cm wide; blade oblong, 1.2–1.7 cm long, 0.8–1.4 cm wide, pale lilac or white with a bright yellow blotch at the side of an entire but wavy, white or partly yellow crest; standards spreading or deflexed, 2–2.5 cm long, rhombic, obovate or 3-lobed, acute or obtuse. Style branches with lobes 1.1–2 cm long, 0.4–0.6 cm wide, obliquely triangular, subacute. **Capsule** and **seeds** not known. **Flowering** March–April. $2n = ?$

Habitat
Stony slopes in the lower mountain zone.

Cultivation
Although rare in cultivation, this is fairly easily cultivated in the bulb frame or alpine house and would probably do well planted out in a well-drained sunny bed.

Observations
For many years this was confused with *I. magnifica* although there is not much similarity. The latter has a very wide wing on the haft of the falls which immediately distinguishes the two species.

I. vvedenskyi Nevski ex Voron.

Synonym
Juno vvedenskyi (Nevski ex Voron.) Nevski.

Distribution
C Asia: Pamir–Alai Mountains.

Description
Bulb tunics papery; storage roots much swollen. **Leaves:** lower 0.4–0.5 cm wide at the base, separated up the short stem and showing short internodes, falcate, tapering gradually to the apex; margin distinct, scabrid. **Stem** 3–5 cm long, with short internodes visible, unbranched with 1(–2) flowers. Bract and bracteole details unknown. **Flowers** pale yellow with an orange crest; tube about 2.5–3 cm long; falls 2.5–3 cm long, the claw without a wing, with almost parallel margins, 0.5 cm wide, the blade obovate, about 1.2 cm long, 0.7 cm wide, pale yellow with an orange crenate crest; standards spreading, about 0.6 cm long, acute, entire or bluntly 3-lobed. Style branches with lobes about 0.7 cm long, obliquely triangular, acute. **Capsule** and **seeds** not known. **Flowering** May. $2n = ?$

Habitat
Not recorded.

Cultivation
No information is available, but it will probably require the standard juno treatment in an alpine house or cold frame.

Observations
This is a poorly known species, probably not in cultivation.

I. warleyensis Foster

Synonym
Juno warleyensis (Foster) Vved.

Distribution
C Asia: W Pamir–Alai Mountains.

Description
Bulb tunics papery; storage roots long and thick, not markedly swollen. **Leaves:** lower 1.5–3 cm wide at the base, spaced out on the stem, especially towards the end of flowering time, falcate, tapering gradually to the apex, light green; margin distinct, scabrid. **Stem** 20–40 cm long, with clearly visible internodes, unbranched, with 2–5 flowers. Bract and bracteole details unknown. **Flowers** varying shade of lilac with a darker violet blade to the falls and yellow zones on each side of an undulate white crest; tube about 4.5–5 cm long; falls about 4–5.5 cm long, the claw not winged, 0.7–1.2 cm wide; blade obovate or oblong, 1.5–2 cm long, 1–1.5 cm wide, pale to deep lilac with a violet apex and a yellow area around an entire or toothed white or yellow crest; standards spreading or deflexed, 1.2–2 cm long, 3-lobed, obtuse. Style branches with lobes 1.1–5 cm long, 0.4–0.7 cm wide, obliquely triangular, subacute. **Capsule** and seeds not known. **Flowering** March–April. $2n = ?$

Habitat
Stony slopes in the lower mountain zone.

Cultivation
This attractive species is relatively easy to cultivate in a bulb frame or alpine house and will also do reasonably well planted out in a sheltered well-drained sunny bed.

Observations
An easily recognised species with its colourful flowers, dark-blotched at the apex of the falls.

I. wendelboi Grey-Wilson & Mathew

Distribution
SW Afghanistan.

Description
Bulb tunics brown; storage roots short and swollen. **Leaves** 3–4, lower ones 15–18 cm long and 0.7–0.8 cm wide at the base, clustered together and well-developed at flowering time, falcate or twisted and lying on the ground, long-tapering, greyish-green, densely papillose between the veins; margins distinctly silvery-white. **Stem** hidden by leaves or with very short internodes visible, unbranched, with 1–2 flowers. Bract and bracteole 4.5–5.5 cm long, green with a transparent margin, papillose. **Flowers** 4–5 cm in diameter, deep bluish-violet with a yellow crest; tube 3.2 cm long; falls 3.2 cm long, the claw without a wing, 0.8 cm wide, the blade lingulate, about 0.6–0.8 cm long, 0.5 cm wide, with a fimbriate, bright golden-yellow crest; standards spreading or slightly ascending, 0.4–0.5 cm long and 0.1 cm wide, linear or 3-lobed. Style branches 2.8–3 cm long, the lobes about 0.8 cm long, narrowly triangular, subacute. Anthers 1.3 cm long, filaments 0.7 cm long. **Capsule** and seeds not known. **Flowering** March–April. $2n = ?$

Habitat
Dry sandy hills, 1700 m.

Cultivation
Little can be said about the cultivation requirements of *I. wendelboi* except that it was presumably rather tricky to grow. Its dry habitat in southern Afghanistan suggests that it might well be one of the more difficult species to keep for long.

Observations
This is an interesting and attractive species, which is probably not now in cultivation. It was introduced to Britain by Chris Grey-Wilson and Tom Hewer in the early 1970s, but did not persist for long. Unfortunately, the fruiting stage is unknown, so we are unable to determine whether the seeds are with or without an aril. The question of its relationship with other species thus cannot be resolved.

I. willmottiana Foster

Synonym
Juno willmottiana (Foster) Vved.

Distribution
C Asia: Pamir–Alai Mountains.

Description
Bulb tunics papery, storage roots long and thick, not markedly swollen. **Leaves:** lower 3–4 cm wide at the base, clustered together, sometimes slightly spaced out on the stem in cultivation, falcate, shiny green; margin distinct, smooth. **Stem** 10–20 cm long, completely hidden at flowering time, sometimes with the internodes becoming visible later on, unbranched, with 4–6 flowers. Bract and bracteole green, membranous. **Flowers** pale blue (or white in 'Alba') with dark violet-blue blotches and lines on a white signal patch and a white crest; tube about 3.5 cm long; falls about 4–5 cm long, the claw winged but with a gradual transition from the claw to the blade, the wing 1.5 cm wide; blade oblong, about 1.7–2 cm long, 1 cm wide, pale blue with a white area around an entire white crest, with dark violet-blue blotches and streaks in the centre; standards spreading or deflexed, about 1.5 cm long, rhombic or 3-lobed, acute. Style branches with lobes about 1.2 cm long, 0.5 cm wide, obliquely triangular, subacute. **Capsule** cylindrical; seeds without an aril. **Flowering** May. $2n = ?$

Habitat
Stony slopes in foothills.

Cultivation
This is an attractive species and is not difficult to cultivate in a bulb frame or alpine house. The albino version, 'Alba', requires further study to determine whether or not it is actually a form of this species: some plants in cultivation appear to be forms of *I. bucharica*.

Observations
A very attractive species, named after Ellen Willmott of Warley Place. It is a stocky plant with broad leaves amid which nestle the quite large blue flowers, so it is rather distinctive, although it and *I. kuschakewiczii* are very similar. The main difference is said to be in the width of the wing of the falls, but this may not be sufficient to distinguish between them; unfortunately the range of variation of both of these is not well known in the wild.

I. xanthochlora Wendelbo

Distribution
NE Afghanistan.

Description
Bulb tunics brown; storage roots swollen. **Leaves** (3–)4–5, up to 20 cm long and 2 cm wide, well developed at flowering time, clustered together, falcate, densely papillose-puberulent, glossy green above, dull green beneath; margin

whitish. **Stem** hidden by the leaves, unbranched with 1–2(–3) flowers. Bract and bracteole green with a transparent margin, 5–7.5 cm long, minutely papillose. **Flowers** 4–5 cm in diameter, greenish-yellow with a deeper crest; tube 5–6 cm long; falls about 4 cm long, the claw about 2.5 cm long, not winged, about 0.6 cm wide at the base, widening gradually to about 1 cm at the apex, the blade elliptic, about 1.5 cm long, 0.8 cm wide, with a very prominent deep yellow crest; standards spreading to deflexed, linear–oblanceolate, acuminate, 1–1.5 cm long. Style branches 3.5–3.8 cm long, the lobes subacute, 1–1.2 cm long, 0.5–0.6 cm wide. Anthers 1.2–1.6 cm long, filaments 1.2–1.6 cm long. **Capsule** ellipsoid. **Seeds** red-brown, with conspicuous aril. **Flowering** (May–)June(–July). $2n = 14+1B$.

Habitat
Dry slopes, subalpine belt, 2600–3100 m.

Cultivation
Very little information is available about the cultivation of this species, but it seems likely that it would not be too difficult to cultivate in an alpine house or bulb frame. It is not known to be in cultivation at present.

Observations
I. xanthochlora is related to *I. kopetdagensis*, but has falls in which the blade is about as wide as the claw, or sometimes narrower, whereas in the latter the reverse is true. It appears that the plants do not become elongated in the fruiting stage.

I. zaprjagajewii (N.V. Abramov)

Synonym
Juno zaprjagajewii N.V. Abramov.

Distribution
C Asia: Pamir Mountains.

Description
Bulb tunics brownish-grey, papery; storage roots much swollen. **Leaves:** lower up to 4 cm wide, clustered together, sheathing the stem, not fully developed at flowering time and overtopped by the flowers, greyish-green (shiny according to original collection), falcate; margin distinct. **Stem** very short, carrying 1–3 flowers. Bract and bracteole membranous, sheathing the tube. **Flowers** white with a yellow crest; tube 6–9 cm long; falls about 4 cm long, the claw not widely winged but with down-turned margins, 2 cm wide, the blade about 1.5 cm wide, white

with a yellow crest; standards subrhomboidal, acute, about 1 cm long, 0.5 cm wide. Style branch lobes 1–1.3 cm long, 0.5–0.6 cm wide, obliquely triangular. **Capsule** and seeds not known. **Flowering** April–May. $2n = ?$

Habitat
In mountains at about 2200 m, but precise habitat not noted.

Cultivation
Although this is very rare in cultivation and there has been little scope for trying out cultivation methods, it seems to require similar treatment to *I. nicolai*.

Observations
This is very closely related to *I. nicolai* and *I. rosenbachiana* and their allies in having down-turned margins to the haft of the falls, a more or less stemless habit with broad sheathing leaves, and very swollen roots. It differs most obviously in the pure white flowers. The specimen which I cultivated was undoubtedly this species but differed from the original description in having grey-green leaves; another plant that flowered in Britain under this name, and which was beautifully illustrated in colour in the *Bulletin of the Alpine Garden Society* (Hulme 1989), is certainly not *I. zaprjagajewii*. With its widely winged falls it is rather more like *I. orchioides,* although much whiter in colour than usual.

I. zenaidae (Vved.)

Synonym
Juno zenaidae Vved.

Distribution
Kirgizya, Tien Shan Mountains.

Description
Bulb tunics greyish-brown; storage roots much thickened in the middle. **Leaves:** lower up to 2 cm wide at the base, clustered together at flowering time, later becoming separated out on the stem, falcate. **Stem** hidden at first, internodes becoming visible later on, carrying 1–3 flowers. Bract details not known. **Flowers** primarily violet-blue with spotted falls; tube 4.5–5.5 cm long; falls 4.5–5.5 cm long, the claw with a conspicuous wing about 2.2–2.8 cm wide, whitish veined violet, the blade violet-blue with a white slightly crinkled edge, the crenations violet; standards 2–2.8 cm long, paler than the blade, broadly oblong and narrowed to a claw, which is slightly longer than the blade. Other details unknown. **Flowering** April. $2n = ?$

Habitat
Rocky places.

Cultivation
As far as I know, *I. zenaidae* has not been introduced into cultivation, but judging from the distribution and habitat it seem likely that it will require similar treatment to *I. magnifica*.

Observations
When describing *I. zenaidae*, A.I. Vvedensky likened it to *I. magnifica*, but noted that it differed in having violet-blue flowers with spotted falls. Unfortunately, little is known of this species and many of the features are not recorded.

Hybrids
Natural hybrids have been recorded by A.I. Vvedensky, although a certain amount of guesswork was involved since the species themselves had not been thoroughly studied at the time and this position has not changed. Possible crosses were *I. narbutii* × *I. maracandica*; *I. narbutii* × *I. orchioides*; *I. subdecolorata* × *I. narbutii*; and *I. vicaria* × *I. bucharica*. In Turkey I have seen plants that could be hybrids between *I. caucasica* and *I. persica*, and there are certainly plants that are intermediate between *I. persica* and *I. galatica* in central Anatolia. In cultivation, several crosses have been successful and it seems likely that there are few barriers between the species, although some of the hybrids may well be sterile. The following species have been crossed artificially.

I. warleyensis × *I. aucheri* = *I.* 'Warlsind'

I. aucheri × *I. galatica* = *I.* 'Sindpur'

I. aucheri × *I. persica* = *I.* 'Sindpers'

I. warleyensis × *I. bucharica*

I. warleyensis × *I. persica*

I. graeberiana × *I. persica*

Recent work by Mr Alan McMurtrie in Toronto suggests that many other crosses are possible. It is essential that all experimental hybridisation has a purpose and is recorded accurately; it would be very unfortunate to lose track of the true species in a mass of garden hybrids long before the species themselves are adequately studied and documented.

Subgenus Hermodactyloides Spach (the Reticulata Irises)

WILLIAM R. KILLENS

Map 27, Figure 27.

I. reticulata was first described by Marschall von Bieberstein in 1808 and remained the only known member of this popular group of irises until 1872 when *I. histrio* Reichb. fil. was included. Several more species had been named and added by 1900. Only three more have been described since that date.

In 1846, Spach first placed this distinct group of irises in the monotypic subgenus Hermodactyloides based on the quadrangular cross section of the leaves, which is similar to that of *Hermodactylus tuberosus* (L.) Salisb. Later re-classification grouped some species in the subgenus Xiphion (Miller) with *I. danfordiae* included in subgenus Juno (Scorpiris). In the most recent classification (Mathew 1989) this group of bulbous irises are all included in subgenus Hermodactyloides.

This subgenus consists of a small group of iris species having a bulbous rootstock with a reticulate-fibrous (rarely membranous) tunic, fibrous roots not usually thickened (except *I. pamphylica*), leaves usually quadrangular in section (except *I. bakeriana* and *I. kolpakowskiana*) and flowers with, normally, a long perianth tube and ovary below ground at flowering (except *I. kolpakowskiana* and *I. pamphylica*). Ten species are recognised, grouped in four sections: A Brevituba, B Monolepis, C Hermodactyloides, D Micropogon.

The general distribution of subgenus Hermodactyloides is from mid-Turkey east through Iran to Tashkent and from Israel north to the Russian Caucasus.

General cultivation

I. reticulata, I. histrioides, I. winogradowii and their cultivars appear to grow quite well in the open ground given a sunny aspect in good fertile soil that is well drained and will not become waterlogged. For other members of this subgenus the raised bed or bulb frame would seem preferable. These irises as a whole do not appreciate pot culture; as an alternative, the use of perforated plastic pots, similar to those used for orchids or aquatic plants, plunged to the rim in the raised bed or bulb frame, might be tried. The root system would have access to the surrounding compost and drainage around the bulbs could be controlled. Plants should not be allowed to dry out while in growth, but should be dried off as soon as the foliage shows signs of dying down. Whichever method of cultivation is used, feeding with a high-potash fertiliser is recommended. Various formulas have been suggested in the past, but one of the present-day proprietary brands would seem adequate as they can be obtained in granular or liquid form: the granular should be applied in autumn when root action is commencing and liquid applied when watering during growth. As all the members of this subgenus are quite hardy it is suggested that if frame culture is used the side lights or glasses should be removed to allow maximum ventilation at all times and overhead protection only used during the drying-off period in summer, or in exceptionally severe weather, or to protect from heavy rain when in flower.

With the exception of ink disease, a fungal infection that can decimate a whole group of bulbs, this subgenus seems to suffer few problems. The treatment recommended (Mathew 1989) is to soak the bulbs when lifted in a fungicidal solution or to dust them with the dry powder. If the bulbs are not lifted, the surrounding soil should be thoroughly watered with the solution in the autumn.

Section A: Brevituba   B. Mathew, sect. nov.

Type of section: *I. pamphylica* Hedge.

This group has two exceptional distinguishing characteristics: the extremely sharp, thin basal leaf fibres and the relatively long stem above ground.

I. pamphylica Hedge

Distribution
Antalya province, Turkey; very local, growing around fields and in *Quercus* scrub on the edges of pine forest; Manavgat to Akselki above Fersinuluni; 700–850 m, sometimes to 1500 m.

Description
Plant 15–25 cm tall at flowering. **Bulb** with fine netted tunic producing thick fleshy annual roots and furnished with short, fine, very sharp hairs at base; few bulblets. **Leaves** 1–2 elongating to 55 cm at maturity, rectangular in section. **Stem** up to 20 cm above ground with several green inflated sheathing leaves. Bract and bracteole more or less equal, sheathing the 2 cm perianth tube. **Flower** bicoloured; falls 3.5–4 cm long, haft suberect, 2.5–2.7 cm long, 0.4–0.7 cm wide, green with darker spots, blade 1.5 cm long, 0.6 cm wide, reflexed, elliptic, obtuse, brownish-purple with a yellow, purple-spotted median ridge; standards 4 cm long, 0.6 cm wide, pale to deep blue veined darker, erect, narrowly oblanceolate, acute, shading to green at base; style branches 3.5 cm long, 0.3–0.4 cm wide at base of lobes, oblanceolate; lobes 0.8 cm long, 0.15–0.2 cm wide, oblanceolate, acute;

stigma bilobed; stamens with filaments 1.5 cm long. **Capsule** erect at first, becoming pendulous; tapering at both ends with a long tapering beak. **Seeds** with an appendage. **Flowering** March–April. $2n = 20$.

Cultivation
I. pamphylica appears to grow quite well in a compost based on John Innes formula with the addition of extra sharp sand or grit and lime or ground chalk to ensure very good drainage and avoid any possibility of stagnant moisture. Messrs van Tubergen are reported to grow their stocks in pure dune sand (Mathew 1971). Probably the best situation for this species is the bulb frame or raised bed.

Propagation
This can best be achieved by dividing the bulbs if well grown, or bulbils, which can then be grown on in the normal way. Seed has been produced after hand pollination, which should be tried whenever possible to ensure the continuation of vigorous and healthy stock.

Observations
I. pamphylica, only recently described and known only from this one area, could be in danger of extinction from over-collecting. It is reported to have already been wiped out from one location. It has several features that tend to isolate it from other members of the subgenus; there are similarities with *I. masia* in the underground parts which could form a link with subgenus Limniris, Series Syriacae (Diels) and it is possible that there was a common ancestor.

No hybrids with other members of the subgenus are known.

Section B: Monolepis (Rodionenko) B. Mathew, comb. nov.

Type of section: *I. kolpakowskiana* Regel.

Synonym
Iridodictyum Rodionenko, section Monolepis Rodionenko. *Alatavia kolpakowskianum* (Regel) Rodion.

The outstanding feature of this section is the channelled leaves, which are quite different from those of other members of the subgenus.

I. kolpakowskiana Regel

Synonyms
Iridodictyum kolpakowskiana (Regel) Rodionenko, *Xiphion kolpakowskiana* Regel.

Distribution
Former USSR, Central Asia, Tien Shan Mountains on open grassy–stony slopes 800–1300 m, endemic, flowering near melting snow.

Description
Plant 15–20 cm tall at flowering. **Bulb** covered with densely reticulate-fibrous tunics, few bulblets. **Leaves** 3–4, 3.5–11 cm long at flowering, extending slightly at fruiting stage, about 0.2 cm wide, linear, canaliculate, prominently ribbed on underside; margins glabrous or papillose-scabrid. **Stem** below ground, bract and bracteole green, stiff, perianth tube 5–9 cm long. **Flower** lilac-violet, falls 3.5–4.5 cm long, lanceolate, blade 2–2.5 cm long, 1–1.4 cm wide, darker violet or purple with yellow pubescent median ridge, ovate, acute or obtuse, haft suberect, 1.5–2.5 cm long, 0.6–0.8 cm wide, violet veined white, pubescent; standards 3.5–5 cm long, 0.6–1.3cm wide, erect, obovate or oblanceolate, apex shape variable; style branches 2.5–3.5 cm long, about 0.6 cm wide at base of lobe, oblanceolate; lobes 0.9–1.2 cm long, 0.2–0.3 cm wide; stigma entire; stamens with filaments 0.5–0.9 cm long; anthers 0.7–1 cm long. **Capsule** about 6 cm long, 0.7 cm diameter, cylindrical, with short beak, carried above ground. **Seeds** without appendage. **Flowering** March–April. $2n = 20$.

Cultivation
I. kolpakowskiana has been in cultivation intermittently over the past century. It is not an easy plant to grow, although some specialists succeed. Mr F.V. Kalich of Albuquerque, New Mexico, found that it grew well under conditions favourable to reticulata irises and raised it from seed to flowering in three years and from bulblets in two years; he was successful with self-pollination by hand at temperatures above 40 °F (4–5 °C), but was unsuccessful in crossing it with other members of the subgenus.

In April 1966 Dr John Marr saw *I. kolpakowskiana* growing wild on the Chimgan Ala Tau, 70 miles NE of Tashkent, at 2000 m. The ground was extremely wet either from a torrent of snow-melt or from a very heavy storm, judging by the state of the flowers. Later in the year the bank on which they were growing would become completely dry and the bulbs thoroughly ripened. It would appear that the best situation in the garden would be the raised bed treatment with ample moisture while in active growth.

Observations
I. kolpakowskiana (excluding *I. winkleri*) is quite distinct from all other members of this subgenus. The pattern of the reticulated bulb tunic differs from all other members of the groups. The leaves are channelled, crocus-like, 3–4 enclosed within a tubular sheath at base. The stigmatic lip is entire, the pollen grains are spherical, not elliptical and the seeds have no appendage.

There have been suggestions of possible links with the Central Asiatic members of subgenus Scorpiris and there may be a common ancestry, but as yet insufficient information is available and for the time being it would appear that this species will remain in this subgenus.

I. winkleri Regel

Synonym
Iridodictyum winkleri (Regel) Rodionenko.

Distribution
USSR, Central Asia, Tien Shan; endemic, 3000–4000 m.

Description
Plant 10–20 cm tall at flowering. **Bulb** with brown membranous tunics. **Leaves** 3–4, single sheath surrounding at base, sometimes overtopping flowers, narrowly linear, underside 3-nerved with central forming a narrow keel. Bract and bracteole green, acuminate. Perianth tube equal in length to limb. **Flower** bluish-violet; falls oblanceolate with an obvious lamina; standards erect, oblong, slightly wider than falls; style branches with oblong lobes. No information about capsule or seeds. **Flowering** June. $2n = ?$

Cultivation
No information.

Observations
Not known in cultivation. Little information is available concerning this species apart from the type at Leningrad. It appears to be similar in some respects to *I. kolpakowskiana* in the cross section of the leaf (Rodionenko 1961). The fact that it was described from an area close to that occupied by *I. kolpakowskiana* leaves some doubt as to the existence of *I. winkleri* until further fresh material is available.

Section C: Hermodactyloides

Type of section: *I. reticulata* M. Bieb.

Synonym
Iridodictyum section *Iridodictyum* Rodionenko.

These are the classic reticulata irises with quadrangular or subcircular octagonal leaves and with tall standards. With 6 species, this is the largest group in the subgenus and they all appear to hybridise, but not with the *Brevituba* or *Monolepis* groups.

I. bakeriana Foster

Synonyms
I. reticulata M. Bieb. var. *bakeriana* (Foster) Mathew & Wendelbo, *Iridodictyum bakerianum* (Foster) Rodionenko.

Distribution
SE Turkey, W Iran, NE Iraq. On open stony or rocky mountain slopes, clearings in *Quercus* scrub, often flowering near snowline, 850–2000 m.

Description
Plant 8–15 cm tall at flowering. **Bulb** with coarsely fibrous-reticulate tunic, with few bulblets at base. **Leaves** usually 2, varying from just visible at flowering to equalling the height of the flower, elongating to 25 cm after flowering, more or less round in section, with 8 ribs. **Stem** subterranean. Bract green, rather stiff and closely sheathing the perianth tube; bracteole membranous. Perianth tube 4–7 cm long. **Flower** usually pale blue. Falls 4–4.8 cm long, narrowly panduriform with yellow or sometimes white median ridge; haft 2.5–3.3 cm long, 0.6–1.0 cm wide, white with deep blue-violet veining or spotting, suberect, blade 1.0–1.2 cm long, 1.0–1.2 cm wide, with dark blue-violet apex, horizontal to reflexed, standards 3.5–4.5 cm long, 0.6 cm wide, bluish-lilac, erect, oblanceolate; style branches 3.5–4 cm long, 0.4–0.6 cm wide at base of lobes, bluish-lilac, oblanceolate or cuneate; lobes 0.6–0.9 cm long, 0.2–0.3 cm wide at base, obliquely obtuse, irregularly toothed; stigma bilobed; stamens with filaments 1.5–1.9 cm long; anthers 0.8–1 cm long, blue. **Capsule** erect, carried at or just above ground level, ellipsoid or fusiform, with short beak at apex. **Seeds** with an appendage. **Flowering** (Feb–)March–April(–early May). $2n = 20$.

Cultivation
Although often found growing naturally in heavy, rocky and stony soils, *I. bakeriana* is not a difficult plant to grow in conditions similar to those used for *I. reticulata*, but probably the best situation would be the bulb frame or a raised bed. This species is quite hardy, but as flowering takes place early in the year overhead cover from the elements will protect and prolong the flowering period.

Observations
I. bakeriana has a restricted distribution within the much wider area covered by *I. reticulata*. The cross sectional shape of the leaves with eight ribs makes it readily distinguishable from the other members of the group. In Iran, west of Shiraz, Dr Grey-Wilson found plants close to *I. bakeriana* in general appearance, but with leaves having 4, 5 or 7 ribs; this suggests the possibility of a natural hybridisation with *I. reticulata* at some time. Dykes refers to bulbs received as *I. bakeriana* var. *metaina,* which produced leaves with six ribs, and suggests the possibility of *I. bakeriana* crossed with pollen of *I. reticulata*. There are forms, particularly among those in cultivation, which are without the yellow marking on the median ridge of the falls.

I. histrio Reichb.

Synonyms
Iris libani Reuter ex Baker pro syn., *I. reticulata* M. Bieb. var. *histrio* (Reichb. fil.) Foster, *Iridodictyum histrio* (Reichb. fil.) Rodionenko, *Xiphion histrio* (Reichb. fil.) Hooker fil.,

Distribution
Israel, Lebanon, Syria and S Turkey. Open rocky slopes and in scrub, 500–1700 m.

Description
Plant 10–18 cm tall at flowering. **Bulb** with finely fibrous-reticulate tunic similar to *I. reticulata*, but somewhat smaller, with numerous basal bulblets. **Leaves** 1–2(–3), variably shorter or taller than flowers, elongating to 25–40 cm, rectangular in section. **Stem** subterranean. Bract and bracteole thinly membranous, white or tinged with green. Perianth tube 4–12 cm long. **Flower** mainly pale blue rarely with darker blue ground colour; falls (4–)4.3–5.0(–6) cm long, oblanceolate-oblong, haft suberect, 2.5–3.5 cm long, 0.8–1.4 cm wide, blade (1.5–)2.0–2.2(–2.5) cm long, 0.8–1.6 cm wide, conspicuously blue-blotched on a pale blue or creamy ground with a yellow median ridge, the darker markings concentrated towards the outer edge of the blade, slightly deflexed, broadly ovate, obtuse or subacute; standards 3.5–4.5 cm long, 0.5–0.6 cm wide, unmarked pale blue, erect, oblanceolate; style branches 3.8–5 cm long, 0.8–1.1 cm wide at base of lobes, oblanceolate, or cuneate, unmarked; lobes 1–2 cm long, 0.3–0.8 cm wide at base, obtuse or nearly rounded, usually irregularly toothed; stigma bilobed; stamens with filaments 1.1–1.6 cm long; anthers 1–1.4 cm long. **Capsule** carried erect at or just above ground level, ellipsoid with a short beak at apex. **Seeds** with appendage. **Flowering**: January–March(–April). $2n = 20$.

Two subspecies may be recognised.

subsp. *histrio*

Description
Blade of falls 1.2–1.6 cm wide, sparsely but not conspicuously blue-blotched over whole surface; anthers 1.1–1.4 cm long, relatively wide and compressed in section.

subsp. *aintabensis* (G.P. Baker) B. Mathew stat. nov.

Synonym
I. histrio var. *aintabensis* G.P. Baker.

Distribution
S Turkey, in the neighbourhood of Gaziantep. Stony or rocky areas, often near or in scrub, 610–1100 m.

Description
Overall flower colour pale blue; blade of falls 0.8–1.1 cm wide with darker blue blotches usually concentrated towards the centre around the yellow median ridges; filaments at least 1.4–1.6 cm long, filiform.

Cultivation
Neither subspecies is a difficult plant to grow, but given their early flowering period it is advisable to give the protection of a bulb frame. C. H. Grey (1938) writes of growing *I. histrio* in a well-drained rock garden with little more than 'natural protection' (?) from gales and frosts and it bloomed for him for years, but as an insurance some were kept in a cold house. *I. histrio* subsp. *histrio* appears to be rare in cultivation although subsp. *aintabensis* has been available from specialist bulb growers. The problem of bulbs dividing up into numerous small bulblets after flowering could be avoided to some extent by deeper planting and feeding during growth.

Observations
I. histrio is a lovely plant and is readily identified by the size of the flower being generally larger than in *I. reticulata*. The conspicuous scattered dark blotches on the creamy, or pale blue, blade of the outer segments, which tend to congregate towards the edges to form a darker margin, distinguish subsp. *histrio* from subsp. *aintabensis* in which the dark blotches are stronger in colour around the central yellow ridge and fade to a paler colour towards the edges of the lamina. Dykes refers to a variety *atropurpurea*, which appears to remain unknown except for bulbs received from Marash in 1908: the whole flower was stained with a dark red colour overlaid with a black sheen. He was unable to obtain seed and so was unable to check whether the colour would remain constant, but in all other respects it was similar to *I. histrio* in shape and growth.

I. histrioides (G.F. Wilson) S. Arnott

Synonyms
I. reticulata M. Bieb. var. *histrioides* G.F. Wilson, *Iridodictyum histrioides* (G.F. Wilson) Rodionenko.

Distribution
N Turkey, in the region of Amasya, and Rize province. Mountain slopes in turf or in open coniferous woods, 1300–1750 m.

Description
Plant 6–15 cm tall at flowering. **Bulb** with coarsely fibrous-reticulate tunic, with a few bulblets at base. **Leaves** 2, only just visible at flowering, elongating later to 20–30 cm, rectangular in section. **Stem** subterranean. Bract and bracteole membranous, white or greenish with green cross-veins. Perianth tube 4–7 cm long. **Flower** blue, falls 4.2–5 cm long, broadly oblanceolate, haft 2.2–3 cm long, 1–1.2(–1.4) cm wide, clearly or slightly inclined; blade 1.7–2.5 cm long, 1.4–2(–2.3) cm wide, with blue-blotched lighter area around yellow median ridge, reflexed, broadly ovate, obtuse or rounded; standards 4–4.5(–5.6) cm long, 0.6–0.8(–1) cm wide, oblanceolate, obtuse or rounded; style branches 3.6–5(–5.7) cm long, 1.3–1.6 cm wide at base of lobes, strongly cuneate; lobes obtuse to acute, 1.1–1.7(–2.2) cm long, 0.6–1 cm wide at the base; stigma bilobed; stamens with filaments 1.3–1.5 cm long; anthers 0.9–1 cm long. **Capsule** erect, carried at ground level, ellipsoid, without a prominent beak. **Seeds** with an appendage. **Flowering** March–April. $2n = 16$.

Cultivation
I. histrioides is an excellent garden plant given the conditions previously recommended. It appears perfectly hardy having been seen with the flowers appearing above a 5–8 cm blanket of snow. It has become very popular commercially and numerous named clones are available; there is some variation in the depth of colour and marking on the falls, but this is not as great as in the wild populations. *I. histrioides* is a most valuable addition to any spring display in the garden.

Observations
The area of distribution covered by *I. histrioides* appears small, limited as it is to two small locations in N Turkey; and, as a consequence, it could be vulnerable in its native habitat. *I. histrioides*, apart from the colour of its flowers, would appear to have a close affinity to *I. winogradowii*: its cytology is similar, as is the habit of growth with leaves just emerging at flowering time. Hybrids between the two have been produced.

var. *sophensis* (Foster) Dykes
Originally collected by Mrs Barnum near Kharput, the flowers appear similar to *I. reticulata*, but the leaves are only just visible at flowering as in *I. histrioides*. The flower segments are narrow and overall dull violet-blue in colour with a yellow median ridge on the falls. It is quite distinct, but cannot compare with the types.

I. reticulata M. Bieb.

Synonyms
I. hyrcana Woron., *Iridodictyum reticulatum* (M. Bieb.) Rodionenko, *Neubeckia reticulata* (M. Bieb.) Alefeld, *Xiphion reticulatum* (M. Bieb.) Klatt.

Distribution
NE, E and SE Turkey; NE Iraq; N, NW and W Iran; former USSR, Caucasus and Transcaucasus. Mountain grassland, stony slopes, in scrub, rarely in woods, 600–2700 m.

Description
Plant 7–15 cm tall at flowering. **Bulb** with fine to coarse fibrous-reticulate tunics, with bulblets at base. **Leaves** 2, varying in length at flowering, later to 30 cm, four-ribbed in section. **Stem** subterranean. Bract green, rigid, closely sheathing the perianth tube, bracteole membranous. Perianth tube 4–7 cm long, **Flower** very variable in colour, blue, violet or purple; falls 3.2–4.8 cm long, often bicoloured with yellow or orange median ridge, narrowly panduriform, haft suberect, 2.1–3 cm long, 0.5–1.2 cm wide, dark-veined, blade 1.1–1.8 cm long, 0.7–1.3 cm wide, ovate, obtuse to subacute, darker than rest of flower; standards 3.5–5 cm long, 0.4–0.6(–1) cm wide, erect, oblanceolate; style branches 3–4 cm long, 0.5–0.9 cm wide at base of lobes, oblanceolate–cuneate; lobes 0.6–1.4 cm long, 0.2–0.5 cm wide at the base, obliquely obtuse or acute, occasionally acuminate, often irregularly toothed; stigma bilobed; stamens with filament 1–2 cm long, anthers 0.6–1.1 cm long. **Capsule** 3.5–5.5 cm long, carried at or just above ground level, ellipsoid, beak short. **Seeds** with an appendage. **Flowering** March–June. $2n = 20$.

Cultivation
As previously stated, *I. reticulata* is not a difficult plant to grow. Given the conditions suggested it makes a very valuable addition to the display of spring flowers in the open garden, is an ideal subject for the rock garden or alpine house and may be used for indoor decoration providing care is taken and the bulbs are not over-forced. If one is fortunate enough to obtain bulbs or seed of wild species of known provenance, it would be advisable to use the bulb frame or raised bed treatment and endeavour to propagate and keep the stock pure as an aid to conservation.

Observations
This is the most widely distributed and variable species in the subgenus. The shape of the bulb can vary from almost globose to pear-shaped, the bulb tunics from slender, finely reticulated fibres to coarsely fibred tunics. The number of bulbils formed at the base of bulbs may vary from a very few large to numerous small, rice-grain-sized bulblets. Flower colour is from reddish-purple, as in the var. *krelagei*, deep violet-blue, the form usual in commerce, to a clear pale blue in var. *hyrcana* without a deeper marking on the lamina. This variability within *I. reticulata* has made it an important plant commercially as it has enabled a large number of cultivars to be created and, with the ability to propagate by means of bulbils, it has become a very popular garden plant.

I. vartanii Foster

Synonym
Iridodictyum vartanii (Foster) Rodionenko.

Distribution
Israel, Lebanon, S Syria, NW Jordan. Open rocky places and clearings in scrub, 300–1000 m.

Description
Plant 10–15 cm tall at flowering. **Bulb** with strongly reticulate-fibrous tunic, few small bulblets at base. **Leaves** usually 2, shorter, sometimes taller, than flower, elongating to 15–30 cm, rectangular in section. **Stem** subterranean. Bract and bracteole thin and membranous, usually greenish-veined towards apex. Perianth tube 5–8 cm long. **Flower** pale or slaty-blue, rarely white; falls (3.5–)4.0–4.5(–5) cm long, oblanceolate with raised yellow median ridge within a white zone with blue veins radiating outwards onto the bluish blade, haft 2–2.7 cm long, 0.4–0.5 cm wide, suberect, blade (1.5–)1.8–2.2 (–2.3) cm long, 1.5 cm wide, horizontal or slightly deflexed, oblong–obtuse or subacute; standards 3.0–3.8 cm long, 0.5 cm wide, erect, narrowly oblanceolate, obtuse or subacute; style branches 3.2–4 cm long, 0.5 cm wide at base of lobes, narrowly wedge-shaped, shallowly and irregularly toothed; lobes 2–2.5 cm long, 0.3–0.4 cm wide, gradually tapering to an acuminate apex; stigma bilobed; stamens with filaments 1.1–1.8 cm long, anthers 0.6–1.2 cm long. **Capsule** erect, carried at ground level, beak short. **Seeds** with a small appendage. **Flowering**: December–February. $2n = 20$.

Cultivation
I. vartanii comes into growth earlier in the year than any other species in the subgenus and is therefore not an easy plant to grow in the colder northern European climate: the leaves, which may be well developed at flowering, are damaged by adverse weather, resulting in weakening of the bulbs. It is therefore advisable if bulbs or seed should become available that they be given the protection of the alpine house or bulb frame.

Observations
I. vartanii is easily recognised by its shape and slaty-blue colour together with long, narrow style branch lobes. It appears to be fairly rare in some of its natural habitats, possibly from over-collecting in the past. A white form referred to as *I. vartanii* var. *alba,* found near Beersheba and Hebron was in commerce, but the blue type is said to be the more attractive. Both Dykes (1924) and Grey (1937) refer to it having a strong or delicious almond scent. Ofer Cohen says that he has not found a white form in the wild during his recent work on Israeli wild plants.

I. winogradowii Fomin

Synonym
Iridodictyum winogradowii (Fomin) Rodionenko.

Distribution
Former USSR, W Caucasus, mountains above Gagrami, and E Transcaucasia, Mount Lomis. Endemic.

Description
Plant 8–15 cm tall at flowering. **Bulb** with coarsely fibrous-reticulate tunic, with few bulblets at the base. **Leaves** 2–3(–4), just visible or shorter than flower, elongating to 15–30 cm, rectangular in section. Stem subterranean. Bract and bracteole thin, membranous, often suffused or veined green towards the tips. Perianth tube 5–7 cm long. **Flower** pale yellow; falls 4.3–5.2 cm long, with yellow-orange median rib, oblanceolate, haft 2.5–2.6 cm long, 0.8–1.2 cm wide, suberect, blade 1.7–2.6 cm long, 1.2–2 cm wide, sparsely spotted green, reflexed, ovate, obtuse; standards 3.7–5.0 cm long, 0.6–1 cm wide, erect, narrowly oblanceolate, acute; style branches 4–4.3 cm long, 0.6–1.1 cm wide at base of lobes, oblanceolate; lobes 1.2–1.6 cm long, 0.6 cm wide at base, obliquely ovate, obtuse or subacute; stigma bilobed; stamens with filaments 1.1–1.2 cm long; anther 0.9–1 cm long. **Capsule** erect, carried at ground level, ellipsoid, without a prominent beak at apex. **Seeds** with appendage. **Flowering** April–May. $2n = 16$.

Cultivation
It is difficult to understand why *I. winogradowii* is not as widely grown as *I. histrioides*. It is not difficult in cultivation, bearing in mind that it is an alpine plant and should be treated as such. It is quite hardy, needs plenty of moisture while in growth, and should not be over-dried while dormant. It is easily propagated from the bulblets produced at the base of the parent bulbs and will readily set seed with hand pollination. Every effort should be made to keep pure stock as an aid to conservation. It is propagated by the Dutch bulb growers and is available commercially.

Observations
I. winogradowii was first noted by P.Z. Winogradow-Nikitin, after whom it was named. In 1972 Dr Rodionenko noted that only a few hundred plants were known to exist on the slopes of Mount Lomis. As has already been noted under *I. histrioides*, it and *I. winogradowii* are closely related. Apart from the colour of the flowers, there is some difference in the size and shape of the style branches.

Hybrids exist between the two species.

Section D: Micropogon (Baker) Boiss.

Type of section: *I. danfordiae* (Baker) Boiss.

The distinctive characteristic of this group is reduction of the standards to tiny, narrow, almost vertical structures.

I. danfordiae (Baker) Boiss.

Synonyms
I. amasiana Bornm., *I. bornmuelleri* Hausskn., *I. crociformis* Freyn, *Iridodictyum danfordiae* (Baker) Rodionenko, *Xiphion danfordiae* Baker.

Distribution
Central Turkey from Ordu, Amasya and Erzincan in the north to the Taurus Mountains in the south. Stony slopes in exposed places and in sparse coniferous woods; usually flowering near the snowline, 1000-2000 m.

Description
Plant 6–7 cm tall at flowering. **Bulb** with coarsely fibrous-reticulate tunic, with many tiny bulblets at the base. **Leaves** (1–)2(–3), shorter than flower, lengthening later to about 20 cm, rectangular in section. **Stem** subterranean. Bract green, bracteole narrower and semi-transparent. Perianth tube 3–4(–6) cm long. Flowers yellow; falls 2.5–4 cm long, oblanceolate, haft 1.5–2.5 cm long, 0.5–0.7 cm wide, suberect, blade 0.7–1.8 cm long, 0.6–1.3 cm wide, sometimes reflexed, ovate, obtuse, with a few small green spots surrounding a deep orange or yellow median ridge; standards 0.3–0.5 cm long, erect, bluntly needle shaped; style branches 2.2–3.3 cm long, 0.6–1 cm wide, at base of lobes, cuneate; lobes 1–1.3 cm long, 0.4–0.7 cm wide at the base, ovate, usually obtuse, irregularly toothed; stamens with filaments 0.9–1.1 cm long, anthers 0.5–0.6 cm long. **Capsule** 2.5 cm long, 1 cm wide, erect, carried at ground level, ellipsoid tapering to a point. **Seeds** with an appendage. **Flowering**: March–April(–May). $2n = 18, 27$.

Cultivation
I. danfordiae is not a difficult plant to grow, but the bulbs have the annoying habit of breaking up into small rice-grain-sized bulblets after the first season. It is suggested that deep planting may help to overcome this tendency. Brian Mathew reports that the wild plant, although smaller, is a much more attractive form and the flowers are scented.

Observations
I. danfordiae is easily recognised by the colour and size of the flower, being somewhat smaller than *I. winogradowii* which is the only other yellow-flowered species

in this subgenus, and by the much reduced bristle-like standards. Cytologically, *I. danfordiae* is distinct from the other members of the subgenus.

I. danfordiae has been in cultivation for a long time: it is extensively grown by the Dutch bulb trade and is readily available in commerce. This clone is somewhat larger than the wild plant and is a sterile triploid.

I. pariensis Welsh 1986

J.W. WADDICK

Distribution
Unknown; original plant found Kane County, Utah.

Description
Rhizome under 1 cm in diameter, carrying fibrous remnants of old leaf bases. **Leaves** numerous at both current and new growing points, 7–24 cm long, 0.2–0.5 cm wide, yellowish to brown or purple at bases. **Stem** 4 cm long, with several sheathing leaves attached; flower single; bracts 5–6 cm long, 0.2–0.3 cm wide, bases close together, tight sheathing, herbaceous; ovary about 1.2 cm long. **Flower** pale in colour, probably white, tube 1.5 cm long, falls about 6 cm long, 1 cm wide; standards about 6 cm long, 0.8 cm wide, narrowly oblanceolate; styles about 2.8 cm long, crests about 7 cm long, anthers about 1.3 cm long. **Capsule** unknown. **Flowering** May? **2n** = ?

Comments
The single specimen of this putative species was found in sandy ground in a grass–shrub area of semi-desert at 1403 m by Vane O. Campbell in Kane County, Utah. It was described by Stanley L. Welsh in the Great Basin Naturalist (1986). Since Dr Norlan Henderson first drew attention to this species there have been extensive and on-going searches for further material, but without success. (In the spring of 1993, following a cold, wet winter after a prolonged drought, an ornithologist in search of an owl believed he might have seen this plant in flower. Sadly, he did not take any photographs.) Dr Welsh thought it might have some affinity to the irises of the Pacific Coast in spite of similarities to *I. missouriensis*. Dr Norlan Henderson has recently reviewed the materials of *I. missouriensis* assembled by the late Dr Homer Metcalf and concluded that *I. pariensis* may be a variant of *I. missouriensis*. Further specimens are needed to confirm identity.

References

Anderson, E. 1928 The problem of species in the northern blue flag. *Ann. Missouri Bot. Gard.* **15**, 241–332.
Anderson, E. 1936 The species problem in iris. *Ann. Missouri Bot. Gard.* **23**, 457–509.
Bastow, K. 1968 Oncocyclus Iris, part 2. *Iris Year Book* 1968, pp. 77–91.
Bennet, R. & Arnold, M.L. 1989 A preliminary report on the genetics of Louisiana irises. *Bull. Am. Iris Soc.* **273**, 22–5.
Bini Maleci, L. & Maugini, L. 1977 Karyotypes of male-fertile and male-sterile *Iris pallida* Lam. growing in Tuscany. *Caryologia* **30** (2), 237–45.
Bini Maleci, L. & Maugini, L. 1981 *Iris pseudopumila* Tin and *Iris attica* Boiss. & Heldr.: two very similar karyotypes. *Bol. Soc. Broteriana* **LII**, 921.
Boissier, E. 1882 *Flora Orientalis*, vol. 5. Geneva: Apud H. Georg.
Chaudhary, S.A., Kirkwood, G. & Weymouth, C. 1975 *Iris* subgenus Susiana in Lebanon and Syria. *Bot. Not.* **128**.
Chimphamba, B.B. 1971 Cytogenetic studies in the genus *Iris*. Ph.D. thesis, University of London.
Clarkson, Q.D. 1958 *Iris* Subsection Oregonae. *Madroño* **14**, 246–7.
Clarkson, Q.D. 1959 Series Oregonae: *I. tenuis*. *Baileya* **6**, 187.
Collins, J. 1986 "Holden Clough": one of its mysteries solved. *Iris Year Book* 1986, pp. 109–10.
Davies, P.H. 1988 *Flora of Turkey*, supplement. Edinburgh University Press.
Davis, A.P. & Jury, S.L. 1990 A taxonomic review of *Iris* L. series *Unguiculares* (Diels) Lawrence. *Bot. J. Linn. Soc.* **103**, 281–300.
Dykes, W.R. 1913a *The Genus Iris*. Cambridge University Press.
Dykes, W.R. 1913b A Dalmatian Iris Hunt. *Gard. Chron.* May 1913, pp. 321–2.
Dykes, W.R. 1915a Notes on Irises. *Gard. Chron.* Sept. 1915, p. 163.
Dykes, W.R. 1915b Notes on Irises. *Gard. Chron.* Feb. 1915, pp. 95–6.
Dykes, W.R. 1915c What is a species? *Gard. Chron.* June 1915, pp. 341–2.
Dykes, W.R. 1924 *A Handbook of Garden Irises*. London: Martin Hopkinson.
Ellis, J.R. 1975 *Hybridisation and the Flora of the British Isles*, ed. C.A. Stace, pp. 471–3. London: Academic Press/Botanical Society of the British Isles.
Ellis, J.R. 1979 Cultivated *I. wattii*. *British Iris Society Year Book* 1979, pp. 62–6.
Ellis, J.R. 1980 A note on the discovery of *I.* "Nova". *British Iris Society Species Group Bulletin*, Dec. 1980, p. 4.
Feinbrun-Dothan, N. 1986 Iridaceae. In *Flora Palaestina*, part 4, pp. 112–38. Jerusalem: Israel Academy of Sciences and Humanities.
Furse, P. 1968 Iris in Turkey, Iran and Afghanistan. *British Iris Society Year Book* 1968, pp. 64–74.

Gabrielson, I.N. 1932 *Western American Alpines*. New York: Macmillan.
Gavrilenko, B.D. 1962 Variability in *Iris iberica*. *Trans. Tblisi Bot, Inst.*, vol. 22.
Grey, C.H. 1937 *Hardy Bulbs*, 1st edn, vol. 1, *Iridaceae*. London: Williams & Norgate.
Gustafsson, M. & Wendelbo, P. 1975 Karyotype analysis and taxonomic comments on irises from SW and C Asia. *Bot. Not.* 1975, pp. 380–407.
Hulme, J. 1989 Some interesting plants at the shows, 1987–88. *Bull. Alpine Gard. Soc.* **57**, 55.
Inariyama, S. 1929 Kanyological studies of *I. kempferi. Jap. J. Bot.* **4**, 405–26.
Kazao, N. 1928 Cytological studies of *Iris. Bot. Mag., Tokyo* **42**, 262–6.
Knowles, G.B. & Westcott, F. 1838 *The Floral Cabinet*, vol. 2, p. 19. London: William Smith.
Köhlein, F. 1981 *Iris*. Stuttgart: Ulmer.
Komarov, V.L. (ed.) 1935 *Flora of the USSR*, vol. 4. Leningrad: Botanical Institute of the Academy of Sciences. (Translated 1968 under Israel Programme for Scientific Translations, Jerusalem.)
Lawrence, G.H.M. 1953 A reclassification of the genus *Iris. Gentes Herbarum* **8** (4), 346–71. Ithaca, New York: L.H. Bailey Hortorium, Cornell University.
Lenz, L.W. 1959 Hybridization and speciation in the Pacific Coast Irises. *Alisonia* **4** (2), 311.
Lin, T.-S. & Ling, S.-S. (eds) 1978 *Flora of Taiwan*. Taipeh, Taiwan: Epoch Publishing Co.
Longley, A.E. 1928 Chromosomes in iris species. *Bull. Am. Iris Soc.* **29**, 43–55.
Lynch, R.I. 1923 *The Book of the Iris*. London: John Lane, The Bodley Head.
Mahan, C. 1988 *Iris florentina*. *SIGNA* **40**, 1443. Minnesota, USA: Species Iris Group of North America.
Marchant, A. 1968 Spurias. *British Iris Society Species Group Bulletin*, July 1968, p. 6.
Mathew, B. 1971 Plant portraits: *I. pamphylica. Bull. Alpine Gard. Soc.* **39**, 217–18.
Mathew, B. 1984 Iridaceae. In Davis, P.H. (ed.) *Flora of Turkey*, pp. 392–3. Edinburgh University Press.
Mathew, B. 1989 *The Iris*, 2nd edn. London: Batsford.
Maugini, L. & Bini Maleci, L. 1981 Le specie nane di *Iris* in Toscana e il loro problema tassonomico. *Webbia (Raccolta di Scritti Botanici, Florence)* **35** (1), 145–86.
Nicholson, G. & Upcott Gill, L. (eds) 1889 *An Illustrated Dictionary of Gardening*, vol. 5 and supplement. London.
Nicholson, G. & Upcott Gill, L. (eds) 1900 *An Illustrated Dictionary of Gardening*, supplement 2. London.
Noltie, H. 1995 New irises from Yunnan. *New Plantsman* **2**, 136–7.
Prodan, J. 1964 Die Iris-Arten Roumänien. *Bull. Bot. Gdn Mus. Univ. Cluj, Romania* **15**, 68–71.
Randolph, L.F. 1934 Chromosome numbers in native American and introduced species and cultivated varieties of Iris. *Bull. Am. Iris Soc.* **52**, 61–6.
Randolph, L.F. (ed.) 1959 *Garden Irises*. St Louis, Missouri: The American Iris Society.
Randolph, L.F. 1966 A new species of Louisiana iris of hybrid origin. *Baileya* **14**, 143–69.

Randolph, L.F. & Mitra, J. 1959a Dwarf bearded species hybrids. In *Garden Irises*, ed. L.F. Randolph, pp. 408–12. St Louis, Missouri: The American Iris Society.

Randolph, L.F. & Mitra, J. 1959b Karyotypes of *Iris pumila* and related species. *Am. J. Bot.* **46** (2), 93.

Randolph, L.F. & Mitra, J. 1961 Karyotypes of iris species indigenous to the USSR. *Am. J. Bot.* **48** (10), 862–70.

Rodionenko, G.I. 1961 *The Genus Iris*. Leningrad: Academy of Sciences of the USSR. (English edition 1984, transl. T. A. Blanco White. London: The British Iris Society.)

Rodionenko, G.I. 1967 Wild iris species of the USSR, part 2. *Iris Year Book* 1967, pp. 96–102.

Salisbury, R.A. 1812 The cultivation of rare plants. *Trans. Hort. Soc. Lond.* **1**, 303.

Sani, G.L. 1963 Technical and economic aspects of *Iris* growing for perfumery. *Atti del 1° Simposio Internazionale dell'Iris*, ed. Italian Iris Society, p. 581. Florence: Typographia Giuntina.

Shevchenko, G.T. 1979 *Iris* L. In Galushko, A.I. (ed.) *Flora Severnogo Kavkaza*, vol. 3, p. 79. Rostov: University of Rostov.

Simonet, M. 1932a Recherches cytologiques et génétiques chez les iris. *Bull. Biol. France Belg.* **105**, 255–444.

Simonet, M. 1932b Recherches cytologiques et génétiques chez les iris. *Bull. Biol. France Belg.* **78**, 696–707.

Simonet, M. 1934 Nouvelles recherches cytologiques et genetiques chez les Iris. *Annls Sci. Nat. (Bot.), Paris* **16**, 229–383.

Snoad, B. 1952 Chromosome counts of species and varieties of garden plants. Report, John Innes Horticultural Institute, pp. 47–50.

Stojanov, N. 1966 Iridaceae. In *Flora Bulgarica*, 2nd edn, p. 247. Sofia University Press.

Webb, D.A. & Chater, A.O. 1976 Notes on the genus *Iris*. In Heywood, V. (ed.), *Flora Europaea, Notulae Systematicae*, No. 20. *Bot. J. Linn. Soc.* **76** (40), 314–16.

Welsh, S.L. 1986. New taxa and combinations in the Utah flora. *Great Basin Nat.* **56** (2), 256.

Wendelbo, P. 1977 *Tulips and Irises of Iran*. Tehran: Botanical Institute of Iran.

Werckmeister, P. 1957 In the homeland of the Oncocyclus. *Iris Year Book* 1957, pp. 86–92.

Whitehouse, B. & Warburton, B. 1978 Miniature Dwarf Beardeds. In Warburton, B. & Hamblen, M. (ed.), *The World of Irises*, pp. 145–55. Wichita, Kansas: American Iris Society.

Wu, Q. J. & Cutler, D.F. 1985 Taxonomic, evolutionary and ecological implications of the leaf anatomy of rhizomatous *Iris* species. *Bot. J. Linn. Soc.* **90**, 253–303.

Yasui, K. 1939 Karyological studies on *I. japonica* Thunb. and its allies. *Caryologia* **10**, 180–8.

Zhao, Y.-T. 1985 *Flora Reipublicae Popularis Sinicae*, vol. 16, part 1, pp. 48–60.

Select Bibliography

Caillet, M. & Mertzweiller, J.K. (eds.) 1988 Waco, Texas: *The Louisiana Iris.* Society for Louisana Irises, Texas Gardener Press.
Cassidy, G.E. & Linnegar, S. 1982 *Growing Irises.* London: Croom Helm.
Cohen, V.A. 1967 *A Guide to the Pacific Coast Irises.* British Iris Society.
Dykes, W.R. 1913 *The Genus Iris.* Cambridge University Press.
Dykes, W.R. 1924 A Handbook of Garden Irises. London: Martin Hopkinson.
Dykes, W.R. 1928 Irises, Present-Day Gardening. London: T.C. & E.C. Jack.
Horinaka, A. 1990 *Iris Laevigata.* Ohuna, Japan: ABOCSHA Co.
Innes, C. 1985 The World of Iridaceae. Ashington, Sussex: Holly Gate International.
Köhlein, F. 1981 *Iris.* Stuttgart: Ulmer.
Lynch, R.I. 1923 *The Book of the Iris.* London: John Lane, The Bodley Head.
Mathew, B. 1989 *The Iris*, 2nd edn. London: Batsford.
McEwen, C. 1990 *The Japanese Iris.* Hanover: The Society for Japanese Irises, U.P. of New England.
McEwen, C. 1995 *The Siberian Iris.* Portland, Oregon: Timber Press.
Randolph, L.F. (ed.) 1959 *Garden Irises.* St Louis, Missouri: The American Iris Society.
Rodionenko, G.I. 1961 *The Genus Iris.* Leningrad: Academy of Sciences of the USSR. English edition 1984, transl. T.A. Blanco White. London: The British Iris Society.
Waddick, J.W. & Zhao, Y.-T. 1992 *Iris of China.* Portland, Oregon: Timber Press.
Warburton, B. & Hamblen, M. (ed.) 1978 *The World of Irises.* Wichita, Kansas: The American Iris Society.
Wendelbo, P. 1977 *Tulips and Irises of Iran.* Tehran: Botanical Institute of Iran.
Wendelbo, P. & Mathew, B. 1975 *Iridaceae, Flora of the Iranian Highlands.* Graz, Austria: Akademische Druck-u. Verlaganstalt.
Werckmeister, P. 1967 *Catalogus Iridis.* Leonberg bei Stuttgart: Deutsche Iris-und Liliengesellschaft.

Publications of:
American Iris Society
British Iris Society
Canadian Iris Society
Cornell University (*Baileya*)
Gesellschaft der Stadenfreude, Germany
Iris Society of Australia
New York Botanic Garden (*Addisonia*)
New Zealand Iris Society
Royal Botanic Gardens, Kew

Bibliography

Royal Horticultural Society
Schweizer Iris & Lilien-freunde
Societa Italiana del Iris
Société Francaise des Iris et Plantes Bulbeuses

British Iris Society Species Group
New Zealand Iris Society Species Section
Species Iris Group of North America

Curtis's Botanical Magazine
The Kew Bulletin

Flora Bulgarica 4th edn. (ed. N. Stojanov, B. Stevanov & B. Kitanov), 1966. Sophia.

Flora Europaea (ed. T.G. Tutin, V.H. Heywood & N.A. Burgess), vol. 5, 1976. Cambridge University Press.

Flora Severnogo Kavkaza (ed. A.I. Galushko), 1980. University of Rostov.

Flora of Taiwan – Iridaceae (ed. T.-S. Lin & Sh.-S. Ling), 1978. Taipej, Taiwan: Epoch Publishing Co.

Flora of Turkey (ed. P.H. Davis)., vol.8, 1984. Edinburgh University Press.

Flora of Turkey supplement (ed. P.H. Davis, R.R. Mill & K. Tan), 1988. Edinburgh University Press.

Flora of the USSR, vol.4 (ed. V.L. Komarov), 1935. Leningrad: Botanical Institute of the Academy of Sciences. Translated 1968 under Israel Programme for Scientific Translations, Jerusalem.

Glossary

Acid of soil, with a pH less than 7
Acuminate long pointed
Aggregate of plants, a group of closely related species known by the same specific name
Alkaline of soil, with a pH of more than 7, not necessarily calcareous
Allopolyploid (= alloploid, amphiploid) a polyploid arising after an interspecific hybridisation
Amphidiploid the original term for an allotetraploid in which the progenital species were strictly diploid
Apex tip or point, end of leaf, perianth segment or capsule
Apiculate a short point on a rounded apex
Appendage an extra attachment to an organ, often of no apparent use
Aril part of the material attaching a seed to the capsule, usually fleshy when fresh
Auricle a small growth, usually at the base of a leaf, which bears a slight resemblance to an ear
Auto- occurring from within
Axil where leaf and stem join
Axillary growing from an axil

Basal in this context, usually the base of a stem or bulb
Beard in irises, usually a narrow hairy area found on the middle of the falls and occasionally the standards
Bifid divided into two parts from the midpoint
Binomial a two part name in Latin indicating genus and species for a living organism
Blade in irises, the outer part of a perianth segment
Bract a modified leaf usually protecting the flower and its parts, sometimes found quite a long way down the stem.
Bracteole a subsidiary bract, sometimes smaller
Bulbil small bulbs occurring round a parent bulb
Bulblet see bulbil

Calcareous of soil, containing chalk
Canaliculate grooved or channelled
Capsule ripe seed pod
Caudate with a tail
Cauline related to the stem, *e.g.* cauline leaf
Chromosome a small self perpetuating body in cell nuclei and which carries the heredity factors (genes)

Glossary

Cilia of plants, short, fine hairs covering a surface hence ciliate; finer than papillae
Claw haft of fall or outer perianth segment
Clone any number of plants which have been derived by vegetative propagation from one original seedling
Colony a group of plants of the same species, probably isolated
Concolor same colour all over
Crest 1 the lobe(s) of the style branch; 2 a narrow raised ridge on the fall
Cucullate of an iris leaf, the tip curved and contracted so that it resembles a hood
Cultigen a cultivated variety of either a species or a hybrid
Cultivar in irises, now usually a result of more than one generation of inter-breeding; in the past, often a selected form or cross given a latinised name
Cuneate ends more or less parallel; sides longer and converging to one end
Cytology study of living cells, but generally used of the study of chromosomes

Deflexed bent sharply downwards
Deltoid similar to the Greek delta, squat triangle
Dilated swollen or expanded
Diploid having two sets of chromosomes
Distichous in one plane as with the leaf fan of a bearded iris
Divergent moving away from each other more or less diagonally

Ellipsoid longer than it is wide; ellipse-like
Elliptic narrow oval with rounded ends
Emarginate notched at the apex
Endemic confined to, and usually fairly common in, one area
Ensiform sword shaped, parallel sides narrowing to a pointed apex
Entire unlobed, untoothed and undivided
Erose irregularly notched

Falcate curved, traditionally like a sickle
Falls the outer perianth segments
Filament stalk of the stamen
Filiform thread-like
Foliaceous leaf-like
Fruit mature seed-bearing organ
Fusiform spindle shaped, long, slender

Gamete the male and female cells which unite in sexual reproduction
Genus related plants having the same name and divided into species
Glabrous entirely smooth
Gland a part of the flower, sometimes only hair-like, which exudes sticky matter
Glaucous with a waxy, greyish covering

Glossary

Habitat the natural conditions in which the plant grows
Haft the inner part of falls or standards, usually narrower than the blade
Herbaceous 1. meaning green in reference to bracts. 2. non-woody. 3. leaf material dying down in winter
Hybrid a cross either between plants of the same or another species

Idiogram diagrammatic representation of the somatic chromosome complement
Inflorescence stem, bract(s) and flowers of the plant
Internode length of stem between two joints
Inter-specific (in hybridisation) involving individuals of two distinct species
Intra-specific (in hybridisation) involving individuals within a species

Karyotype a linear arrangement of the chromosomes from a cell to show their number and morphology
Keel in irises, a sharp ridge on the outer side of a bract

Laciniate of iris parts, with the edges divided into many slender thread-like segments
Lamina blade of fall or standard
Lanceolate like a lance blade, narrow at base, widening more or less abruptly and then narrowing gradually to a point
Lateral arising from the side
Lax loose, floppy
Linear edges parallel, narrow
Linear-lanceolate very narrowly lanceolate, almost linear
Lingulate tongue-shaped
Lobe usually a rounded projection on some part of the perianth
Local where a species is restricted to a very limited area, but may be present in quite large quantities
Locules iris capsules are divided into three chambers or locules

Membranous paper-like, dry and flexible, translucent
Mesophyte plant from an area neither desert nor very wet, most irises are this
Monograph a written review of a single group of plants
Morphology details of appearance of external or internal features

Naturalised a plant established in an area well separated from where it originated
Nectary the source of nectar in a plant, usually at the base of the perianth segments
Nerve prominent vein in leaf or petal

Oblanceolate reverse of lanceolate, widest just above centre
Oblong sides more or less parallel, one pair longer
Obovate egg shaped with broad end as apex
Obtuse blunt

Glossary

Offset vegetatively produced growth from rhizome or bulb
Opposite usually 2 leaves or stalks arising from the same node
Orbicular more or less round
Ovary the part of the flower which will develop into the capsule after fertilisation

Panduriform constricted at mid-length like a violin
Papillose minutely hair-like protuberances covering an area, coarser than cilia
Pedicel the stalk of a single flower between ovary and main stem
Peduncle the main flower stalk of the plant
Perianth in iris, the falls and standards together with the style arms
Perianth tube between perianth and ovary, formed from the bases of the perianth segments
Pollination the transfer of pollen from an anther to a receptive stigma
Progenitor an original parental type; were once strictly diploid
Puberulent thinly or minutely pubescent
Pubescent hairy
Pyriform like a stylised pear with rounded ends and one half markedly wider than the other

Raphe a conspicuous growth on a seed which is not an aril
Recurved curving downwards and backwards toward the stem
Reflexed bending sharply downwards
Reticulate netted
Rhizome the conspicuous 'root' of many irises carrying growing points at the ends and along the sides together with true roots; actually a modified stem
Rhombic more or less diamond shaped with very obtuse lateral angles
Rootstock the basal part of a plant, bulb or rhizome

Saline of soils – this does not mean a high proportion of sodium chloride, but an accumulation of other insoluble 'salts' resulting from dehydration rather than drainage of the soils
Scabrid rather rough,
Scarious membranous, dry, not green
Segment a single part of the perianth
Sessile without a stem
Somatic relating to the body of an organism (as opposed to the germ cells)
Spathe modified leaf, often papery, which protects the sexual parts of the flower
Spathulate broad with a rounded tip and narrowed to the base
Species a convenient unit for classification of a group of plants with many characteristics in common
Spike in irises, the inflorescence from rootstock upwards
Standard inner perianth segment usually more or less vertical
Sterile usually flowers which cannot produce seeds because of a genetic fault in ovules or pollen grains; self-sterile requiring pollination by a plant from a different seed

Glossary

Stigma in irises, a 'lip' on the underside of the style arm
Stolon a horizontal stem, usually slender, possibly with roots at nodes and giving rise to another plant at the end.
Striated with fine, longitudinal grooves or ridges particularly on leaves
Style in irises, mostly a fairly long, flattened growth lying on the upper surface of the fall
Subequal nearly equal
Subspecies a classification group of plants usually distinguished from the main group by one characteristic such as flower colour
Subulate slender with a sharp point
Synonyms various names which have been given to a species, but which should not be used and cannot be neglected

Taxon a morphologically distinct group at any level of a classification scheme
Terminal produced at, or referring to, the end of a leaf or stem
Tetraploid having four sets of chromosomes
Toothed an irregular edge usually to the style crests
Trigonal pyramidal, but not necessarily with flat sides
Triploid having three sets of chromosomes
Truncate ending abruptly, almost as if broken or cut short
Tube see perianth tube
Tuberous of roots which have a storage function and are necessary for the good growth of the plant
Tunic the outer coat of a bulb
Type the original specimen from which a plant and so a species was described
Type locality where the type was collected

Unguiculate said of a petal which is narrowed sharply at the haft to look like a claw

Variety a classification group below the subspecies possibly only distinguished by flower colour or location
Ventricose swollen centrally, to one side or round the base

Widespread the plant is found over a large area, but may be thin on the ground

$2n = \ldots + ?B$ it is frequently found that there are small pieces of chromosomal material which are not attached to a chromosome and which are not chromosomes in their own right. They are accounted for separately in this fashion

Maps

WILLIAM R. KILLENS

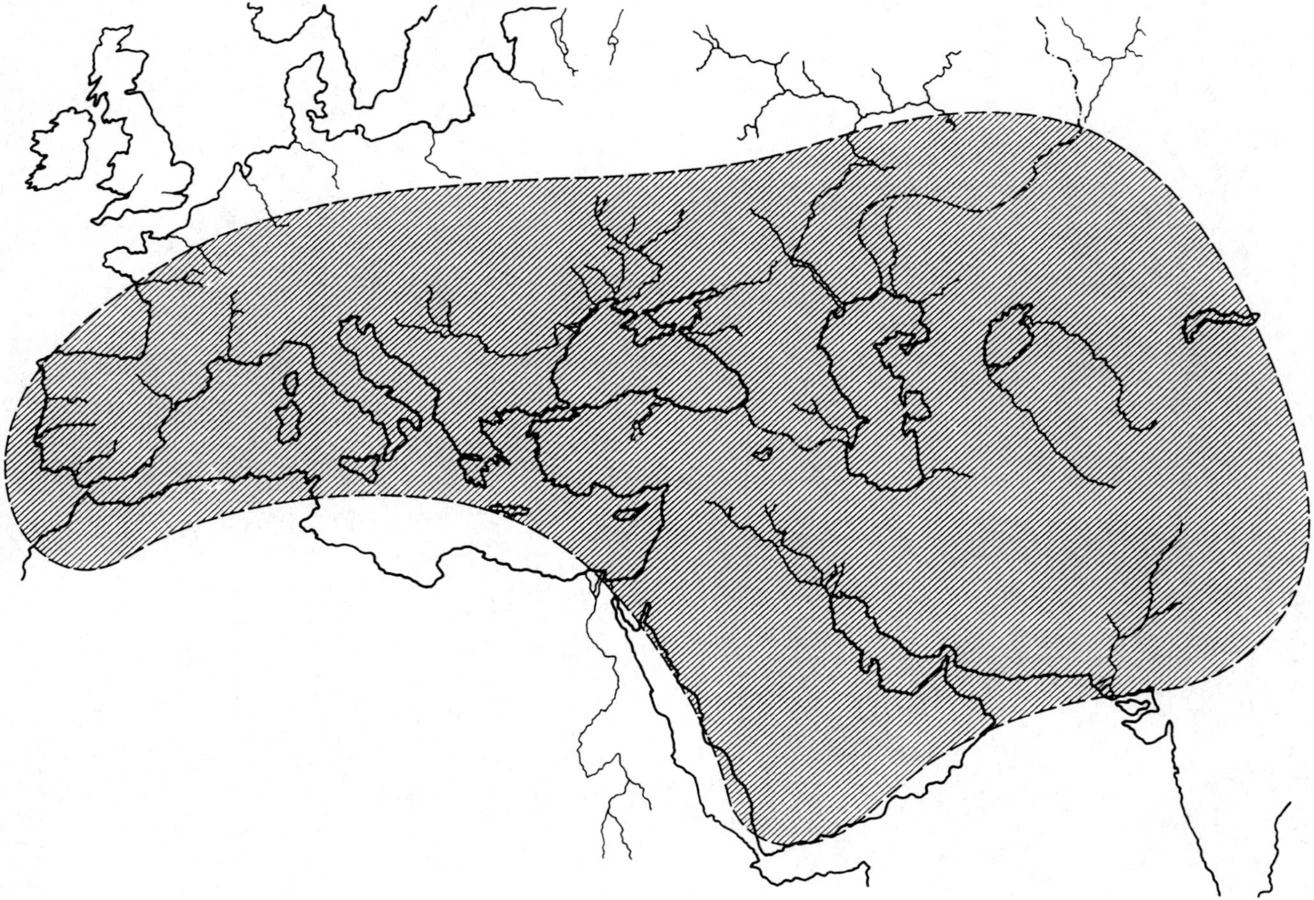

MAP I SECTION IRIS

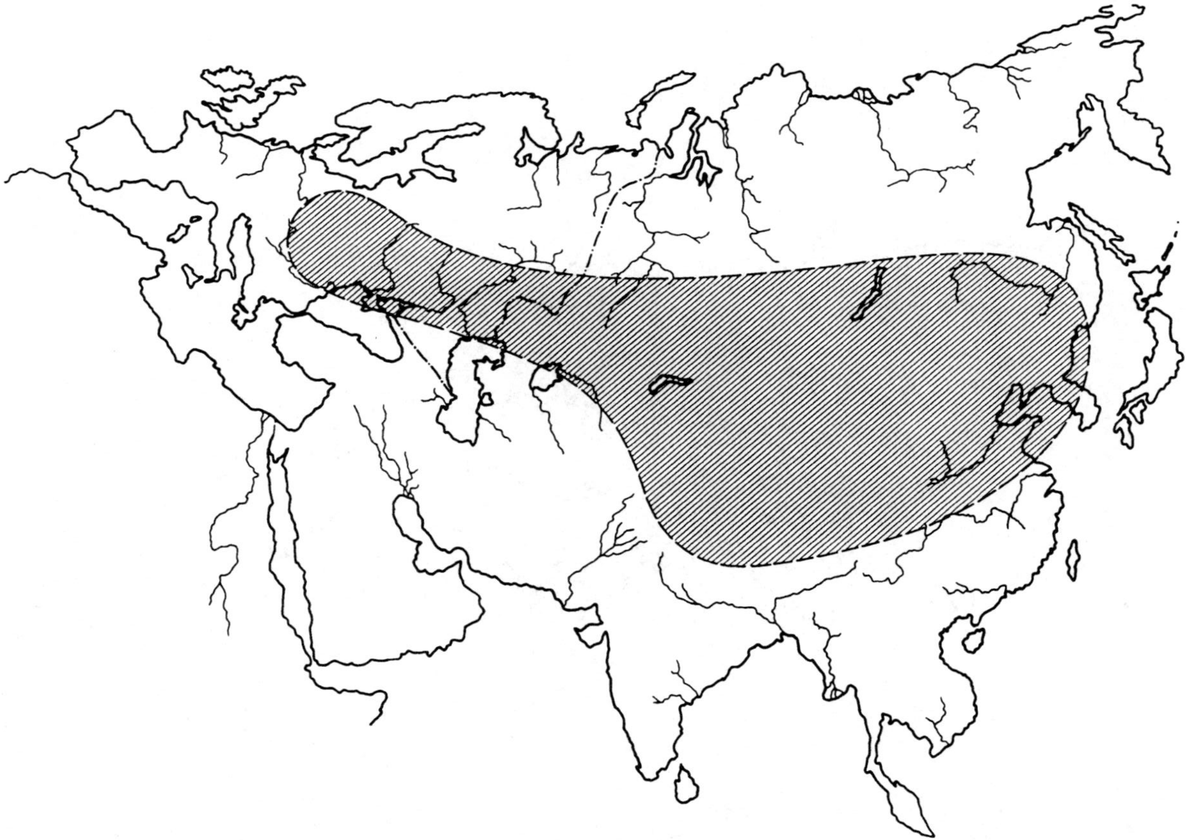

MAP 2 SECTION PSAMMIRIS

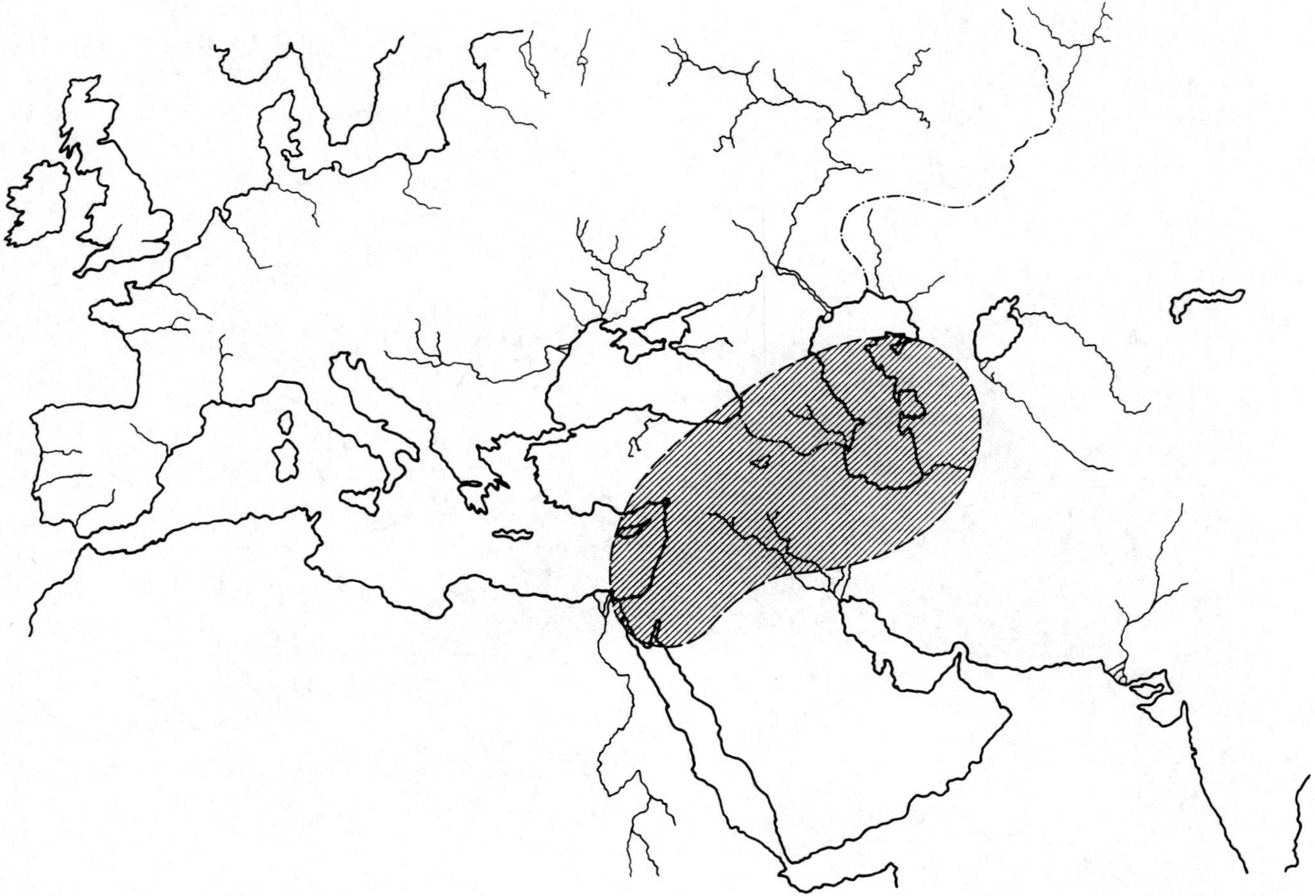

MAP 3 SECTION ONCOCYCLUS

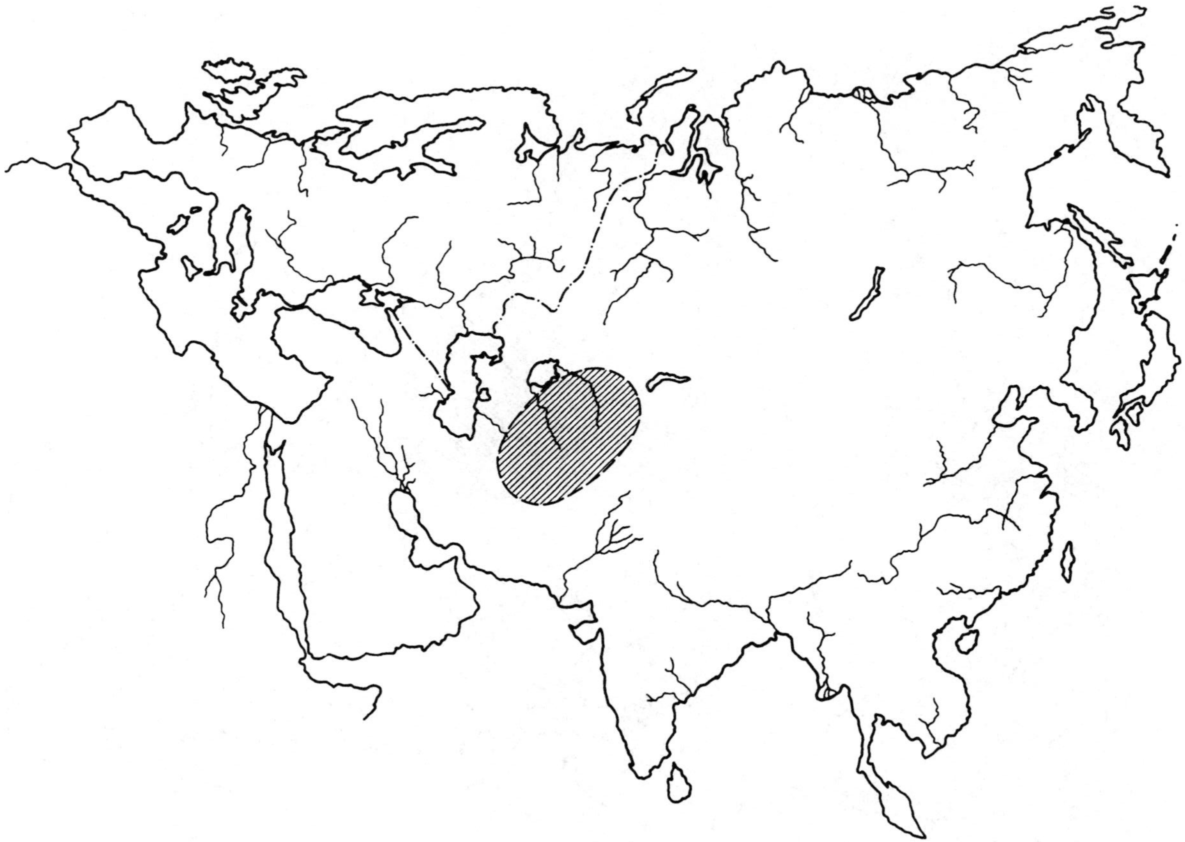

MAP 4 SECTION REGELIA

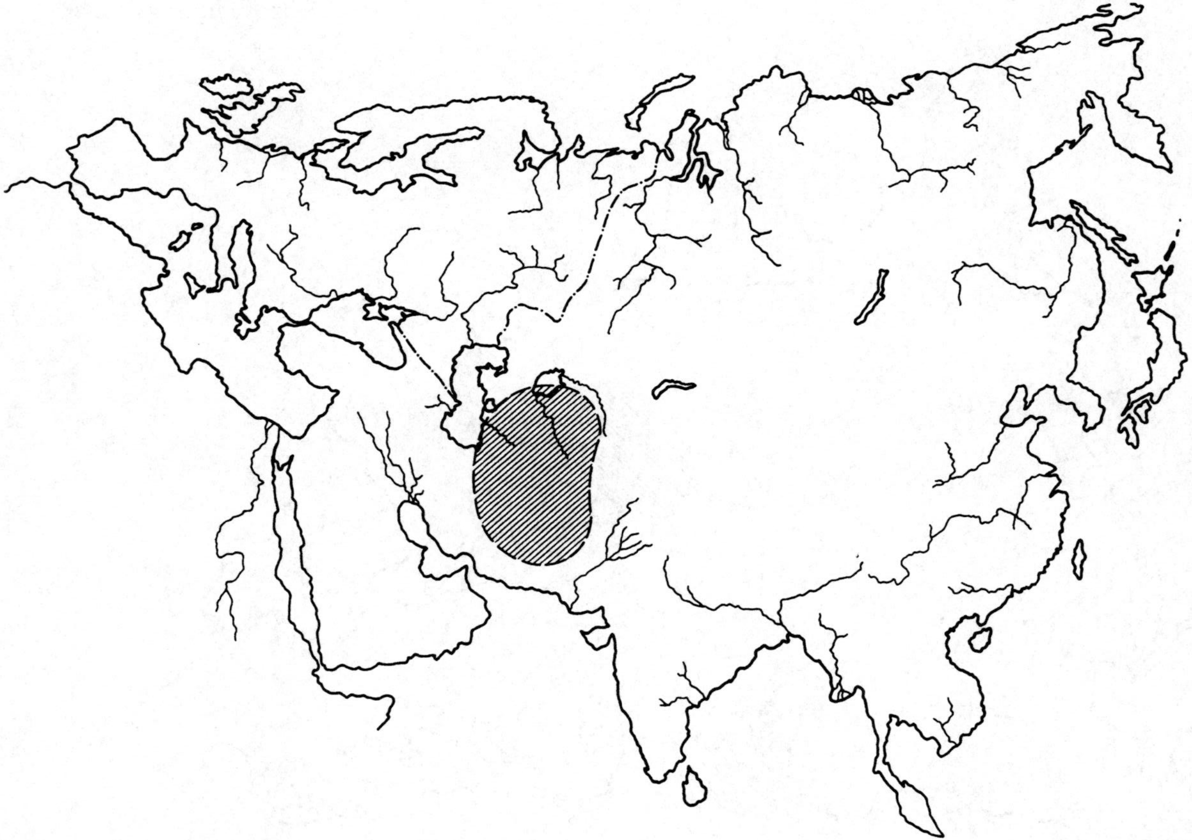

MAP 5 SECTION HEXAPOGON

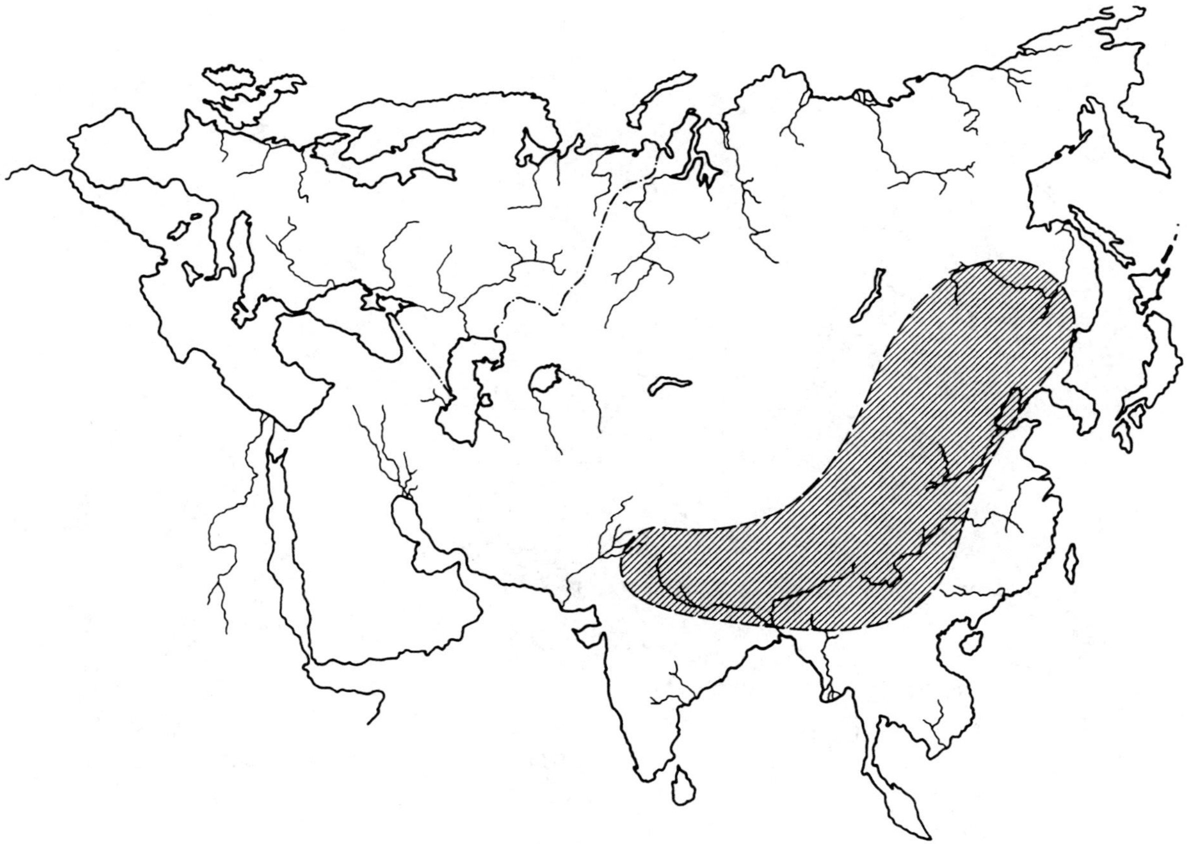

MAP 6 SECTION PSEUDOREGELIA

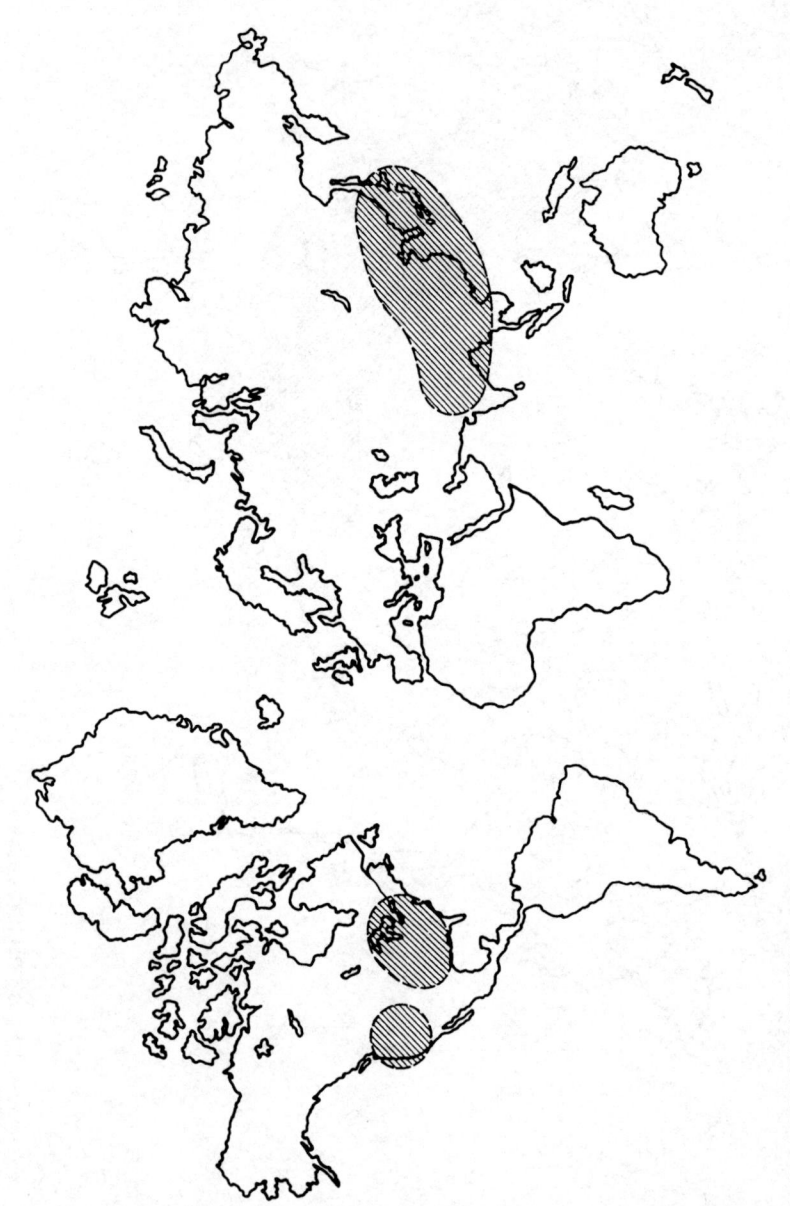

MAP 7 SECTION LOPHIRIS

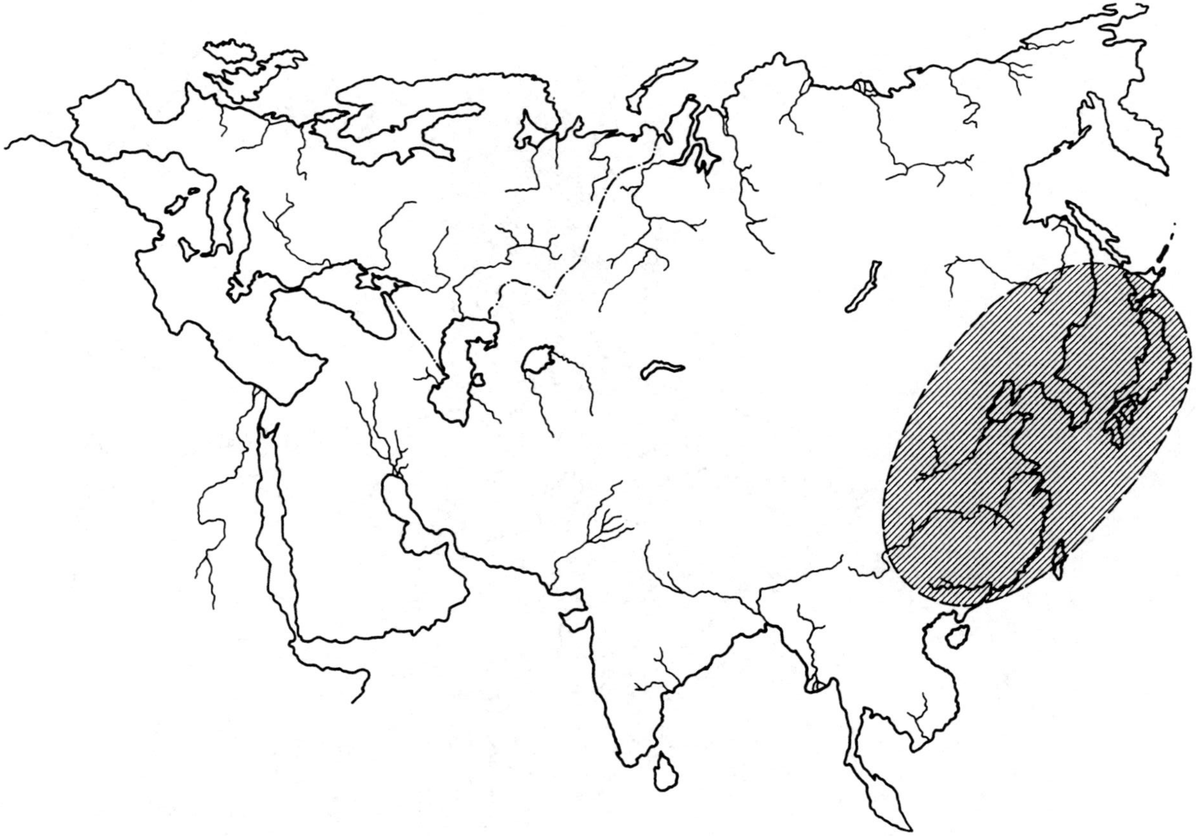

MAP 8 SERIES CHINENSES

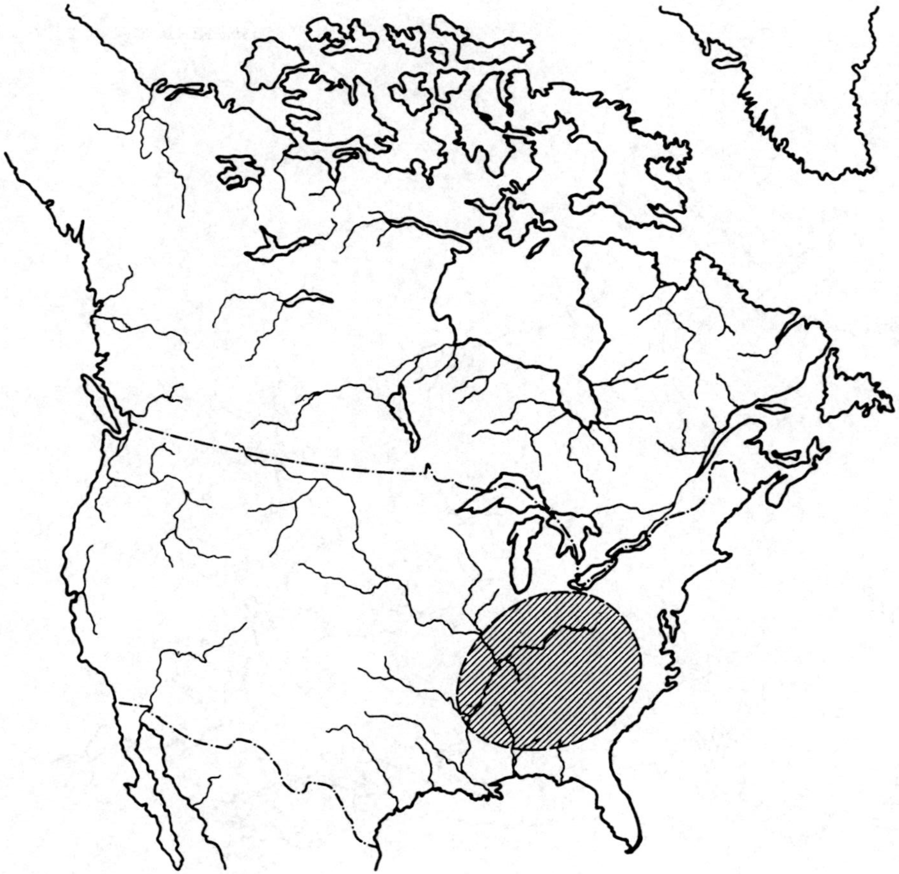

MAP 9 SERIES VERNAE

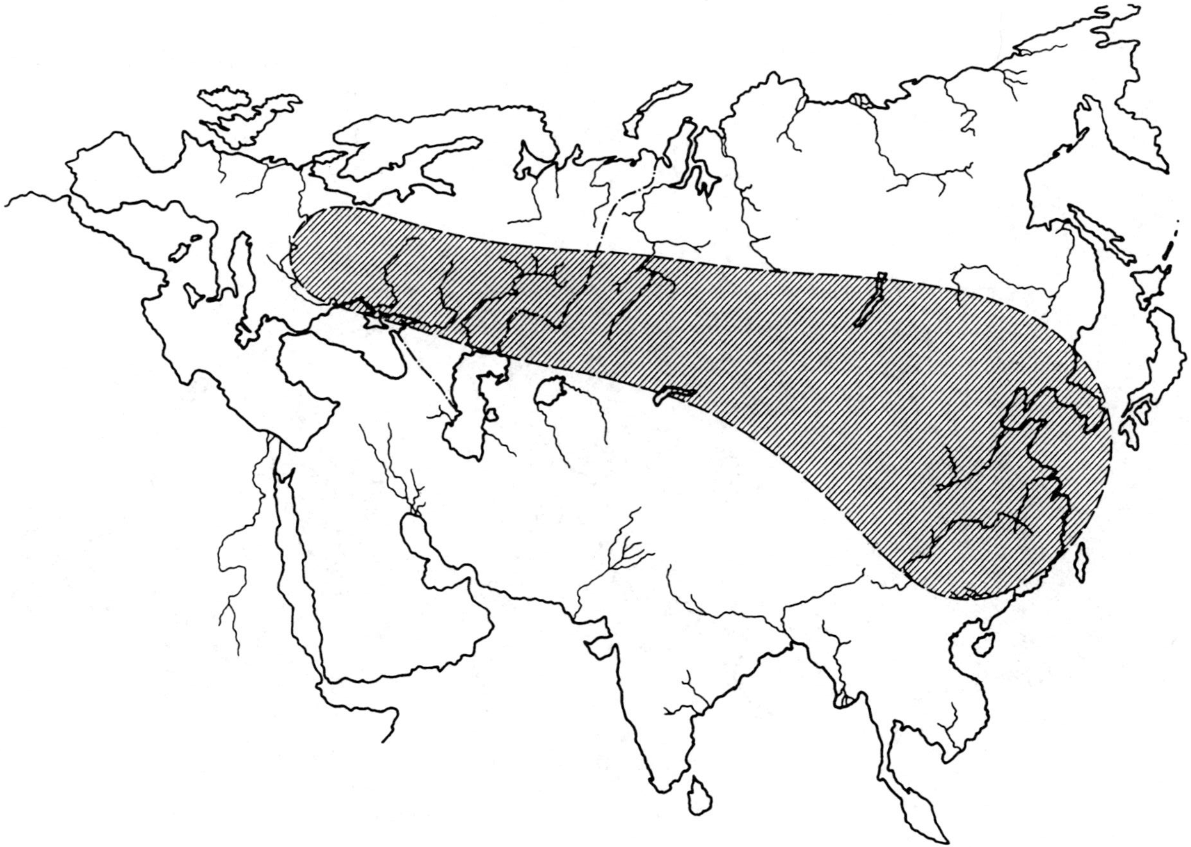

MAP 10 SERIES RUTHENICAE

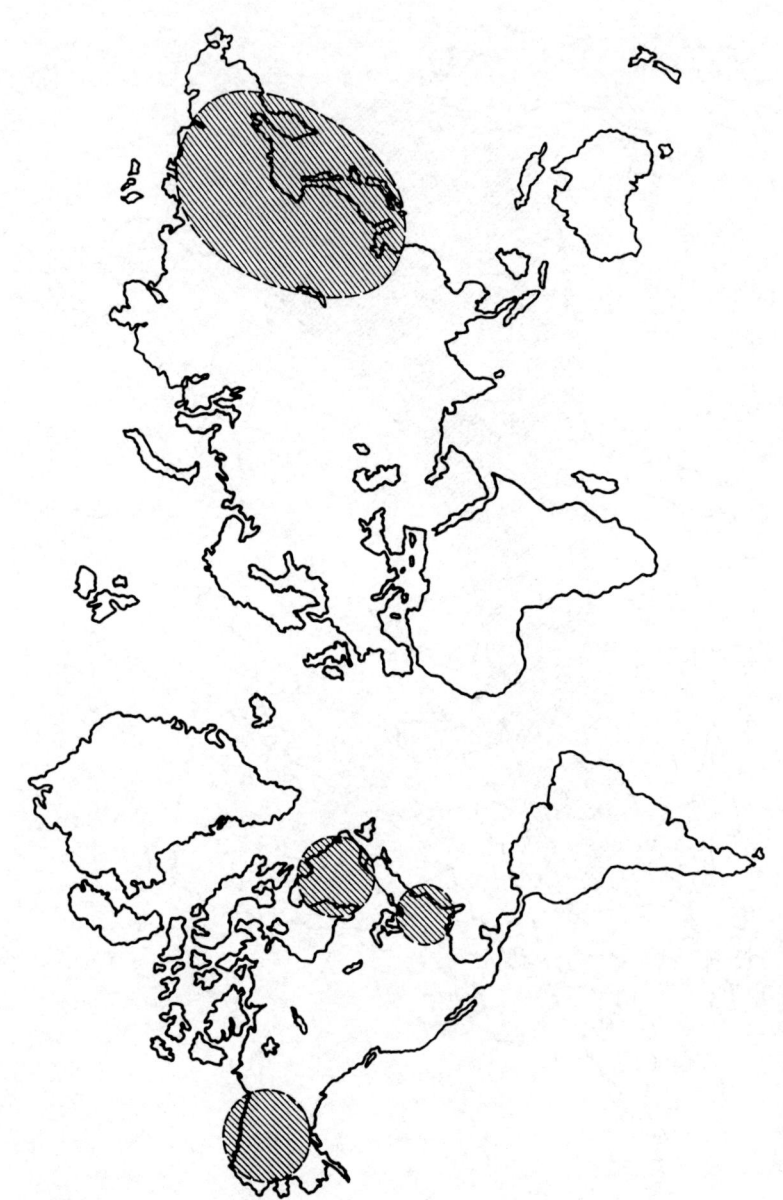

MAP 11 SERIES TRIPETALAE

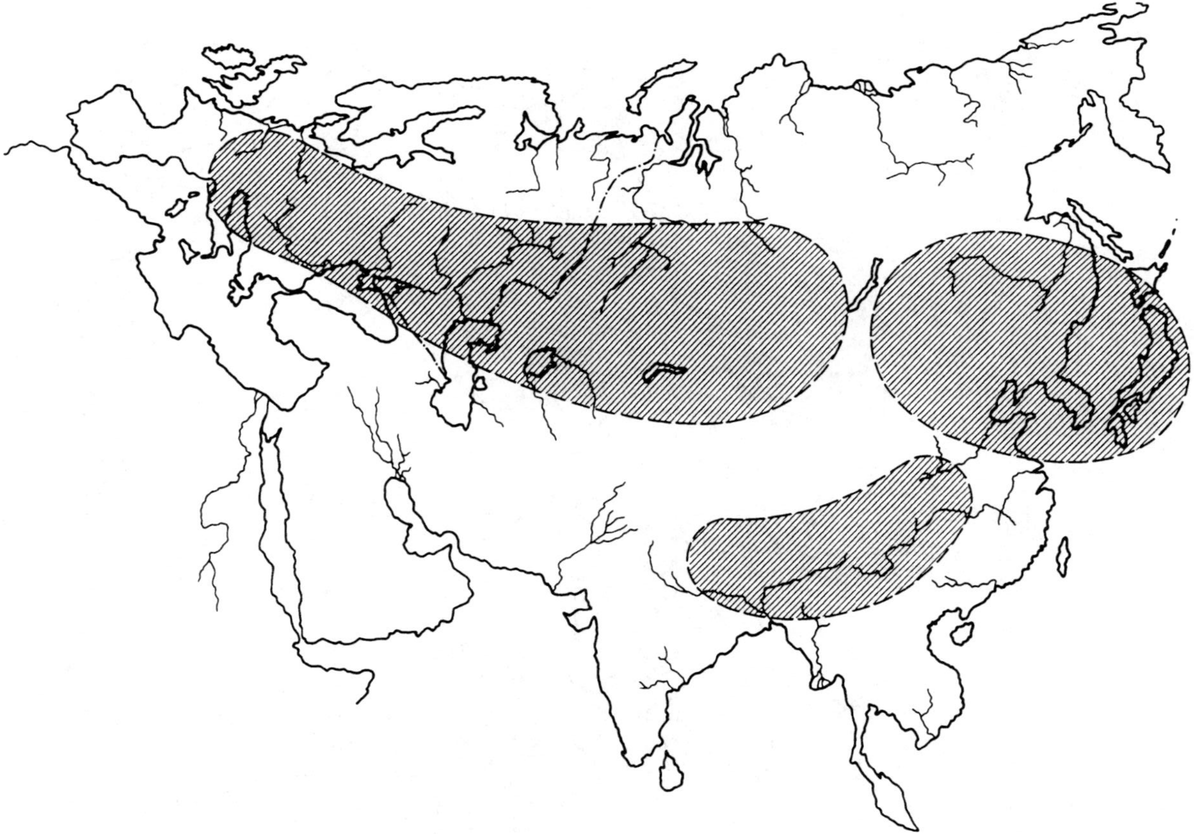

MAP 12 SERIES SIBERICAE

MAP 13 SERIES CALIFORNICAE

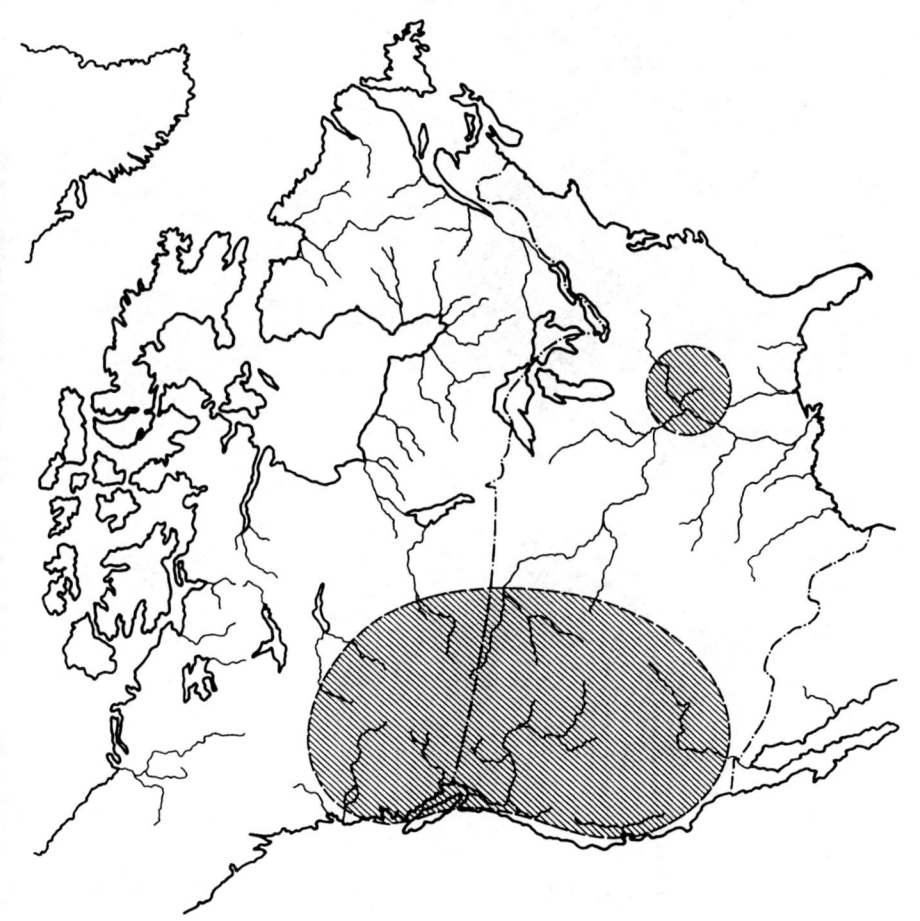

MAP 14 SERIES LONGIPETALAE

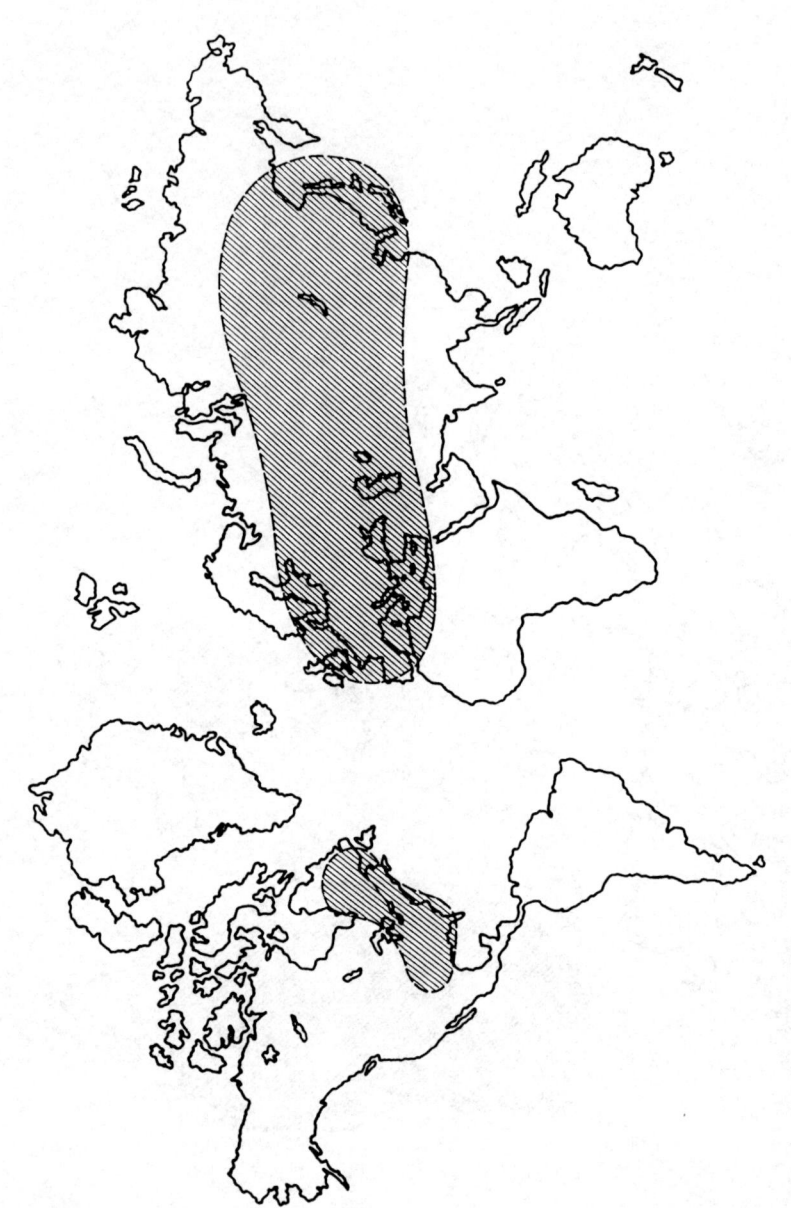

MAP 15 SERIES LAEVIGATAE

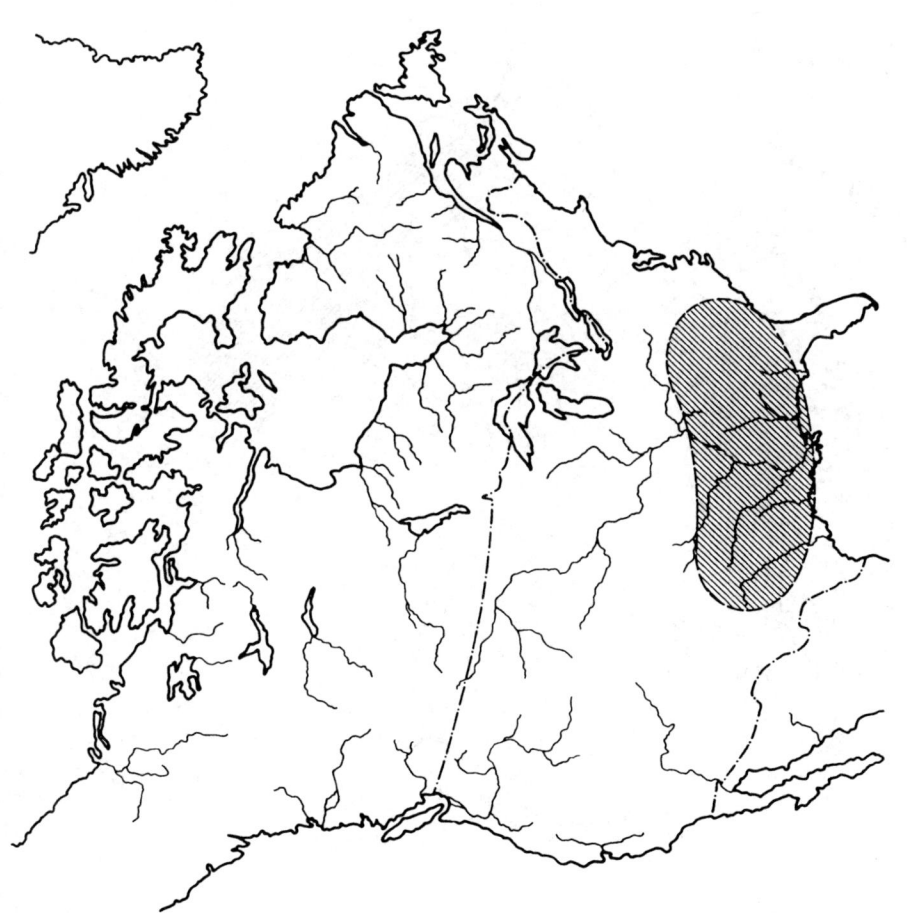

MAP 16 SERIES HEXAGONAE

MAP 17 SERIES PRISMATICAE

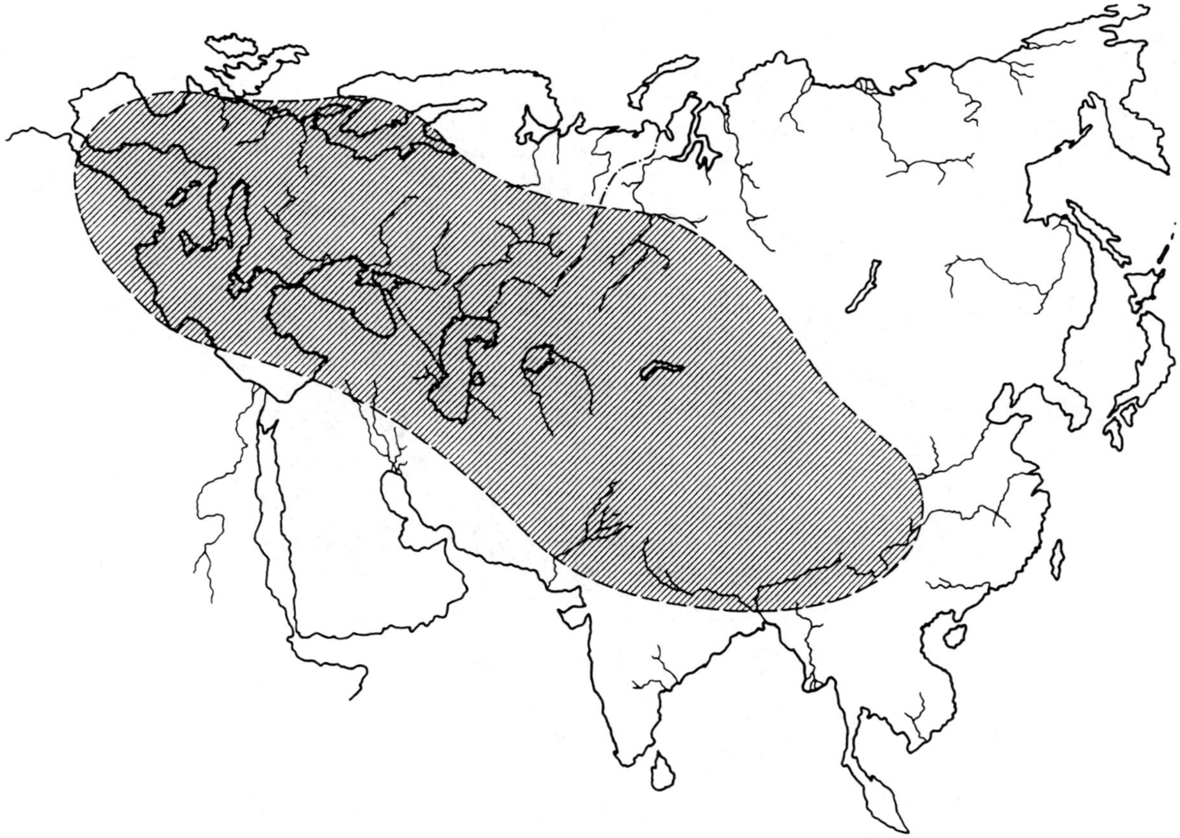

MAP 18 SERIES SPURIAE

MAP 19 SERIES FOETIDISSIMAE

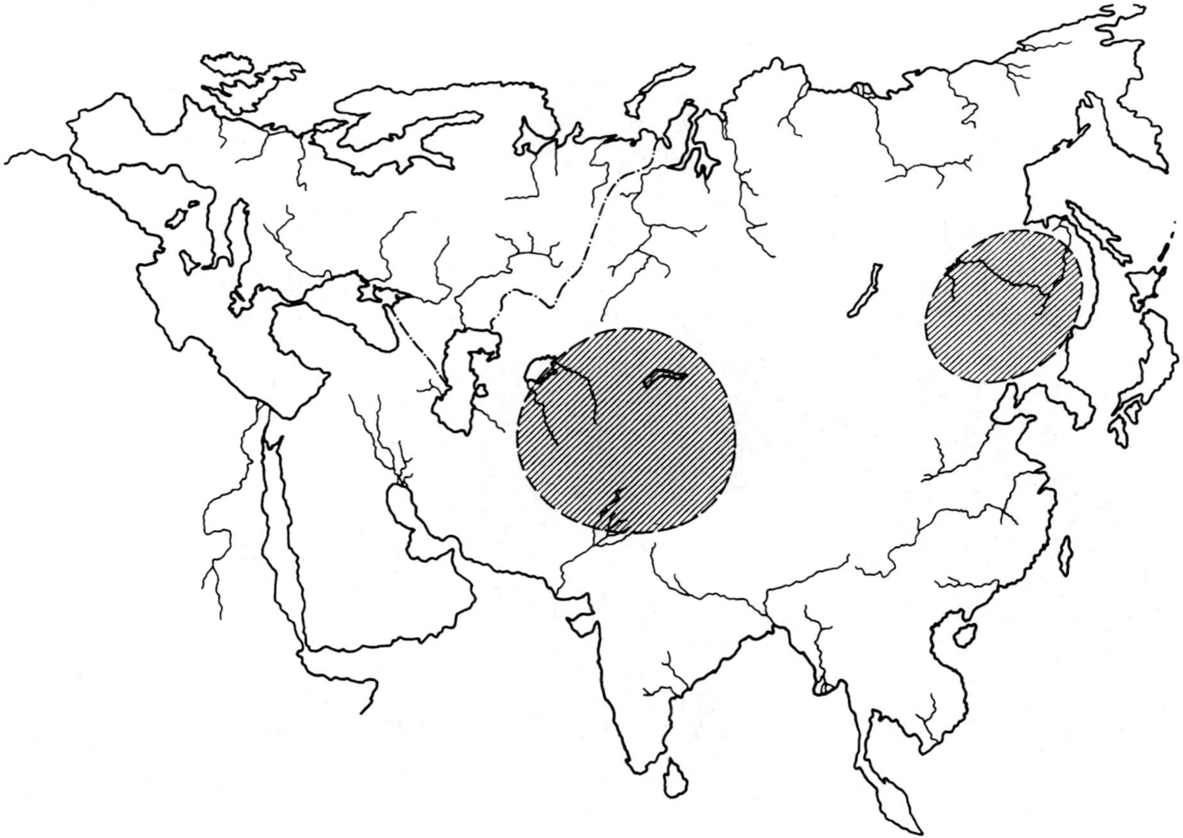

MAP 20 SERIES TENUIFOLIAE

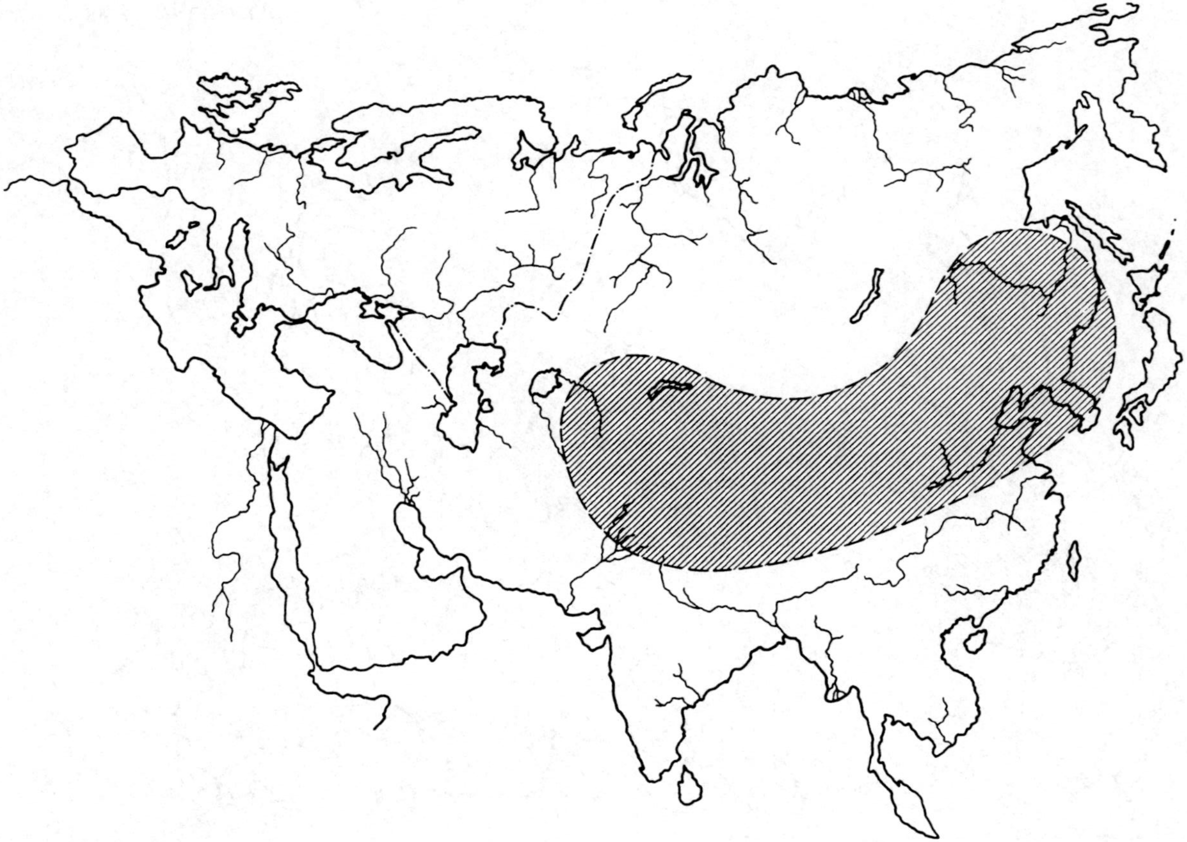

MAP 21 SERIES ENSATAE

MAP 22 SERIES SYRIACAE

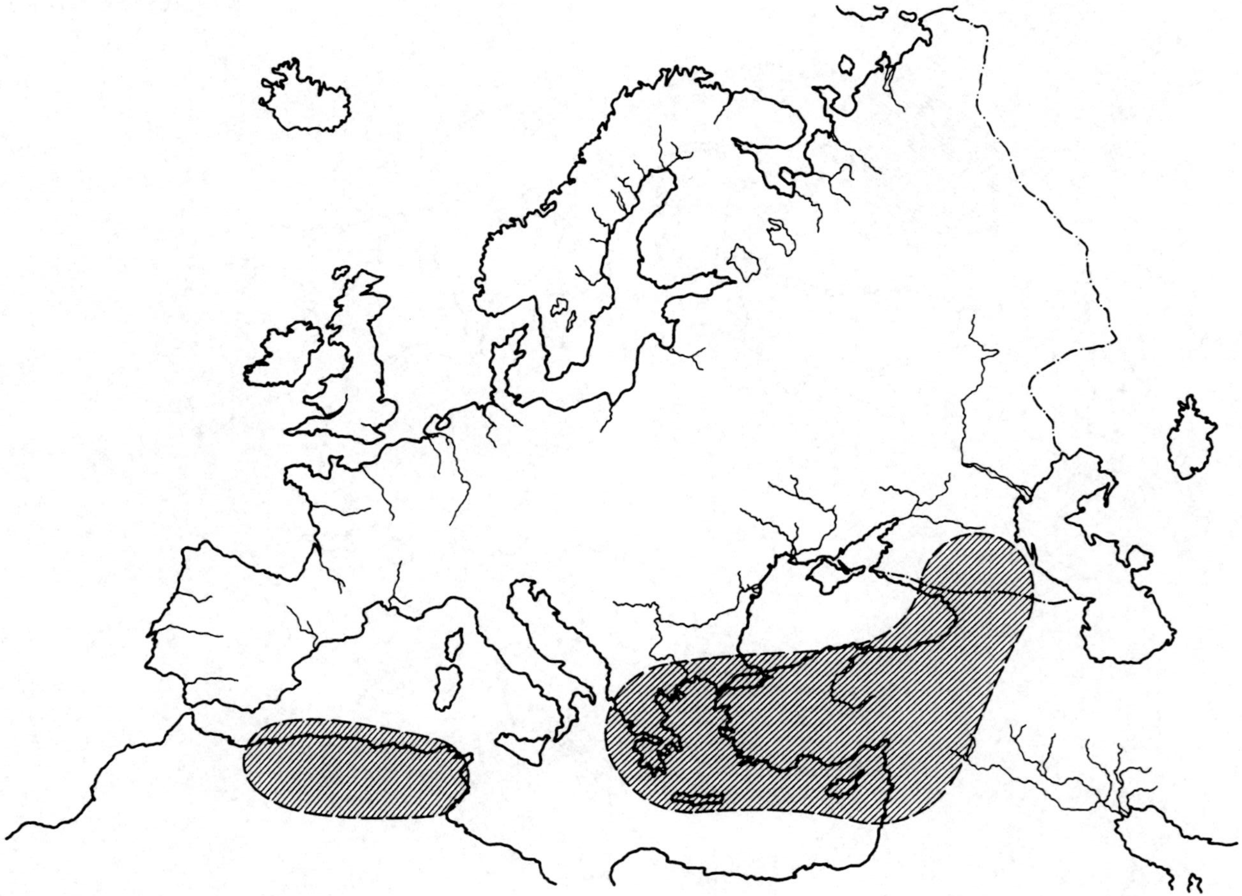

MAP 23 SERIES UNGUICULARES

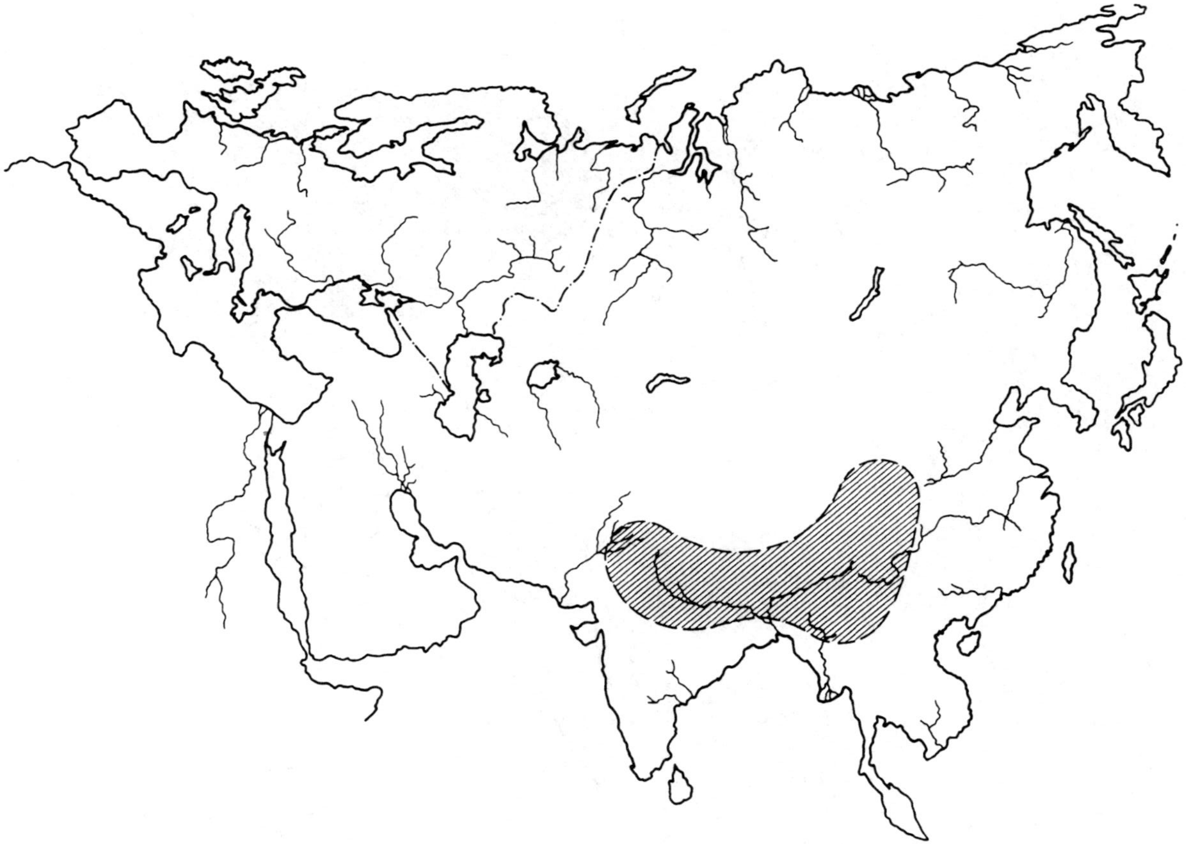

MAP 24 SUBGENUS NEPALENSIS

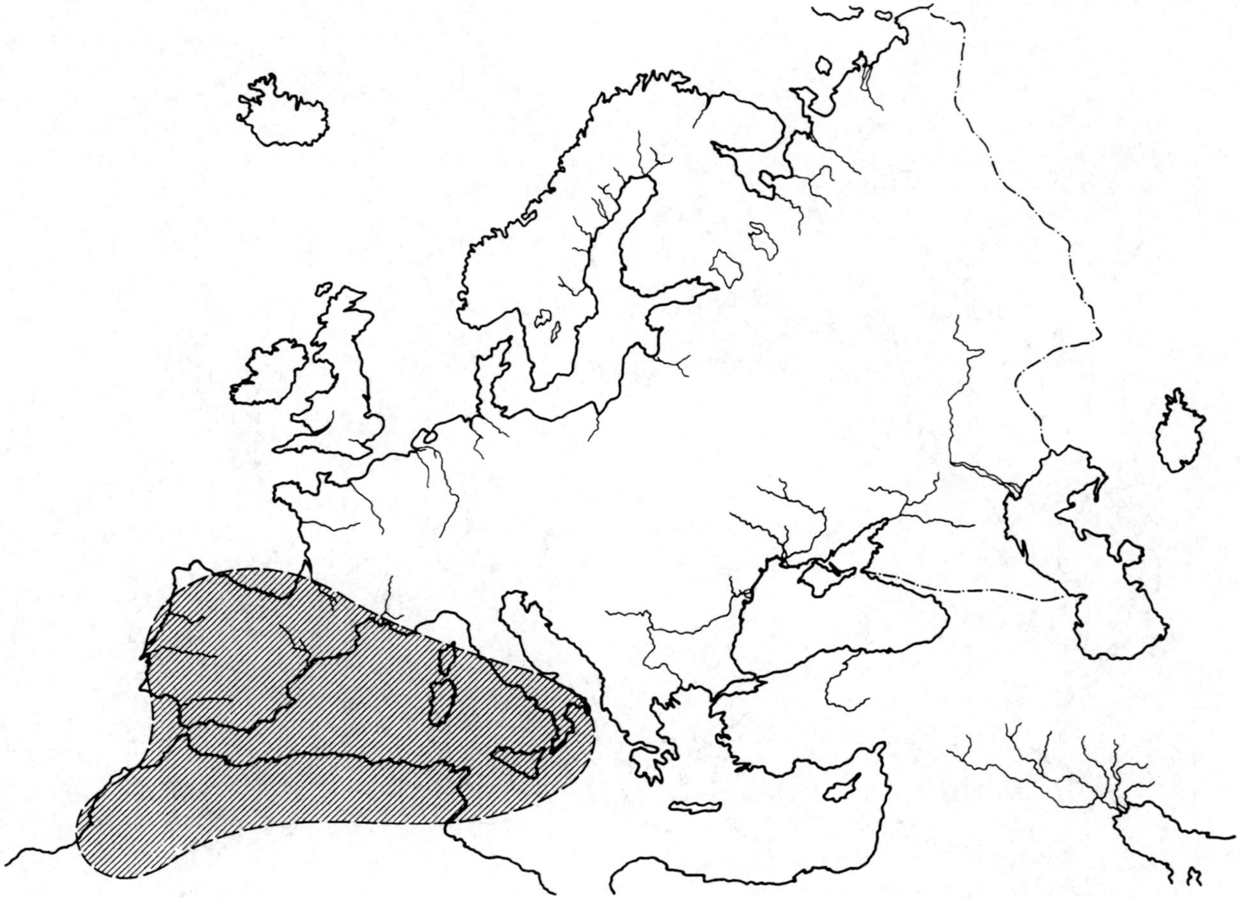

MAP 25 SUBGENUS XIPHIUM

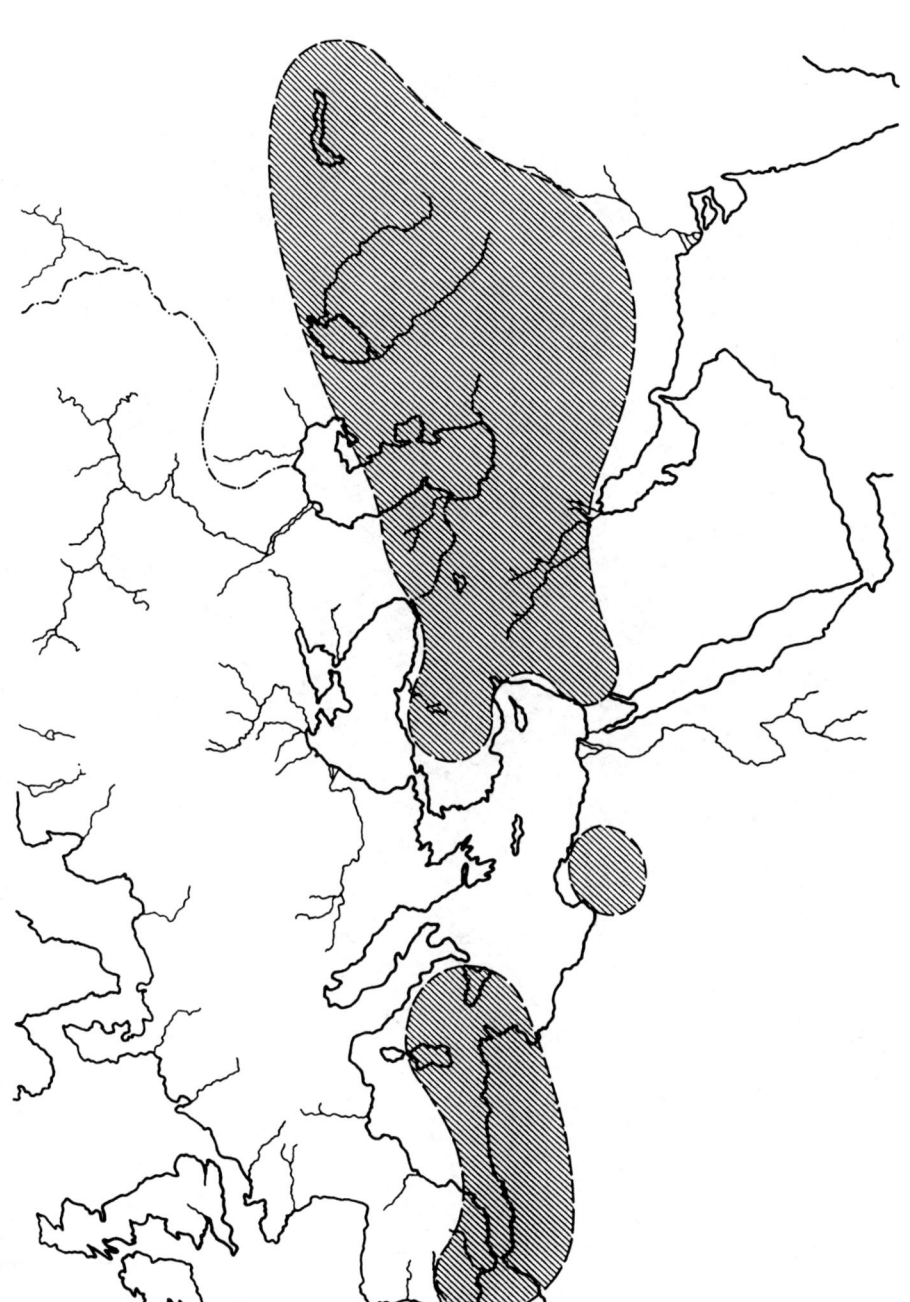

MAP 26 SUBGENUS SCORPIRIS

MAP 27 SUBGENUS HERMODACTYLOIDES

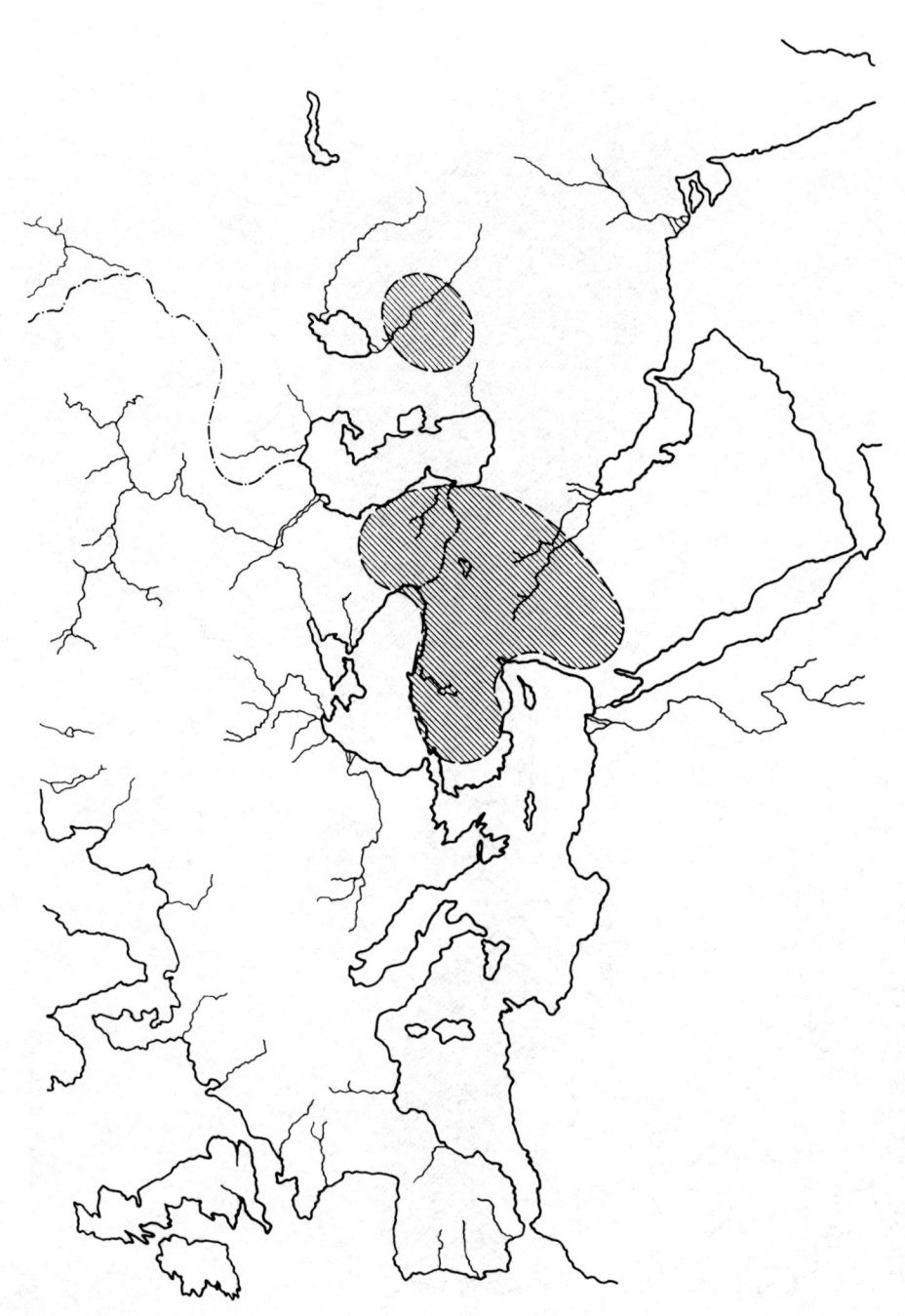

Line drawings

CHRISTABEL KING

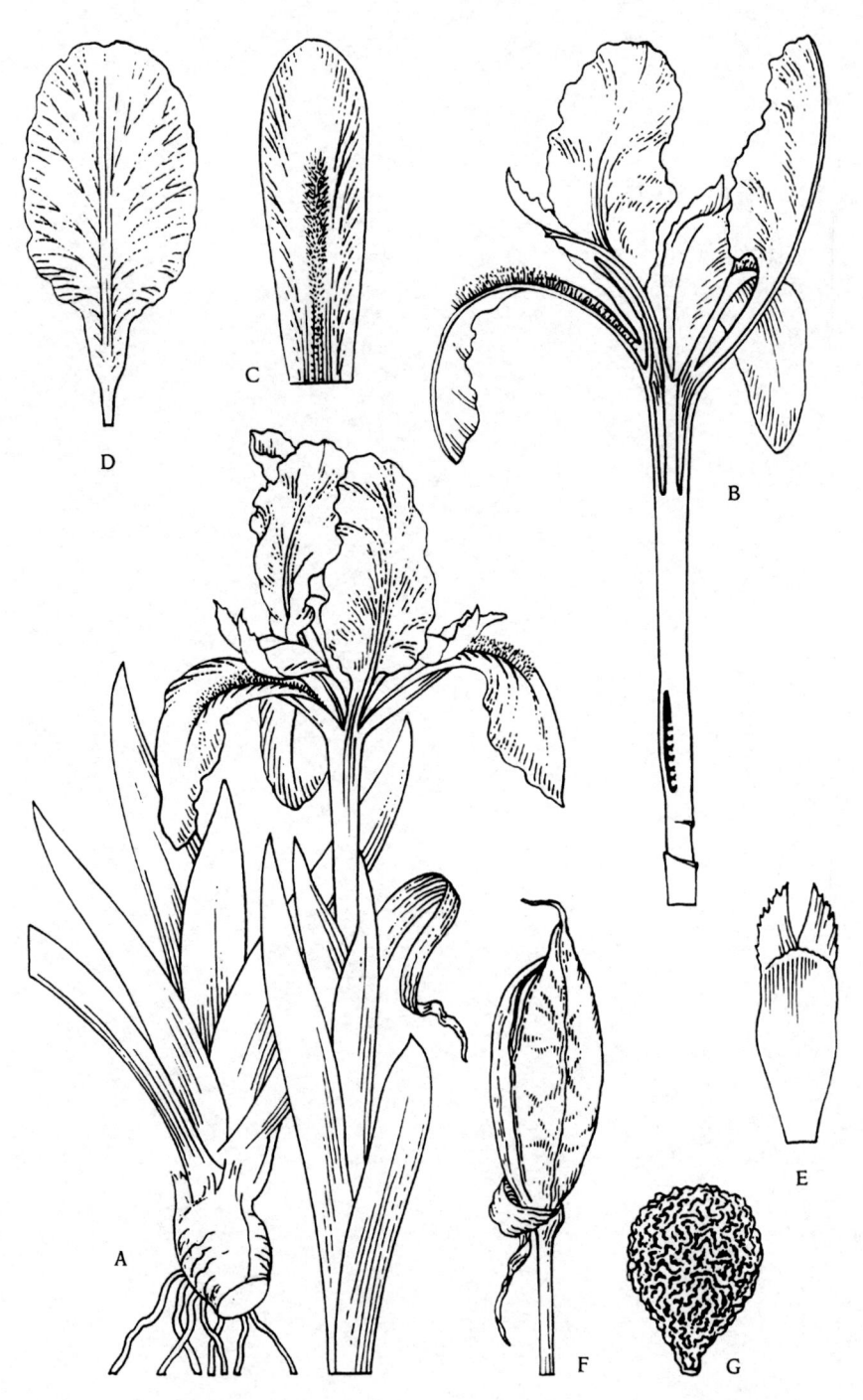

Figure 1 Section Iris. *I. lutescens*. A: habit and flower, × 3/4; B: 1/2-flower, × 3/4; C: outer perianth segment, × 3/4; D: inner perianth segment, × 3/4; E: style arm, × 3/4; F: capsule, × 3/4; G: seed, × 9.

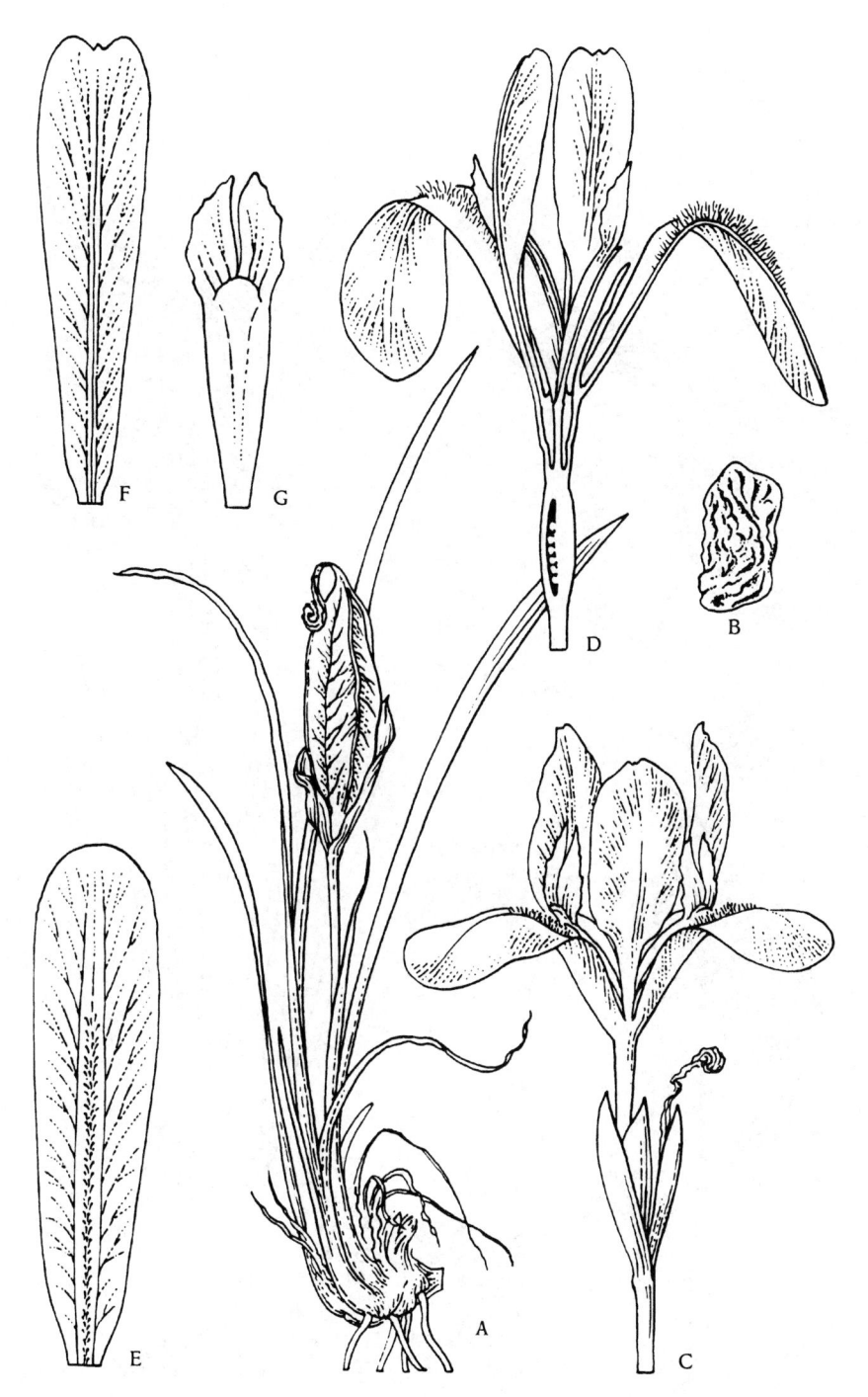

Figure 2 Section Psammiris. *I. bloudowii.* A: habit and capsule, × 1; B: seed, × 30. *I. humilis:* C: flower, × 1 1/2; D: 1/2-flower, × 1 1/2; E: outer perianth segment, × 2; F: inner perianth segment, × 2; G: style arm, × 2.

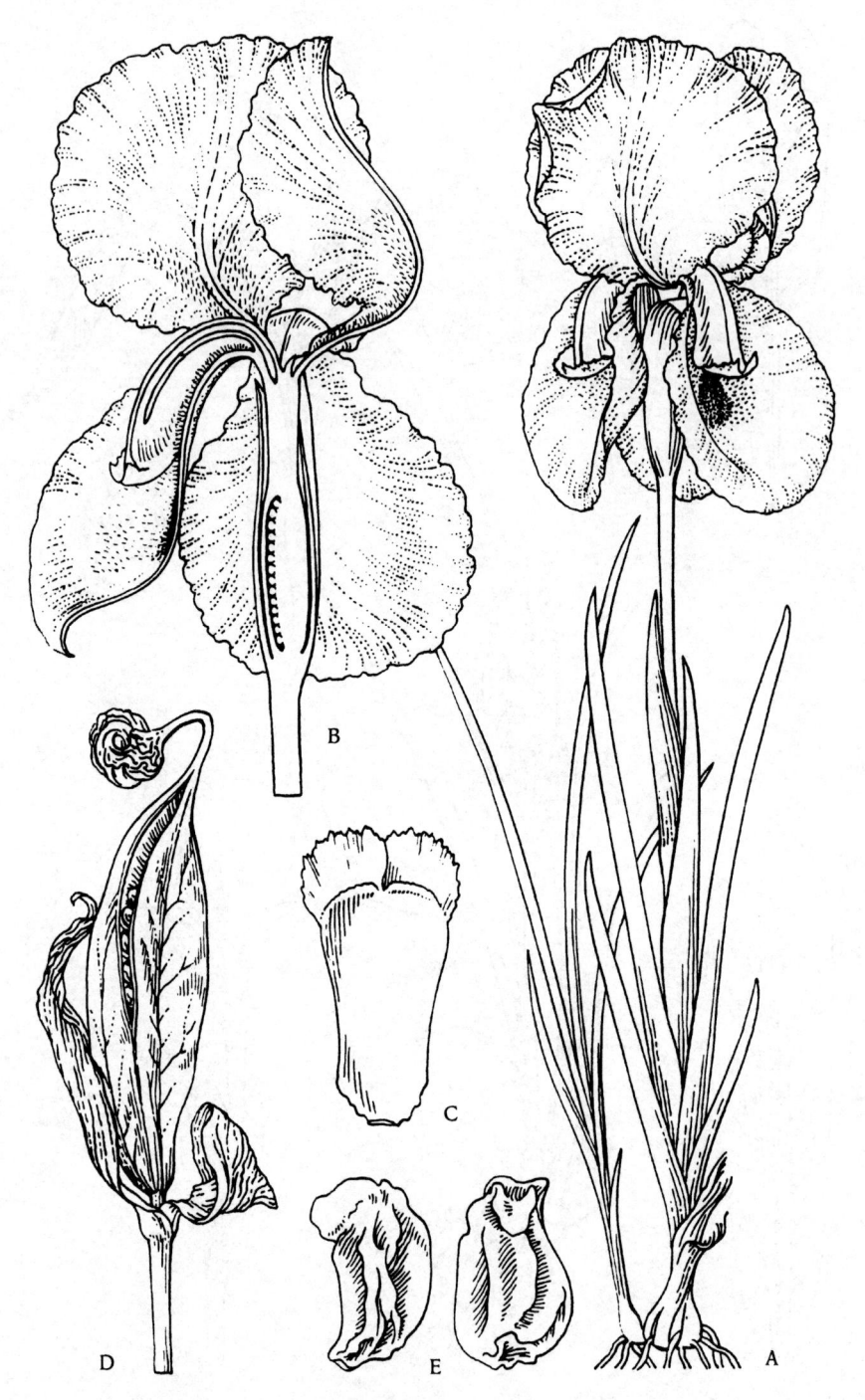

Figure 3 Section Oncocyclus. *I. iberica* ssp. *elegantissima*. A: habit, × 2/3; B: 1/2-flower, × 3/4; C: style arm, × 3/4; *I. iberica*: D: capsule, × 1; E: seed, 2 views, × 6.

Figure 4 Section Regelia. *I. afghanica*. A: habit, × 1; B: 1/2-flower, × 1; C: style arm, × 1 1/2; D: capsule, × 1; E: seed, × 10.

Figure 5 Section Hexapogon. *I. longiscapa*. A: habit in flower, × 1; B: habit and capsule, × 1; C: seed, × 6

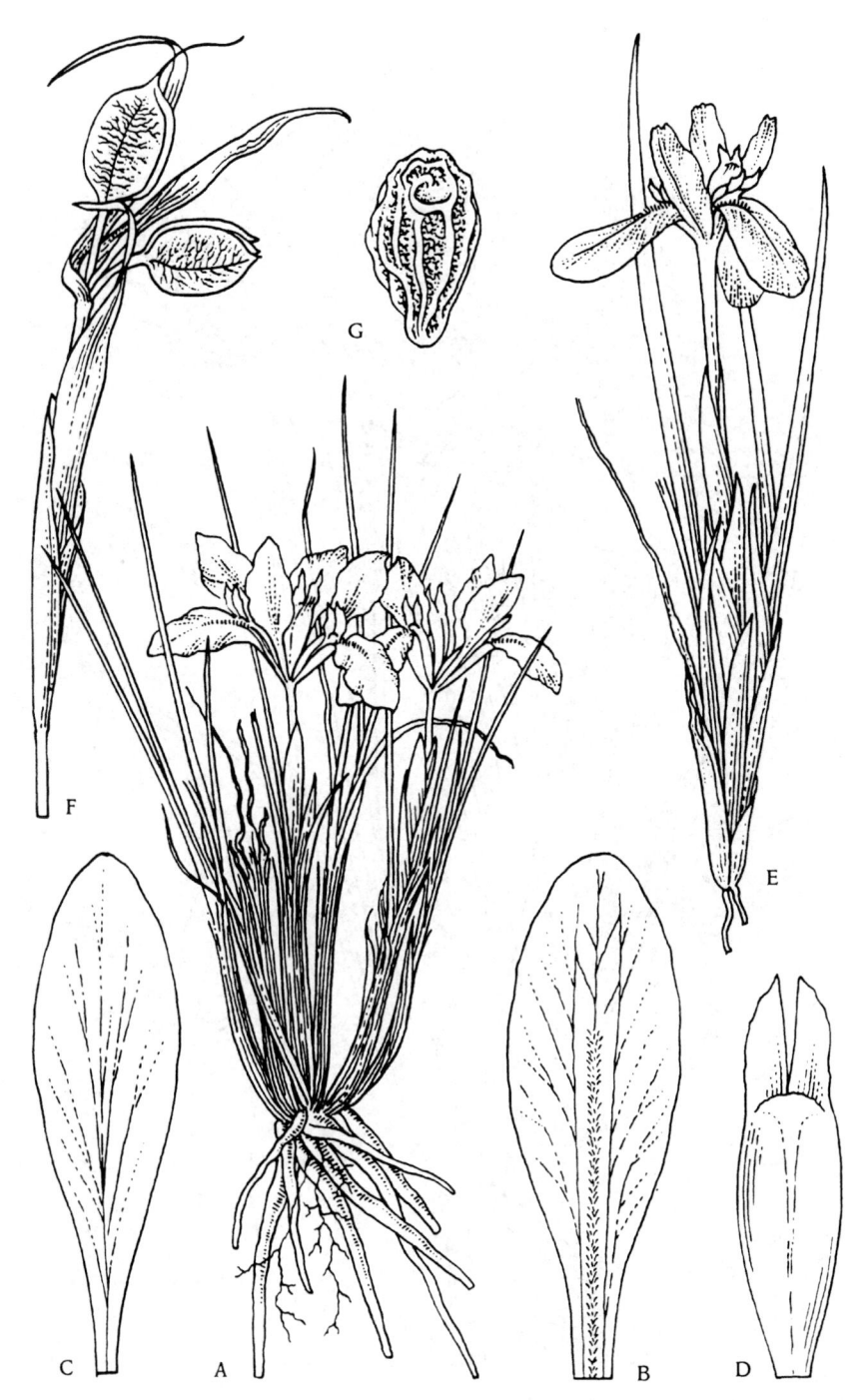

Figure 6 Section Pseudoregelia. *I. tigridia*. A: habit, × 1; B: outer perianth segment, × 3; C: inner perianth segment, × 3; D: style arm, × 3; *I. kemaonensis*: E: habit, × 1; F: capsules, × 1; G: seed, × 6.

Figure 7 Section Lophiris. *I. tectorum*. A: habit and inflorescence, × 2/3; B: crest from outer perianth segment, × 2; C: style arm, × 1 1/2; D: gynoecium and stamen, L.S. × 2; E: capsules, × 1; F: seed, × 4.

Figure 8 Series Chinenses. *I. minutoaurea*. A: habit in late summer, × 1; B: habit and flowers, × 1; C: inner perianth segment, × 3; D: stamens, × 10; E: stigmatic surface and part of style arm, × 3; F: capsule, × 1; G: seed, × 3.

341

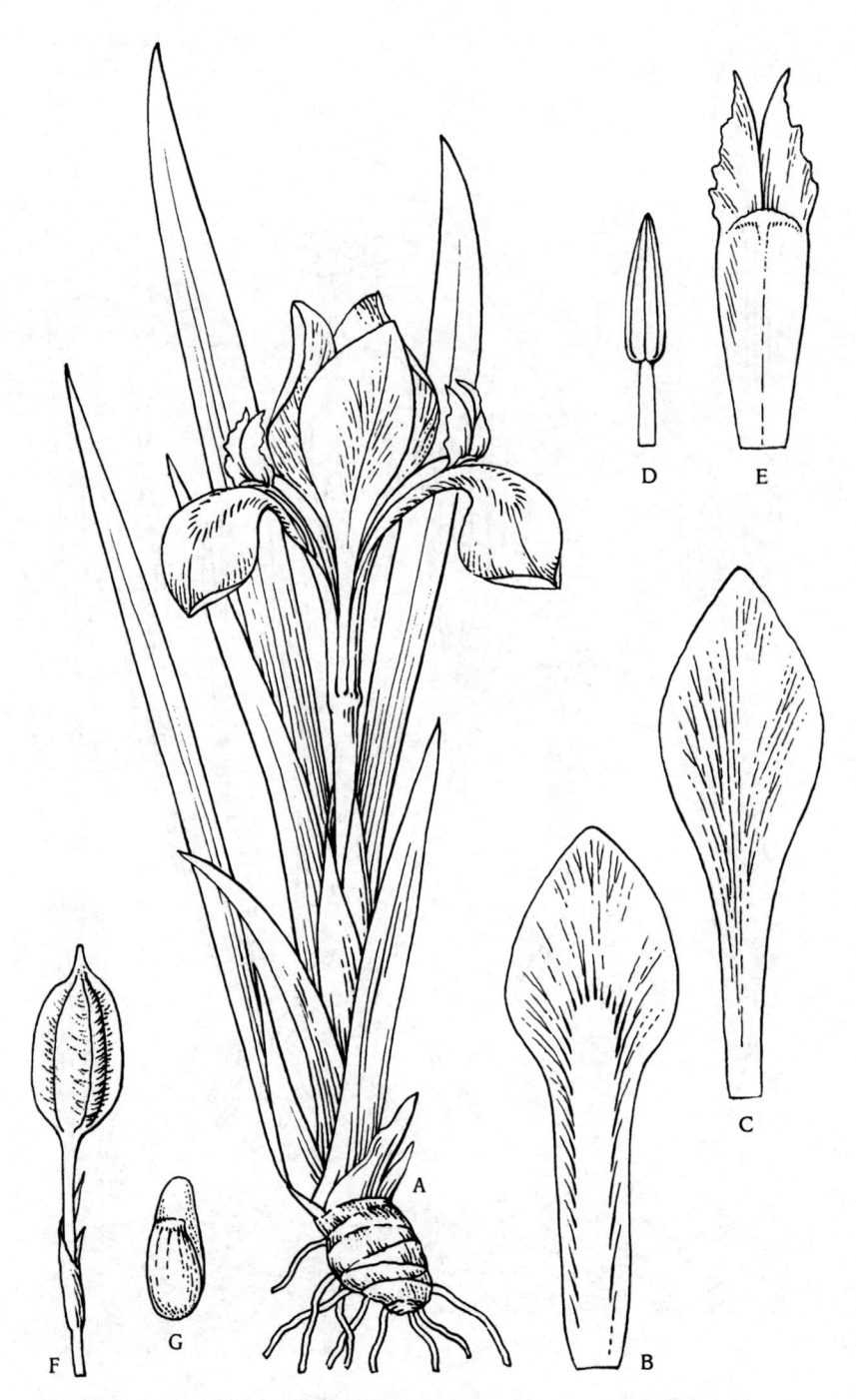

Figure 9 Series Vernae. *I. verna*. A: habit, × 1; B: outer perianth segment, × 1 1/2; C: inner perianth segment, × 1 1/2; D: stamen, × 1 1/2; E: style arm, × 1 1/2; F: capsule × 1; G: seed, × 4.

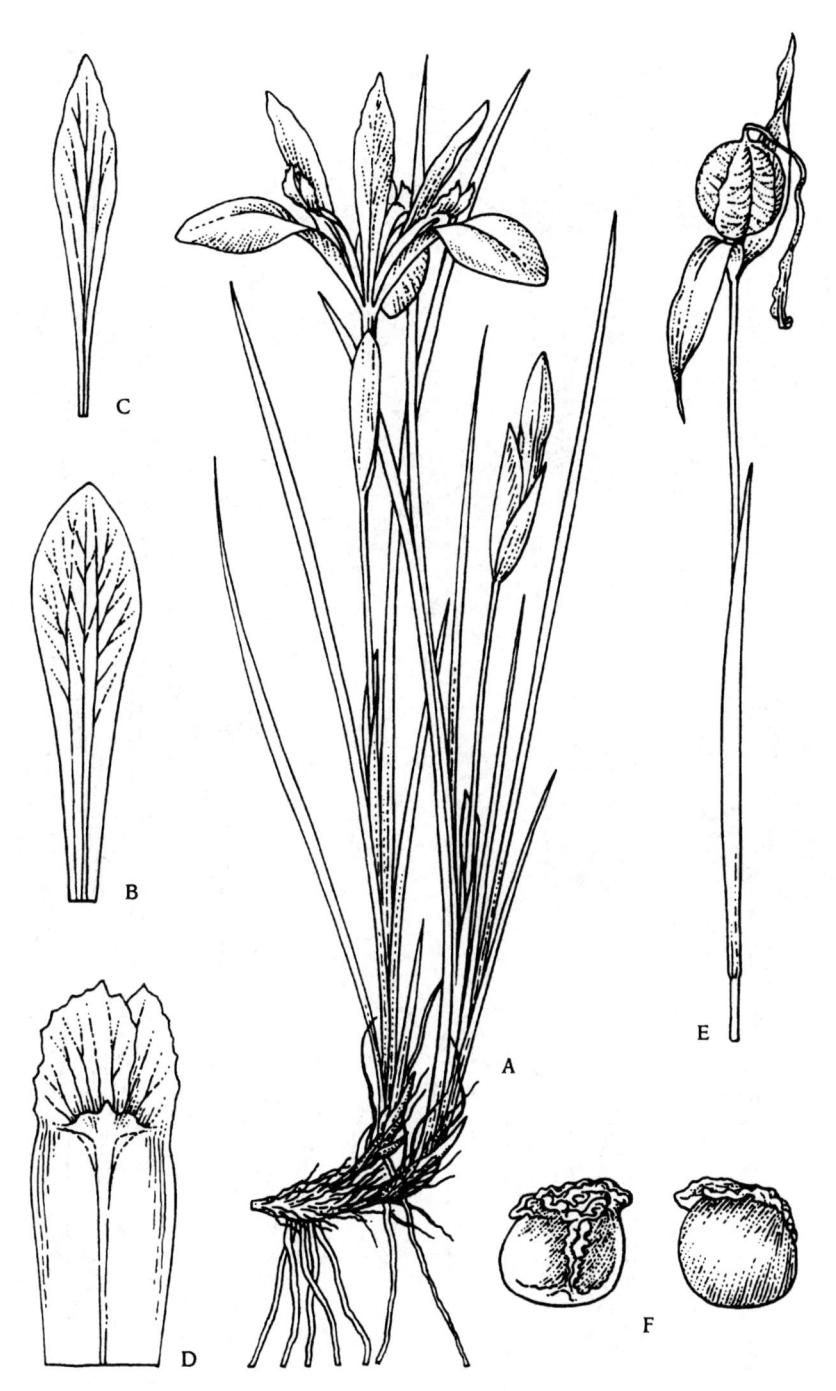

Figure 10 Series Ruthenicae. *I. ruthenica*. A: habit and flower, × 1; B: outer perianth segment, × 1 1/2; C: inner perianth segment, × 1 1/2; D: part of style arm, × 3; E: capsule, × 1; F: seed, 2 views, × 6.

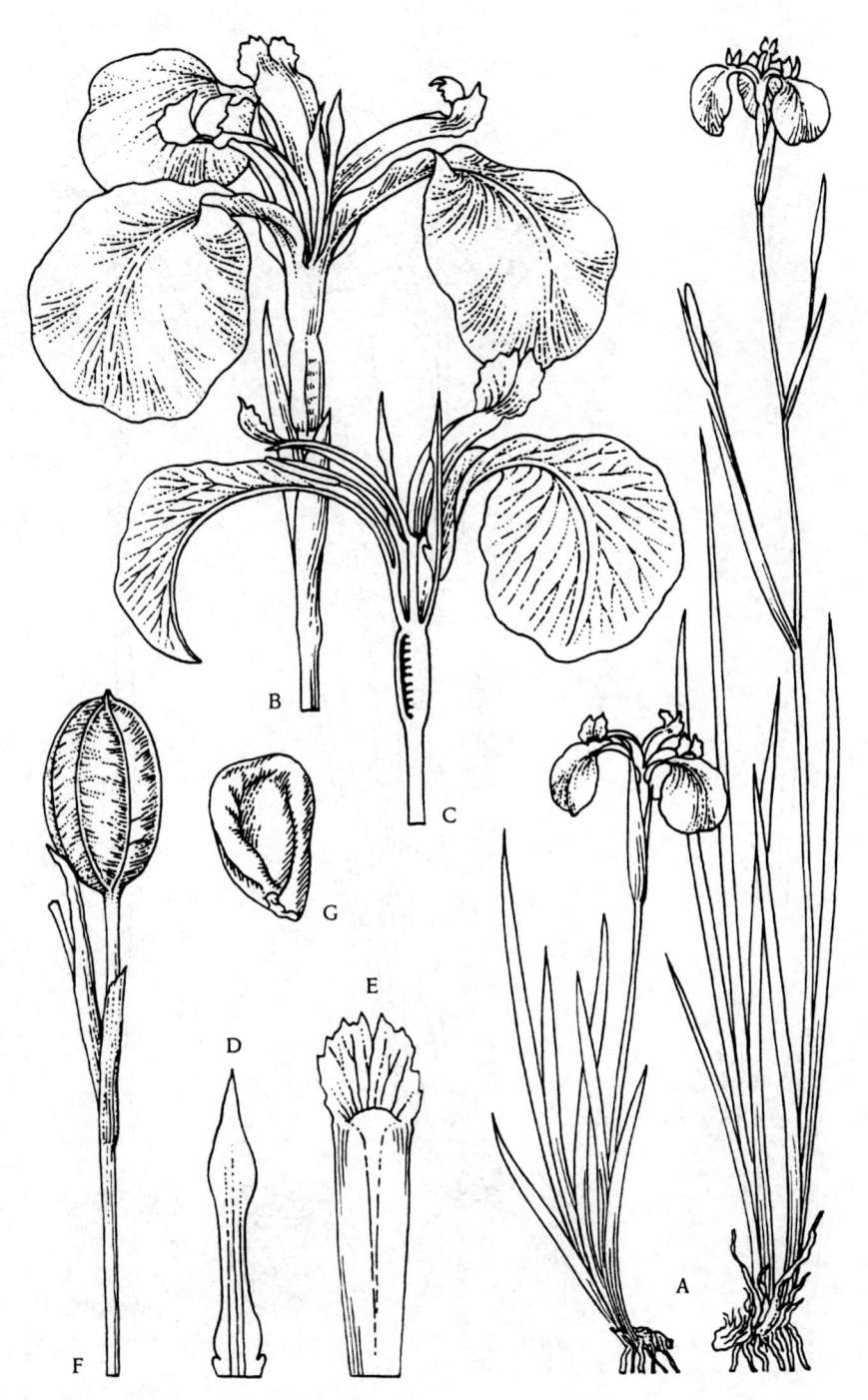

Figure 11 Series Tripetalae. *I. setosa*. A: habit of tall and short forms, × 1/2; B: flower, × 1; C: 1/2-flower, × 1; D: inner perianth segment, × 1 1/2; E: style arm, × 1 1/2; F: capsule, × 1; G: seed, × 6.

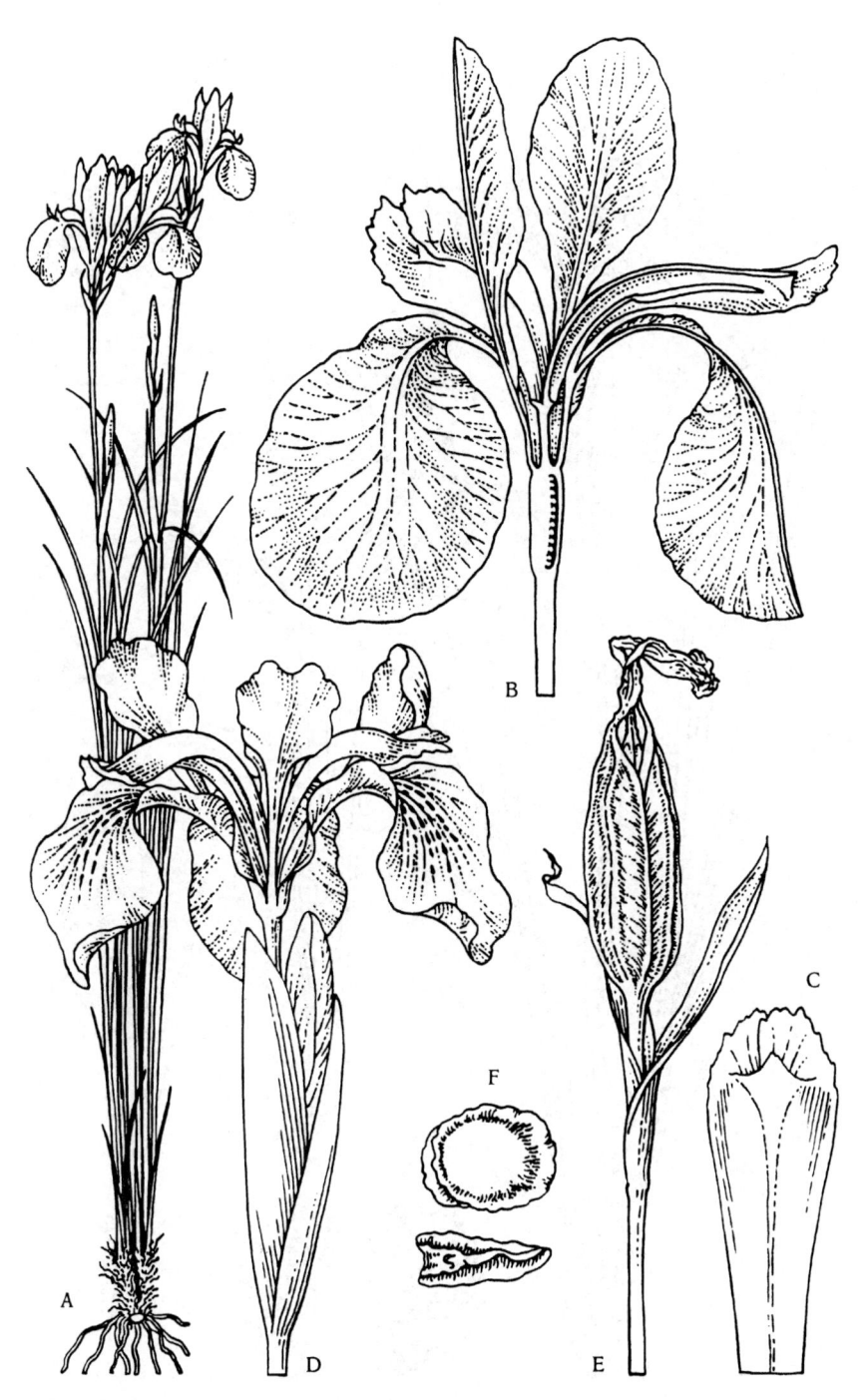

Figure 12 Series Sibericae. *I. sibirica.* A: habit × 1/4; B: 1/2-flower, × 1 1/2; C: style arm, × 1 1/2.
I. wilsonii: D: flower, × 1; E: capsule, × 1; F: seed, 2 views, × 4.

Figure 13 Series Californicae. *I. innominata*. A: habit, × 1; B: 1/2-flower, × 1 1/2; C: stigmatic surface and part of style arm, × 2. *I. tenax*: D: capsule, × 1; E: seed, 2 views, × 9.

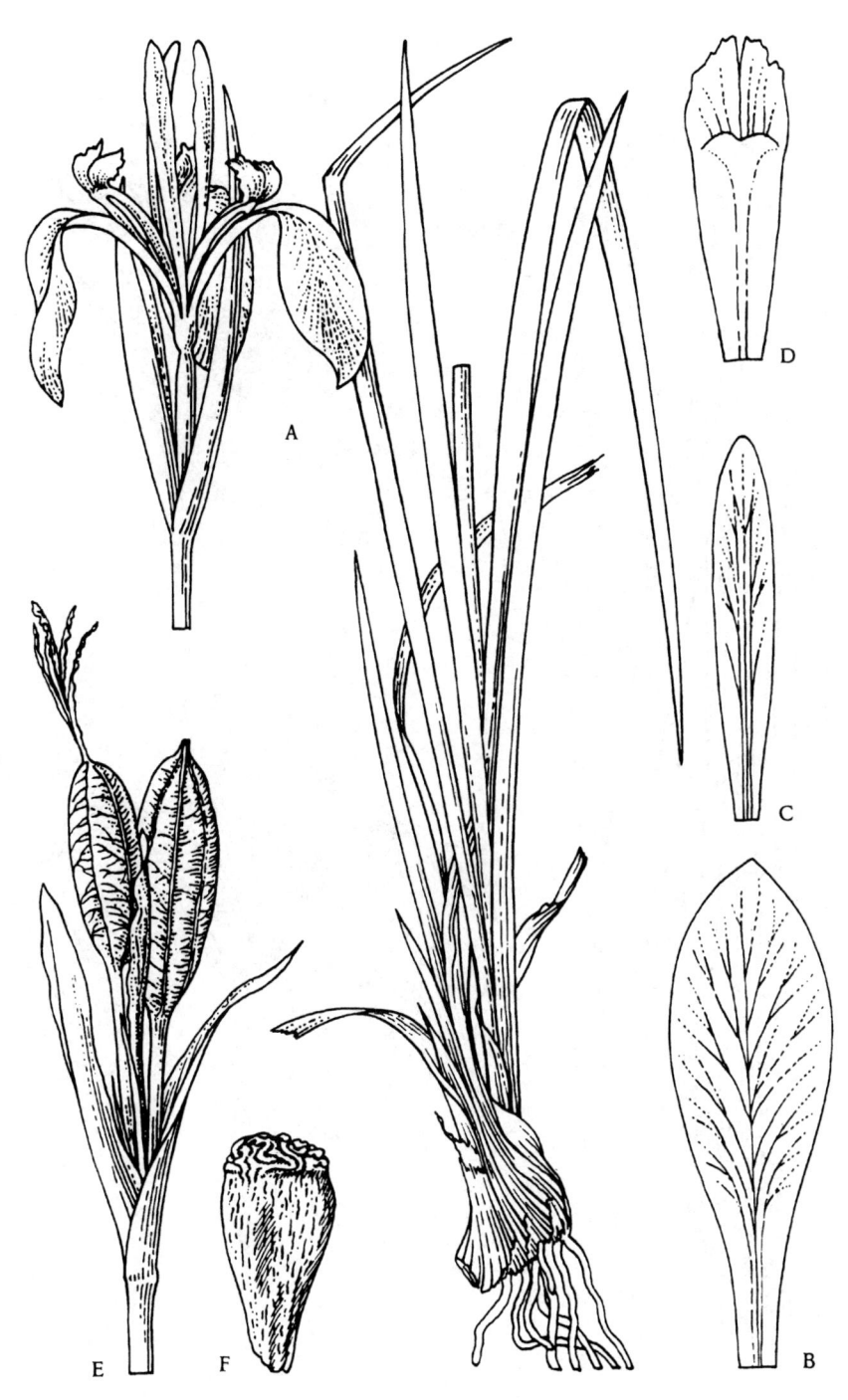

Figure 14 Series Longipetalae. *I. missouriensis* (Longipetala form). A: habit and flower, × 2/3; B: outer perianth segment, × 1; C: inner perianth segment, × 1; D: style arm, × 1. *I. missouriensis*. E: capsules, × 1; F: seed, × 9.

Figure 15 Series Laevigatae. *I. laevigata*. A: habit, × 1/2; B: inflorescence, × 1/2; C: 1/2-flower, × 1; D: stigmatic surface and part of style arm, × 1; E: capsules, × 1; F: seed, × 4.

Figure 16 Series Hexagonae. *I. brevicaulis*. A: habit, × 1/2; B: capsule, × 1; C: capsule T.S., × 1; D: seed, × 2. *I. fulva*: E: 1/2-flower, × 1; F: style arm, × 2.

Figure 17 Series Prismaticae. *I. prismatica*. A: habit, × 3/4; B: inflorescence, × 1; C: 1/2-flower, × 1 1/2; D: stigmatic surface and part of style arm, × 3; E: capsules, × 1; F: seed, 2 views, × 6.

Figure 18 Series Spuriae. *I. graminea*. A: habit in winter, × 1; B: inflorescence, × 1; C: 1/2-flower, × 1 1/2; D: stigmatic surface and part of style arm, × 3; E: capsule and T.S., × 1; F: seed, × 9.

351

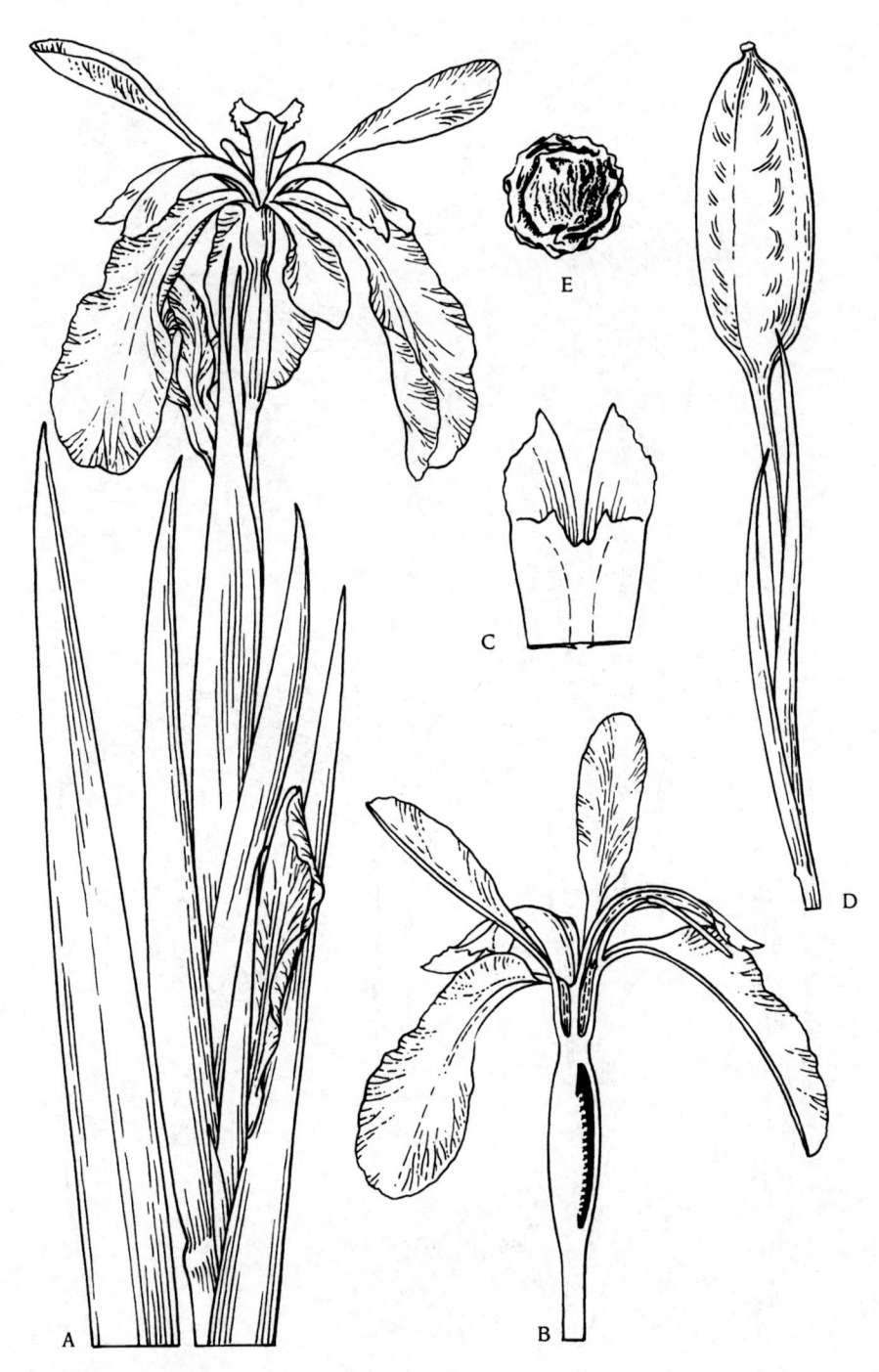

Figure 19 Series Foetidissimae. *I. foetidissima* f. *citrina*. A: inflorescence, × 1; B: 1/2-flower, × 1; C: stigmatic surface and part of style arm, × 2. *I. foetidissima*: D: capsule, × 1; E. seed, × 4.

Figure 20 Series Tenuifoliae. *I. songarica*. A: habit, × 1/2; B: inflorescence, × 1; C: style arm, × 1 1/2; D: capsule, × 1; E: seed, 2 views, × 4.

Figure 21 Series Ensatae. *I. lactea*. A: habit and flower, × 3/4; B: 1/2-flower, × 1 1/2; C: outer perianth segment, × 1 1/2; D: inner perianth segment, × 1 1/2; E: style arm, × 1 1/2; F: capsules, × 1; G: seed, two views, × 6.

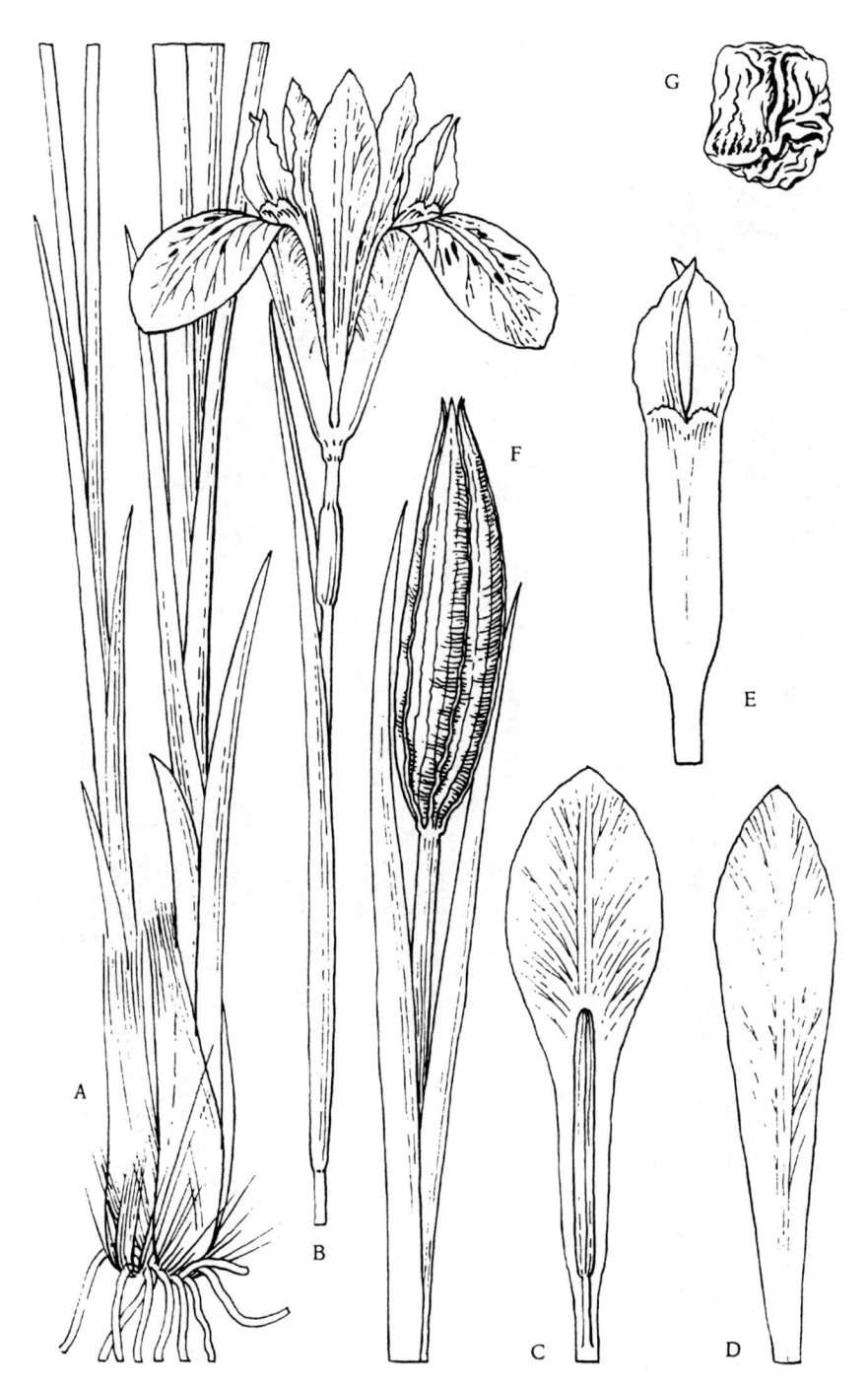

Figure 22 Series Syriacae. *I. grant-duffii*. A: habit, × 1; B: flower × 1; C: outer perianth segment and stamen, × 1 1/2; D: inner perianth segment, × 1 1/2; E: style arm, × 1 1/2; F: capsule × 1; G: seed, × 4.

355

Figure 23 Series Unguiculares. *I. unguicularis*. A: habit and flower, × 2/3; B: longitudinal section of flower and base of shoot with ovaries, × 1; C: stamens, × 1 1/2; D: part of style arm, × 2; E: capsules, × 3/4. *I. lazica*: F: seed, × 8.

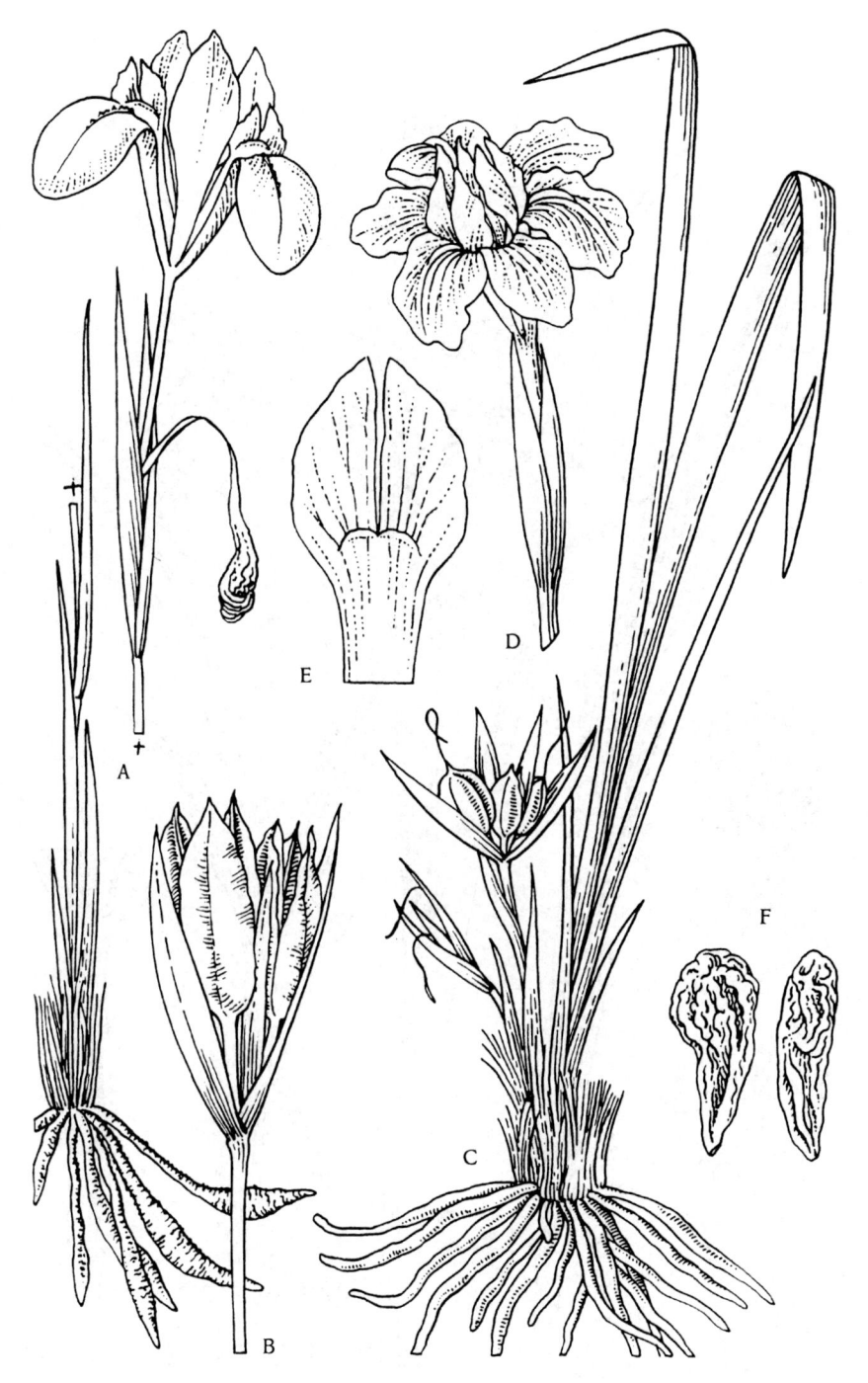

Figure 24 Subgenus Nepalensis. *I. decora.* A: habit and flower × 1; B: capsules, × 1. *I. collettii.* C: habit and capsules, × 1/2; D: flower, × 1; E: upper part of style arm, × 2; F: seed, × 9.

Figure 25 Subgenus Xiphium. *I. xiphium*. A: bulb and leaves, × 3/4; B: flower, × 3/4; C: style arm, × 3/4; D: ovary L.S., × 1; E: capsules, × 3/4; F: seed, × 12.

358

Figure 26 Subgenus Scorpiris. *I. bucharica*. A: bulb roots, × 3/4; B: inflorescence, × 3/4; C: 1/2-flower, × 1; D: inflorescence with ripe capsules, × 3/4; E: capsule, × 1; F: seed, × 4.

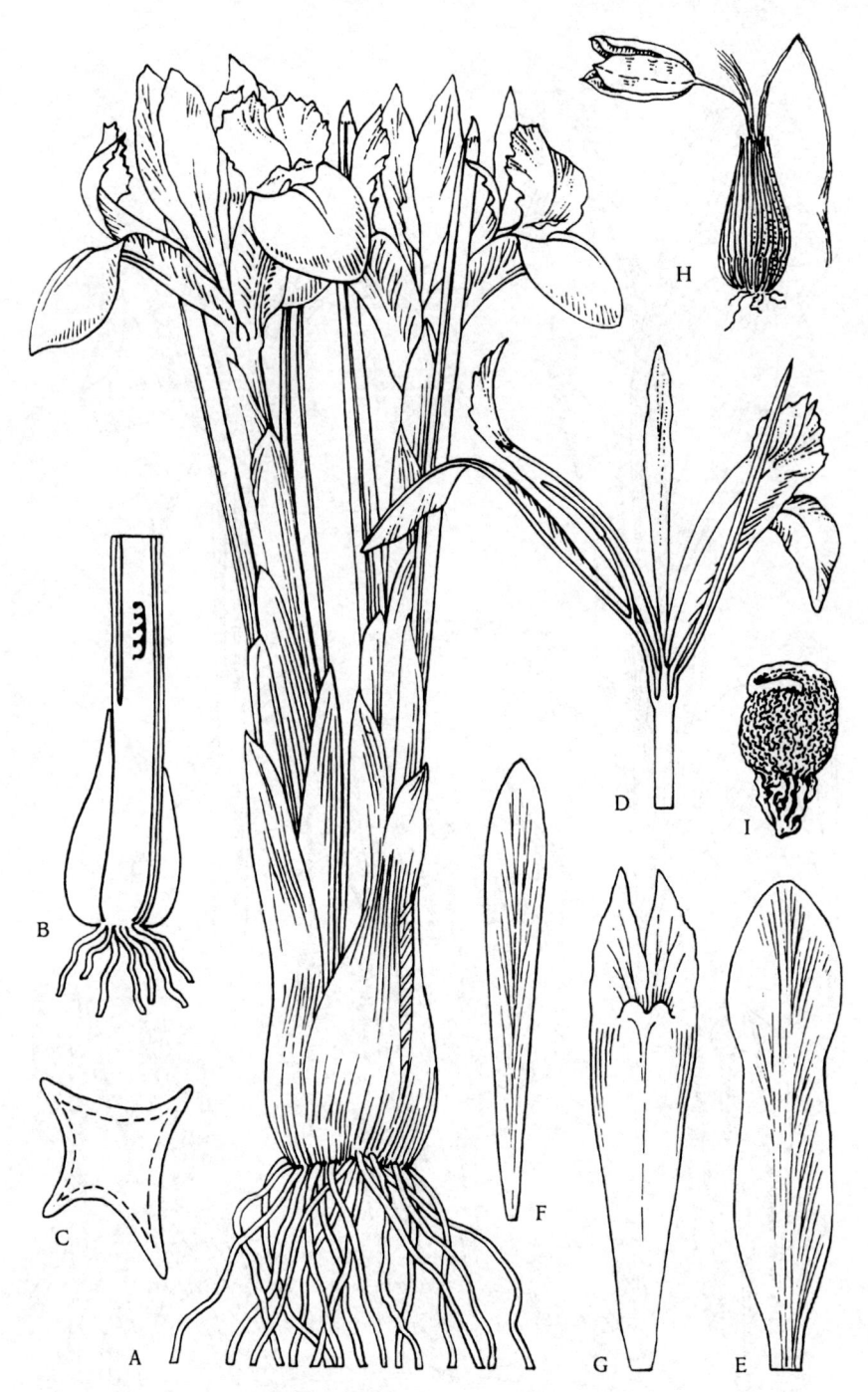

Figure 27 Subgenus Hermodactyloides. *I. reticulata* 'Harmony'. A: whole plant, × 1; B: L.S. of bulb and gynoecium, × 1 1/2. *I. reticulata* coll. Stevens 94: C: leaf, T.S., × 8; D: 1/2-flower, × 1 1/2; E: outer perianth segment, × 1 1/2; F: inner perianth segment, × 1 1/2; G: style arm, × 1 1/2. *I. reticulata* coll. Baytop, Iste: H: bulb and capsule, × 1; I: seed, × 8.

SECTION : IRIS

1 *albertii* Tashkent Botanic Garden

2 *aphylla* (spathes)

3 *belouini* Morocco

4 *biliottii* Turkey

A colour version of these plates is available for download from www.cambridge.org/9780521206433

5 *germanica* 'Alba'

6 *germanica* probably 'Fontarabie'

7 *imbricata*

8 *junonia* Turkey

9 *kashmiriana* 'purpurea' Kashmir

10 *mesopotamica* Turkey

11 *perrieri* (spathes)

12 *pseudopumila*

13 *purpureobractea* (yellow form)

14 *timofejewii*

SECTION : PSAMMIRIS

15 *bloudowii* Borohorasan

16 *humilis*

SECTION : ONCOCYCLUS

17 *auranitica*

18 *barnumae* f. *urmiensis*

19 *demavendica*

20 *elegantissima*

21 *haynei*

22 *mairiae*

23 *nectarifera*

24 *paradoxa* f. *mirabilis*

25 *sofarana*   26 *sprengeri*

SECTION : REGELIA

27 *afghanica* Afghanistan

28 *heweri* Afghanistan

29 *kuschkensis* Afghanistan 30 *stolonifera*

SECTION : PSEUDOREGELIA

31 *dolicosiphon* ssp. *dolicosiphon* 32 *goniocarpa*

33 *hookeriana* Kashmir

34 *pandurata* Gansu

SECTION : LOPHIRIS

35 *tigridia*

36 *confusa* Baoxing

37 *cristata* 'Navy Blue'

38 *formosana*

SERIES : CHINENSES

39 *gracilipes*

40 *minutoaurea*

41 *rossii*

SERIES : VERNAE

42 *speculatrix*

SECTION : RUTHENICAE

44 *ruthenica*

43 *verna*

SERIES : TRIPETALAE

45 *ruthenica* var. *nana* Yunnan

46 *setosa alba* Japan

47 *setosa* var. *nasuensis*

48 *tridentata* Florida

SERIES : SIBERICAE

49 *bulleyana* Yunnan

50 *delavayi* Yunnan

51 *typhifolia*

52 *wilsonii*

SERIES : CALIFORNICAE

53 *fernaldii*

54 *macrosiphon*

55 *tenax*

56 *thompsonii*

SERIES : LONGIPETALAE

SERIES : LAEVIGATAE

57 *longipetala*

58 *ensata*

60 *versicolor* (white)

59 *laevigata variegata* (double form)

61 *virginica* (pale)

62 *virginica* (pink)

63 *virginica* (white)

SERIES : HEXAGONAE

64 *brevicaulis*

65 *fulva* 'Lockett's luck'

66 *giganticaerulea* 'Her Highness'

SERIES : PRISMATICAE

67 *nelsonii*

68 *prismatica* var. *rosea*

69 *prismatica* 'plicata'

70 *prismatica* white

SERIES : SPURIAE

71 *spuria* ?var *lilacina*

72 *spuria* var *maritima*

SERIES : FOETIDISSIMAE

73 *pseudonotha*

74 *foetidissima variegata*

SERIES : TENUIFOLIAE

75 *foetidissima* (seedcoat colours)

76 *anguifuga*

77 *loczyi* Gansu

78 *polysticta* Sichuan

79 *qinghainica* (root system)

80 *songarica* Afghanistan

81 *tianschanica*, Ala Archa;    82 *ventricosa*

SERIES : ENSATAE

83 *lactea* Gansu

84 *lactea* var. *chrysantha* Tibet

SERIES : SYRIACAE

85 *grant-duffii* Israel

86 *masia* Turkey

SERIES : UNGUICULARES

87 *unguicularis* ssp. *carica* var. *carica* S E Turkey

88 *unguicularis* ssp. *carica* var. *angustifolia,* Corfu

89 *unguicularis* ssp. *carica* var. *carica* Kastellorhgo    90 *lazica*

SUBGENUS : NEPALENSES

91 *collettii* Yunnan    92 *leptophylla* Sichuan

SUBGENUS : XIPHIUM

93 *staintonii*

94 *filifolia*

95 *latifolia* (pale form)

96 *tingitana*

SUBGENUS : SCORPIRIS

97 *xiphium*

98 *aitchisonii*

99 *baldschuanica*

100 *carterorum*

101 *caucasica* ssp. *turcica*

102 *drepanophylla*

103 *edomensis*

104 *fosteriana*

105 *kuschakewiczii*    106 *leptorrhiza;*

107 *maracandica*    108 *microglossa*

109 *narbutii*

110 *nicolai*

111 *nusairiensis*

112 *orchioides*

113 *parvula*

114 *platyptera*

115 *postii*

116 *pseudocaucasica*

117 *regis-uzziae* (blue)

118 *regis-uzziae* (yellow)

119 *stenophylla*

120 *wendelboi*

SUBGENUS : HERMODACTYLOIDES

121 *zaprjagajewii*

122 *histrio* Lebanon

123 *kolpakowskiana*

124 *pamphylica*

125 *reticulata*

126 *vartanii* Israel

127 *winogradowii*

I. PARIENSIS

128 *pariensis* (holotype)

Index of *Iris* species

Page numbers of species descriptions are given in bold type.

acutiloba 66, **67**, 89
   subsp. *acutiloba* **67**
   subsp. *lineolata* 67, **90**
aequiloba **46**
aestiva **161**
afghanica 62, **91**
aitchisonii **227**
   var. *chrysantha* **227**
alata **258**
alba 20, 31
albertii 19, **36**
   f. *erythrocarpa* 20
albicans **20**
   f. 'Madonna' 21
albida **183**
albomarginata **183**
alexeenkoi 21, **22**
alpina **128**
amabilis **153**
amankutanica **248**
amasiana **291**
ambertellon **49**
amoena 56
anguifuga **195**, 200
angustifolia **46**
annae **89**
antilibanotica 66, **68**
aphylla **22**, 25, 27, 29, 44, 45, 58
   var. *hungarica* 23
arenaria **60**
aschersonii 203, 206, **207**
assadiana 66, **68**
assyriaca **229**
astrachanica **50**
athoa **49**
atrofusca 67, **69**
atropatana **229**, 234
atropurpurea 34, 67, **69**
   var. *purpurea* 73
atroviolacea **32**
attica 8, 10, 23, **24**
aucheri **229**, 230, 278

aurantiaca **70**
   f. *wilkiana* 70
   var. *unicolor* 70
aurea 169, 175, **176**
australis 32

babadagica **24**
bakeriana 279, 283, **284**
   var. *metaina* 284
baldschuanica 230, 237, **253**, 265
balkana 49
barbatula **215**
barnumae 66, **70**, 71
   subsp. *barnumae* f. *barnumae* 70
   subsp. *demavendica* **72**
      f. *protonyma* 71
      f. *urmiensis* 71
   var. *zenobiae* 68, **87**
basaltica 66, **72**, 89
bastardii **164**
battandieri **224**
beecheyana **147**
belgica 56
belouini **28**
benacensis 22, **40**
biflora 22, 34, 46, **53**
bifurca 22
biggeri **75**
biglumis 202
biliottii **28**, 37
binata 46
bisflorens 22
bismarkiana 66, **73**, 76
bloudowii 58, 60, **62**
bohemica 22
boissieri 220, **221**
bolleana 257, **258**
boltoniana 165, **171**
bornmuelleri **291**
bosniaca 49
bostrensis 66, **73**
brachycuspis **131**

361

Index of *Iris* species

bracteata 145, **146**
brandzae 186
brevicaulis 163, 166, 167, **168**
brevicuspis 131
breviscapa 22
bucharica 231, 255, 275, 278
buiana 57
bulleyana 99, 135, 136, 137
    f. *alba* 136
bungei 196
burnatii 40

cabulica 232
caerulea 32
caeruleo-violacea 204, 208
caespitosa 130, 160, 202
californica 153
camillae 66, **89**
capnoides 232
caricifolia 202
carolina 171
caroliniana 171
carterorum 233
carthaliniae 175, **187**
caspica 163
cathayensis **197**
caucasica 229, 234, 235, 263, 278
    var. *bicolor* 239
    subsp. *caucasica* **234**
    var. *linifolia* 247
    subsp. *turcica* 235
caurina 165
cedretii 65, 66, **74**
chalcidice 49
chameiris 40, 41
chinensis 114, 115
chrysantha 71
chrysographes 9, 134, 135, 136, 137, 140, 144, 145
    var. *rubella* 137
chrysophylla 145, **146**, 147, 157
clarkei 15, 134, 135, **138**
clausii 25
clusiana 22, 46
coelestina 168
coerulea 46, 228
colchica 176
collettii 215, **216**, 219
    var. *acaulis* 217
confusa 109, 110, 111, 113, 114, 115, 119, 120, 121
corygeii 56
cretensis 20

    f. *latifolia* 213
cretica 210, 211, 212
cristata 16, 109, 111, 113, 115, 117, 118, 128
croatica **29**
crocea 161, **175**, 176, 182, 183
crociformis 291
cuniculiformis **99**
cuprea 168
curvifolia **59**, 60
cycloglossa 225, 235, 236
cypriana 27, **29**

dacica 22
daënensis 190
dalmatica 42
damascena 66, **74**
danfordiae 10, 279, **291**, 292
darwasica 92, 95, 96
decora 104, 215, **217**, 219
deflexa 32, 33
delavayi 134, 135, **138**, 139, 163, 169, 171
demavendica 72
demetrii 175, **188**
dengerensis 250
desertorum 188
diantha 46
diversifolia 221
doabensis 236
dolicosiphon 100, 101
    subsp. *orientalis* 101
doniana 160, 202
douglasiana 144, 145, **147**, 148, 151, 152, 155, 158, 160
douglasiana pygmaea 131
dragalz 56
drepanophylla 95, 237, **238**, 246
    subsp. *chlorotica* **238**
    subsp. *drepanophylla* 237
duclouxii 216
duthieii 103
dykesii 135, 139

edomensis 239
eleonorae 240
elizabethae 86
elongata 48, 50
ensata 8, 132, **160**, 162, 163, 164, 166, 167, 176, 182, 194, 202
    var. *grandiflora* 161
    var. *spontanea* 161
erratica 40

Index of *Iris* species

eulefeldii 34, 35
ewbankiana 67, 86
    var. *elizabethae* 86
extrafoliacea 22

falcata 22, 58
falcifolia 93, 97, 98
farreri 194, **197**
fasciculata 217
fernaldii 145, **148**, 149
fibrosa 81
fieberi 22
filifolia 97, 221
    var. *filifolia* 221
    var. *latifolia* 221
fimbriata 114, 117
flaccida 165
flava 163
flavescens 36, 56, 57
flavissima 58, 60
    var. *bloudowii* 58
    subsp. *stolonifera* 60
    var. *umbrosa* 58
florentina 1, 2, 20, 33
foetidissima 3, 6, 9, 14, 15, **194**
foliosa 168
fominii 67
fontanesii 223
formosana 109, 112, 120
forrestii 5, 134, 135, 136, 237, **140**, 144, 145, 160
fosteriana 95, 239
fragrans 53, 202
fulgida 42
fulva 164, 166, 167, **168**, 171, 176
fumosa 229
furcata 22, 25

galatica 241, 178
gatesii 65, 66, 75, 77, 79, 82
georgiana 166
germanica 2, 3, 17, 26, 27, 29, 31, 37, 42, 217
    'Alba' 20, 31, 33
    'Amas' 31, 32
    'Asiatica' 33
    'Askabadensis' 32
    'Atroviolacea' 32
    'Australis' 32
    'Caerulea' 32
    'Cephalonian form' 32
    'Cretan form' 32
    'Deflexa' 32, 33
    'Florentina' 33
    'Fontarabie' 33
    'Gypsea' 33
    'Istria' 33
    'Kharput' 32, 33
    'Metkevic form' 33
    'Mostar form' 33
    'Nepalensis' 27, 32, 33, 34
    'Sivas' 34
    'Veglia' 34
    'Vulgaris' 33, 34
giganticaerulea 167, **169**, 170, 171
gilgitensis 102
glaucescens 34, 35, 39, 51
gmelini 161
goniocarpa 99, 100, 101
    var. *grossa* 99, 102
    var. *tenella* 102
gormanii 155, 156
gracilipes 3, 16, 109, **113**, 121
gracilis 22, 40, 46, 101, 171
graeberiana 230, 241, 178
graminea 174, **176**, 177, 202
    var. *pseudocyperus* 177, 178
graminifolia 178
gransaultii 178
grant-duffii 203, 204, 205, 206, 207, 208
griffithii 35
grisjii 126
grossheimii 89
gueldenstadtiana 188
guertleri 46

haematophylla 202
halophila 174, 175, 188
hartwegii 145, **146**, 150
    subsp. *australis* 145, 150
    var. *australis* 150
    subsp. *columbiana* 144, 145, **151**
    subsp. *pinetorum* 145, **151**, 152
hauranensis 68
haussknechtii 178, 257, 258
haynei 66, 75
heldreichii 266
helena 67
henryi 121, 122, 123
hermona 66, 76
heweri 92
hexagona 167, 170
heylandiana 66, 76, 77, 82
hippolyti 242

Index of *Iris* species

hissarica 250
histrio 279, 285, 286
  subsp. *aintabensis* 285
  var. *aintabensis* 285
  var. *atropurpurea* 286
  subsp. *histrio* 285
histrioides 279, 286, 287, 290
  var. *sophensis* 287
hoogiana 93
hookeri 131
hookeriana 100, 102, 107
hortensis 42
humilis 13, 59, 60, 147, 184, 210
hungarica 22
hymenospatha 243, 244, 258
  subsp. *hymenospatha* 243
  subsp. *leptoneura* 244
hyrcana 287

iberica 66, 77, 89
  subsp. *elegantissima* 78, 83
  subsp. *iberica* 77
  subsp. *lycotis* 77
iliensis 202
illyrica 42
imberbis 222
imbricata 36
inconspicua 244
innominata 145, 152, 157, 158
issica 257, 258
italica 40, 41
itsihatsi 161

japonica 109, 110, 111, 112, 113, **114**, 115, 119, 120
  f. *pallescens* 115
jordana 69
jugoslavica 52
juncea 222
  var. *mermieri* 222
  var. *numidica* 222
  var. *pallida* 222
junonia 27, 29, 37

kaempferi 160, 161
kamaonensis see *kemaonensis*
karategina 95
kashmiriana 31, 38
kasruana 86
kazachensis 89
kemaonensis 100, 103, 107
kermesina 166

kerneriana 178, 180
ketzhowelli 89
kingiana 103
kirkwoodii 65, 66, **79**
  subsp. *calcarea* 79
  var. *macropetala* 79
klattii 190, 191
kobasensis 49
kobayashi 197
kochii 32, 39
koenigi 89
kolpakowskiana 279, **281**, **282**, 283
kopetdagensis 245, 246, 276
koreana 122
korolkowii 10, **94**
  f. *concolor* 94
  f. *leichtliniana* 94
  f. *venosa* 94
  f. *violacea* 94
kumaonensis see *kemaonensis*
kumaonensis caulescens 102
kuschakewiczii 245, **246**, 269, 275
kuschkensis 95

lactea 202
  var. *chinensis* 202, 203
  var. *chrysantha* 203
lacustris 16, 109, 112, **115**
laevigata 8, 14, 132, **161**, 164, 166, 167, 194
  var. *alba* 162
  var. *albo-purpurea* 162
  var. *atro-purpureum* 162
  var. *kaempferi* 160
  var. *plena* 162
  var. *variegata* 162
lamancei 168
latifolia 220, 221, **222**, 224
latistyla 109, 116
lazica 14, 209, 213, 214
leichtlinii 96
lepida 56
leptophylla **104**, 106, 219
leptorrhiza 247
leucographer 56
libani 285
lilacina **179**, 180
limbata 56
lineata 92, **95**
lineolata **67**, 89
linifolia 247, 270
lisbonensis 53

loczyi 198
longiflora 46, 50, 53
longifolia 163, 202
longipedicellata 180, 183
longipetala 148, 158, 159
longiscapa 98
longispatha 160, 202
lortetii 66, 80
 var. *lortetii* 80
 var. *samariae* 80
ludwigii 181
lurida 57
lutea 46, 163
lutescens 17, 40, 41, 45, 46
lycotis 89

macedonica 49
mackii 161
macrantha 32
macrosiphon 145, 153, 155
magnifica 248, 272, 278
mairiae 66, 80
majoricensis 20
mangaliae 56, 57
mandraliscae 42
mandschurica 61, 161
mangaliae 56
manissadjianii 85
maracandica 249, 270, 271, 178
marchesettii 42
maritima 175, 189
maritima hispanica 190
marschalliana 184
marsica 41, 42
masia 204, 205, 207, 208, 281
meda 66, 81, 87
medwedewii 83
melanosticta 203, 206
mellita 52
melzeri 22
mesopotamica 27, 30
microglossa 250
milesii 14, 109, 111, 116, 117, 119, 120
minuta 122
minutoaurea 16, 122
missouriensis 3, 148, 158, 159, 293
monnieri 161, 181, 182, 183, 193
montana 159
moorcroftiana 202
munzii 144, 145, 151, 153, 154
murrayana 165

musulmanica 175, 190
nana 127, 128
napocae 46
narbutii 250, 278
narcissiflora 105
narynensis 251
nazarena 73
nectarifera 66, 77, 81, 82, 87
 var. *nectarifera* 82
 var. *mardinensis* 82
neglecta 40, 57
nelsonii 167, **170**
nepalensis 27, 32, 33, **34**, 217, 218
 f. *depauperata* 216
 var. *letha* 216
nertschinskia 141
nicolai 230, 232, 237, **252**, 253, 260, 261, 265, 277
nigricans 66, **82**
notha 191
'nova' 120, 121
nudicaulis 22, 53
nusairiensis 253
nyaradyana 57

obtusifolia 36
ochridana 23
ochroaurea 176
ochroleuca 183
odaesanensis 123
odontostyla 254, 268
odoratissima 42
olbiensis 40, 41
oncocyclus 1, 34
orchioides 231, **255**, 277, 278
orientalis 133, 134, 163, 174, 175, 176, 179, 180, 182, 183, 193
oxypetala 202

palaestina 255
 var. *caerulea* 262
pallasii 202
pallida 2, 33, **42**, 43, 57, 118, 128
 subsp. *cengialti* **43**, 118
 var. *dalmatica* 44
 var. *loppio* 44
 subsp. *pallida* 42
pallidiflava 163
pallidiflora 163
pallidior 163
pallido-caerulea 42
paludora 163

365

Index of *Iris* species

palustris 163
pamphylica 205, 279, 280, 281
pandurata 105
panormitana 45
paradoxa 66, 82, 89
  f. *atrata* 83
  f. *choschab* 83, 84, 90
  f. *mirabilis* 84
  f. *paradoxa* 84
  f. *vulgaris* 84
pariensis 293
parvula 233, 256
pelogonus 159
perrieri 44
persica 235, 241, 257, 258, 260, 263, 266, 267, 278
  var. *magna* 258
  var. *mardinensis* 258
petrana 66, 84
phragmitetorum 135, 140
picta 165
pineticola 60
planifolia 225, 256, 258
platyptera 259
plicata 42
pluriscapia 46
polakii f. *protonyma* 71
polonica 22
polysticta 194, 199
pontica 181, 184
popovii 260
porphyrochrysa 261, 262
portae 43
postii 239, 262
potaninii 61
prilipkoana 188
prismatica 150, 163, 164, 168, 169, 171, 172
  var. *austrina* 172
proantha 124, 125
  var. *valida* 124, 125
pseudacorus 3, 8, 9, 10, 160, 161, 163, 166, 167, 169, 170, 171, 194
  var. *acoriformis* 164
  var. *alba* 164
  var. *bastardii* 164
  var. *capsica* 164
  var. *flore pleno* 164
  var. *gigantea* 164
  f. *longiacuminata* 164
  var. *mackii* 164
  f. *nyaradyana* 164
pseudocaucasica 263

pseudocyperus 177
pseudonotha 184
pseudopumila 8, 10, 22, 24, 45, 46
pseudopumilaeoides 46
pseudorossii 125
  var. *valida* 124
pulchella 165
pumila 8, 10, 23, 24, 25, 40, 41, 45, 46, 47, 48, 56
  subsp. *attica* 23
  var. *elongata* 48
  subsp. *pumila* 46, 48
  subsp. *taurica* 48
purdyi 145, 154, 155, 158
purpurea 240
purpureobractea 37, 48

reflexa 22
regelii 199
reginae 56
regis-uzziae 264
reichenbachiana 49
reichenbachii 24, 49, 50, 53, 58
repanda 14
reticulata 205, 279, 283, 284, 286, 287, 288
  var. *bakeriana* 283
  var. *histrio* 285
  var. *histrioides* 286
  var. *hyrcana* 288
  var. *krelagii* 288
revoluta 50
rhaetica 57
rigida 22
robusta 9
rosaliae 57
rosenbachiana 230, 232, 253, 261, 265, 277
  var. *albo-violacea* 252
  var. *baldschuanica* 230
rossii 124, 125
  f. *alba* 125
rosthornii 117
rothschildii 57
rubromarginata 52
rudskyi 56
ruthenica 127, 128
  var. *brevituba* 129
  f. *leucantha* 129
  var. *nana* 129

samariae 80
sambucina 57
sanguinea 132, 133, 134, 135, 141, 142, 148, 163

Index of *Iris* species

var. *sanguinea* f. *albiflora* 141, 142
var. *yixingensis* 141
sarajevoensis 46
sari 66, 82, 85
sasha 166
sativa 163
scariosa 22, 34, 35, 50, 51, 56, 60
schachtii 51
schelkownikowii 90
schischkinii 265
schmidtii 22
semperflorens 161
serbica 49
serotina 223
setina 52
setosa 8, 9, 131, 161, 163
   f. *alpina* 132
   var. *arctica* 132
   subsp. *canadensis* 132
   subsp. *hondoensis* 132
   subsp. *interior* 132
   var. *nasuensis* 132
   f. *platyrhyncha* 132
   f. *serotina* 132
sibirica 10, 132, 133, 134, 135, 142, 160, 163, 164, 166, 167, 172, 194
sichuanensis 106
sicula 42
sieheana 257, 258
sikkimensis 106
'Sindpers' 278
'Sindpur' 278
sinistra 90
sindjarensis 229
sintenesii 5, 10, 185, 186, 187
   subsp. *brandzae* 186, 187
   subsp. *brandzae* f. *topae* 187
   f. *constantinopolitan* 185
   var. *urumovii* 186
skamnili 26, 31
skorpilii 49
sofarana 66, 86, 89
   f. *franjieh* 86
   subsp. *kasruana* 86
   f. *qassioumensis* 74
sogdiana 175, 180, 191
songarica 200, 202
   var. *gracilis* 200
sosnowskii 89
spathulata 189
species 58

spectabilis 26
speculatrix 109, 126
sprengeri 66, 86
spuria 161, 164, 169, 171, 175, 182, 185, 187, 189, 190, 192, 193
   subsp. *carthaliniae* 187, 190, 191
   var. *danica* 193
   subsp. *demetrii* 188
   subsp. *halophylla* 179, 188, 191, 192
   var. *lilacina* 179, 180
   subsp. *maritima* 189, 190
   var. *maritima* 189, 193
   subsp. *musulmanica* 179, 190, 191
   subsp. *notha* 188, 190, 191, 193
   var. *reichenbachiana* 190
   subsp. *sogdiana* 180, 191, 192
   var. *sogdiana* 191
   subsp. *spuria* 192, 193
   subsp. *spuria* var. *danica* 193
squalens 56, 57, 114
squamata 57
staintonii 219
statellae 40
steniloba 46
stenogyna 188
stenophylla 266
   subsp. *allisonii* 267
   subsp. *stenophylla* 266, 267
stocksii 254, 263, 267, 268
stolonifera 94, 95, 96
straussii 49, 52
strictum 163
stylosa 210, 212
   var. *angustifolia* 210, 211, 212
suaveolens 26, 52, 53, 58
   var. *jugoslavica* 53
subbiflora 53, 54
   var. *lisbonensis* 54
subdecolorata 268, 278
subtriflora 22
sulcata 217
sulphurea 36
susiana 66, 72, 89
suworowii 92
svetlanae 249, 269, 271
swensoniana 87
sylvatica 176
szovitsii 67

tadshikorum 270
talischii 36

Index of *Iris* species

taochia 55
tatianae 90
tauri 266
taurica 48
tectorum 14, 109, 113, 116, **117**, 118, 120
tenax 144, 145, 147, 150, **155**, 156
    var. *australis* 150
    var. *gormanii* 155
    subsp. *klamathensis* 145, **156**
tenuifolia 49, 198, 199, **200**
    var. *thianschanica* 198
tenuis 109, **118**
tenuissima 157, 158
    subsp. *purdyiformis* 145, 157, 158
thompsonii 145, **158**
thoroldii 61
tianschanica 198
tigridia 62, 107, 108
    var. *fortis* 108
tigrina 103
timofejewii 55, 56
tingitana **223**, 224
    var. *fontanesii* 223
    var. *mellori* 223
tolmeiana 159
tomiolopha 117
transylvanica 46
tridentata 131, **132**
triflora 202
trigonocarpa 171
tripetala 132
tristis 46
trojana 30
tubergeniana 270, 271
'Turkey Yellow' 182
typhifolia 135, **143**

unguicularis 6, 209, 213, 214
    subsp. *carica* 211, 214
    subsp. *carica* var. *angustifolia* 212
    subsp. *carica* var. *carica* 211
    subsp. *carica* var. *syriaca* 212
    subsp. *cretensis* 210, 211, 212, 214
    var. *lazica* 213
    subsp. *unguicularis* 210, 212, 214
    'Abington Purple' 214
    'Alba' 210
    'Angustifolia' 212
    'Ellis's Variety' 214
    'Greek White' 212
    'Marginata' 210

'Mary Barnard' 210
'Oxford Dwarf' 214
'Speciosa' 212
'Starker's Pink' 214
'Walter Butt' 210
'Winter Treasure' 214
uniflora 130
    var. *alba* 130
    var. *caricina* 130
urumovii 186

vaga 96
varbossiana 26
variegata 56, 57
    var. *pontica* 57
vartanii **289**
ventricosa 50, **201**
verna 5, 16, **127**
    var. *smalliana* 127, 128
    var. *verna* 127
versicolor 2, 8, 9, 160, 161, 163, 164, **165**, 166, 167, 168, 169, 171
    var. *arkansensis* 165
    var. *columnae* 165
    var. *kermesina* 166
    var. *rosea* 166
vicaria **271**, 278
violacea 26, 45, 46, 190
virescens 40
virginiana 165
virginica 8, 9, 160, 161, 163, 164, **166**, 169, 170, 171
    var. *shrevei* 166, 167
    var. *shrevei* var. *virginica* 167
vvedenskyi 272

warleyensis 273, 278
'Warlsind' 278
watsoniana 147, **148**
wattii 109, 110, 111, 115, 118, **119**, 120, 121
wendelboi 234, **274**
westii 66, **87**
wilsonii 15, 135, 136, 140, 143, 144, 169, 171
willmottiana 242, 243, 247, **274**
winkleri 282, **283**
winogradowii 279, 287, 290, 291
wittii 121

xanthochlora 275, **276**
xanthospuria 181, 182, 183, **193**
xiphium **224**

Index of *Iris* species

var. *battandieri* 224
var. *lusitanica* 224
var. *praecox* 224
var. *taitii* 224
var. *xiphium* 224
xiphioides 222

yebrudii 66, 88

subsp. *edgecombii* 88
yedoensis 131
yunanensis 217

zaprjagajewii 276
zenaidae 277
zuvandicus 90

Index of species names from other classifications

Alatavia kolpakowskianum 281
   winkleri 283

Evansia chinensis 114
   fimbriata 114
   nepalensis 217

Ioniris biglumis 202
   doniana 160, 202
   fragrans 202
   longispatha 160, 202
   pallassii 202
   ruthenica 128
   stylosa 210
   triflora 202

Iridodictyum bakerianum 283
   danfordiae 291
   histrio 285
   histrioides 286
   kolpakowskianum 281
   libani 285
   reticulatum 287
   vartanii 289
   winkleri 283
   winogradowii 290

Isis fimbriata 114, 117
   fulva 168
   ruthenica 128

Juno aitchisonii 227
   almaatensis 268
   aucheri 229
   baldschuanica 230
   bucharica 231
   cabulica 232
   capnoides 232
   caucasica 234
   coerulea 228
   drepanophylla 237
   fosteriana 239
   hippolyti 242
   inconspicua 244
   kopetdagensis 245
   kuschakewiczii 246
   leptorrhiza 247
   linifolia 247
   magnifica 248
   maracandica 249
   microglossa 250
   narbutii 250
   naryensis 251
   nicolai 252
   orchioides 255
   parvula 256
   persica 257
   popovii 260
   pseudocaucasica 263
   rosenbachiana 265
   stocksii 267
   subdecolorata 268
   svetlanae 269
   tadshikorum 270
   tubergeniana 270
   vicaria 271
   vvedenskyi 272
   warleyensis 273
   willmottiana 274
   zaprjagajewii 276
   zenaidae 277

Junopsis decora 217

Limniris pseudo-acorus 163

Limnirion pseudo-acorus 163

Neubeckia cristata 111
   decora 217
   fulva 168
   reticulata 287
   stylosa 210
   sulcata 217

Pseudo-iris palustris 163

Index of species names from other classifications

Siphonostylis cretensis subsp. *carica* 211
   cretensis subsp. *cretensis* 210
   cretensis subsp. *syriaca* 212
   'Hybridcomplex' 212
   lazica 213
   unguicularis 210

Xiphion acoroides 163
   aitchisonii 227
   alatum 258
   aucheri 229
   brachycuspis 131
   caucasicum 234
   danfordiae 291
   donianum 160, 202
   flaccidum 165
   histrio 285
   kolpakowskianum 281

   palaestinum 255
   pallassii 202
   planifolium 258
   pseudo-acorus 163
   reticulatum 287
   ruthenicum 128
   stocksii 267
   versicolor 165
   virginianum 165

Xiphium pseudacorus 163

Xyridion acoroideum 163
   laevigatum 161
   pseudacorus 163
   setosum 131
   violaceum 161

Lightning Source UK Ltd.
Milton Keynes UK
UKOW031424210212

187698UK00005B/5/P